Applications of Modern Dynamics to Celestial Mechanics and Astrodynamics

NATO ADVANCED STUDY INSTITUTES SERIES

Proceedings of the Advanced Study Institute Programme, which aims at the dissemination of advanced knowledge and the formation of contacts among scientists from different countries

The series is published by an international board of publishers in conjunction with NATO Scientific Affairs Division

A	Life Sciences	Plenum Publishing Corporation
B	Physics	London and New York
C	Mathematical and Physical Sciences	D. Reidel Publishing Company Dordrecht, Boston and London
D	Behavioural and Social Sciences	Sijthoff & Noordhoff International Publishers
E	Applied Sciences	Alphen aan den Rijn and Germantown U.S.A.

Series C – Mathematical and Physical Sciences

Volume 82 – Applications of Modern Dynamics to Celestial Mechanics and Astrodynamics

Applications of Modern Dynamics to Celestial Mechanics and Astrodynamics

Proceedings of the NATO Advanced Study Institute held at Cortina d'Ampezzo, Italy, August 2-14, 1981

edited by

VICTOR SZEBEHELY

L. B. Meaders Professor of Aerospace Engineering, University of Texas, Austin, U.S.A.

D. Reidel Publishing Company

Dordrecht : Holland / Boston : U.S.A. / London : England

Published in cooperation with NATO Scientific Affairs Division

Library of Congress Cataloging in Publication Data

CIP

NATO Advanced Study Institute (1981 : Cortina d'Ampezzo, Italy)
Applications of modern dynamics to celestial mechanics and astrodynamics.
(NATO advanced study institutes series. Series C, Mathematical and physical sciences ; v. 82)
"Published in cooperation with NATO Scientific Affairs Division".
Includes index.
1. Mechanics, Celestial–Congresses. 2. Astrodynamics–Congresses. 3. Dynamics–Congresses. I. Szebehely, Victor G., 1921– II. Title. III. Series.
QB349.N37 1981 521'.1 81-23475
ISBN 90-277-1390-1 AACR2

Published by D. Reidel Publishing Company
P.O. Box 17, 3300 AA Dordrecht, Holland

Sold and distributed in the U.S.A. and Canada
by Kluwer Boston Inc.,
190 Old Derby Street, Hingham, MA 02043, U.S.A.

In all other countries, sold and distributed
by Kluwer Academic Publishers Group,
P.O. Box 322, 3300 AH Dordrecht, Holland

D. Reidel Publishing Company is a member of the Kluwer Group

Printed in The Netherlands

TABLE OF CONTENTS

PART III: CONNECTION WITH STATISTICAL MECHANICS

PART IV: REGULARIZATION AND GEODESICS

PART V: NON-LINEARITY, DETERMINANCY AND MANIFOLDS

PART VI: ABSTRACTS OF SEMINAR CONTRIBUTIONS

PREFACE

This volume contains the detailed text of the major lectures and the abstracts of the lectures delivered during the seminar sessions.

The subject of our NATO Advanced Study Institute in 1981 was the Application of Modern Dynamics to Celestial Mechanics and Astrodynamics. This Preface will first explain the terminology, then it will review shortly the content of the lectures and will outline how all this was made possible and, finally, it will disclose our future aspirations.

Periodicity is an extremely important concept in our field, therefore, it should not be unexpected that our NATO Advanced Study Institute is enjoying a period of three years. Since 1972 we conducted four Institutes with increasing interest and enthusiasm displayed by the participants, lecturers and by this Director.

Celestial Mechanics or Dynamical Astronomy is part of Astronomy dealing mostly with the motion of natural celestial bodies. Astrodynamics or Orbital Mechanics is the application of dynamics to problems of Space Engineering and it treats mostly the dynamical behavior of artificial satellites and space probes. The underlying mathematical and dynamical principles are, of course, the same for Celestial Mechanics and for Astrodynamics.

This Director of the Institute and Editor of the Proceedings was extremely fortunate to have obtained the cooperation of outstanding lecturers who were clear, thorough, understandable, patient to answer questions, but above all, had knowledge of the

V. Szebehely (ed.), Applications of Modern Dynamics to Celestial Mechanics and Astrodynamics, ix–x.

most recent developments in our field of interest. No less was the competence of the members of the audience who contributed to the seminar sessions and discussion periods. In fact, probably the most interesting and outstanding feature of this year's Institute was that lecturers and audience were almost indistinguishable in their technical preparation. Mostly age separated the lectures from the seminar contributors. The fact that the lecturers dared to mention applications to space and astronomy, while the younger seminar contributors emphasized mathematics is not unusual in the Institutes and their mixture creates just the perfect atmosphere.

The problems of three bodies were once again a star (or better, three stars!), stochasticity and statistical mechanics, stability, mappings, chaotic motion, regularization and collisions, invariant manifolds, Lie algebraic methods, hierarchical systems, geodesics, lunar theory, resonance, bifurcation, recently discovered planetary rings, ergodic theory, Trojan asteroids, determinancy in celestial mechanics, etc. -- is just a short list of problems and techniques which were there in Cortina to excite the interest of workers in our field.

We always knew that Sir Isaac Newton did not solve all problems of celestial mechanics, and it was of considerable regret that we found that we could not solve all problems either. This will necessitate conducting another Institute, three years hence, in spite of the fact that many participants felt that from now on our quality can only decline. Our mathematics combined with our increased knowledge of non-linear dynamics and new observational results might, three years from now, allow us to solve more problems and understand better what we did not understand in August 1981.

This Institute was made possible by the support of the Scientific Affairs Division of NATO, of the Italian Research Council and of the University of Texas. The support would not have helped much if the technical level of the participants would not have been unusually high. But let the reader be the judge. If he was fortunate enough to be present in Cortina he already knows what is included in this volume. If the reader, regretfully, was unable to attend, we invite his questions and criticisms. This Director-Editor claims no credit. If the Institute was successful, credit should go to the participants. If anything is to be criticized, the Director should have known better after nine years and four Institutes.

Cortina d'Ampezzo, Italy and
Austin, Texas, USA

Victor Szebehely
Director NATO Advanced
Study Institute and
Editor of the Proceedings

LIST OF SPEAKERS AND PARTICIPANTS

Aksnes, K.	(Norwegian)	Norwegian Defense Research Est. and Univ. of Tromsø, 2007 Kjeller, Norway
Alseda, L.	(Spanish)	Universitat de Barcelona, Barcelona, Spain
Altavista, C.	(Argentinian)	Univ. Observatory of La Plata, La Plata, Argentina
Andersen, P.	(Norwegian)	Norwegian Defense Research Est., Box 25, 2007 Kjeller, Norway
Aubanell, A.	(Spanish)	Dept. Equacions Funcionals, Facultat de Matematiques, Univ. de Barcelona, Barcelona, Spain
Banfi, V.	(Italian)	Osservatorio di Pino Torinese, Torino, Italy
Baumgarte, J.	(West German)	Mechanik Zentrum, Technische Universitat, Postfach 3329, D-3300, Braunschweig, F.R.G.
Benseny, A.	(Spanish)	Facultat de Matemàtiques, Universitat de Barcelona, Barcelona, Spain
Bianchini, L.	(Italian)	University of Padova, Dept. of Mechanics, Padova, Italy
Bozis, G.	(Greek)	Dept. of Theoretical Mechanics, Univ. of Thessaloniki, Thessaloniki, Greece
Bryant, J.	(U.S.A.)	Laboratoire de Mécanique Théorique, Faculté des Sciences, 25030 Besancon, France
Burkhardt, G.	(West German)	Astron. Rechen Institut, D-6900 Heidelburg, F.R.G.

Casasays, J.	(Spanish)	Dept. Teoria de Funcions, Facultad de Matematicas, Universitat de Barcelona, Barcelona, Spain
Coffey, S.	(U.S.A.)	Naval Research Laboratory, Washington, D. C. 20375, USA
Colombo, G.	(Italian)	University of Padova, Dept. of Mathematics, Padova, Italy
Delshams, A.	(Spanish)	Facultat de Matematiques, Univ. de Barcelona, Barcelona, Spain
Dvorak, R.	(Austrian)	Institut f. Astronomie, Universitaetsplatz 5, A8010 Graz, Austria
Easton, R.	(U.S.A.)	Dept. of Math, University of Colorado, Boulder, Colorado 80309, USA
Ferrer, S.	(Spanish)	Departamento de Astronomia, Facultad de Ciencias, Universidad di Zaragoza, Zaragoza, Spain
Ferrini, F.	(Italian)	Dept. of Astronomy, University of Pisa, Pisa, Italy
Froeschle, Ch.	(French)	Observatoire de Nice, 06007 Nice, France
Froeschle, Cl.	(French)	Observatoire de Nice, 06007 Nice, France
Galgani, L.	(Italian)	Istituto di Fisica dell' Università, Via Celoria 16, Milano, Italy
Garfinkel, B.	(U.S.A.)	Dept. of Astronomy, Yale University, New Haven, Connecticut 06520, USA
Gomez, G.	(Spanish)	Dep. D'Equacions Funcionals, Univ. de Barcelona, Barcelona, Spain

Gonczi, R.	(French)	Observatoire de Nice, 06007 Nice, France
Gurel, D.	(Turkish)	New York University, 4 Washington Place, New York, New York 10003, USA
Gurel, O.	(Turkish)	IBM Cambridge Scientific Center, 545 Technology Square, Cambridge, Massachusetts 02139, USA
Hadjidemetriou, J.	(Greek)	University of Thessaloniki, Thessaloniki, Greece
Hanslmeier, A.	(Austrian)	Institut f. Astronomie, Universitaetsplatz 5, A-8010 Graz, Austria
Heggie, D.	(U.K.)	Dept. of Mathematics, Univ. of Edinburgh, King's Buildings, Mayfield Road, Edinburgh EH9 352, U.K.
Helleman, R.	(Dutch)	Theoretical Physics, P. O. Box 217, 7500 AE Enschede, Netherlands
Henrard, J.	(Belgian)	Facultes Universitaires de Namur, 8 Rempart de la Vierge, 5000 Namur, Belgium
Hitzl, D.	(U.S.A.)	Lockheed Research Laboratory, 3460 Hillview Avenue, Palo Alto, California, U.S.A.
Hoots, F.	(U.S.A.)	Directorate of Astrodynamic Applications, ADCOM/DO6, Peterson AFB, Colorado 80914, U.S.A.
Irigoyen, M.	(French)	Institut Henri Poincaré, 11 rue P. et M. Curie, 75005 Paris, France
Kirchgraber, U.	(Swiss)	Sem. f. Angew. Math., ETH-Zentrum, CH-8092, Zürich, Switzerland

Kourogenis, C.	(Greek)	Dekelia Airforce Academy, Greece 6. Katsantom, Athens, Greece
Kovalevsky, J.	(French)	Centre d'Etudes et de Recherches Geodynamiques, Avenue Copernic, 06130 Grasse, France
Lemaitre, A.	(Belgian)	Facultes Universitaires de Namur, Rempart de la Vierge, 5000 Namur, Belgium
Liu, J.	(U.S.A.)	Directorate of Astrodynamic Applications, ADCOM/DO6, Peterson AFB, Colorado 80914, U.S.A.
Llibre, J.	(Spanish)	Secciò de matemàtiques, Universitat de Barcelona, Barcelona, Spain
Lundberg, J.	(U.S.A.)	Dept. of Aerospace Engineering, Univ. of Texas, Austin, Texas 78712, U.S.A.
Magnenat, P.	(Swiss)	Observatoire de Genève, CH-1290 Sauverny, Switzerland
Marchal, C.	(French)	D.E.S.-Onera-92320, Chatillon, France
Martin, F.	(Italian)	Inst. of Applied Mathematics, Univ. of Padova, Padova, Italy
Martinez, J.	(Spanish)	Dept. de Astronomia y Geodesia, Universidad Complutense, Madrid, Spain
Martinez, R.	(Spanish)	Dept. de Matematiques, Universitat de Barcelona, Barcelona, Spain
Message, J.	(U.K.)	Dept. of Applied Math., Liverpool University, Liverpool L693BX, U.K.

Milani, A.	(Italian)	Istituto Matematica "L. Tonelli", Università di Pisa, 56100 Pisa, Italy
Nahon, F.	(French)	Institut Henri Poincaré, 11 Rue P. et M. Curie, 75005 Paris, France
Nobili, A.	(Italian)	Istituto di Matematicà "L. Tonelli", Universita di Pisa, 56100 Pisa, Italy
Oesterwinter, C.	(U.S.A.)	Naval Surface Weapons Center, Dahlgren, Virginia 22448, U.S.A.
Petrovsky, R.	(Belgian)	Service de Chimie Physique II, Code Postal n° 231, Campus Plaine U.L.B., Bvd. du Triomphe, 1050 Bruxelles, Belgium
Rapaport, M.	(French)	Observatoire de l'Université de Bordeaux, 33270 Floirac, France
Renz, W.	(West German)	Dept. of Theoretical Physics, D-5100 Aachen, F.R.G.
Roy, A.	(U.K.)	Dept. of Astronomy, Glasgow University, Glasgow, U.K.
Simo, C.	(Spanish)	Facultat de Matematiques, Universitat de Barcelona, Barcelona, Spain
Sivaramakrishnan, A.	(Indian)	Dept. of Astronomy, Univ. of Texas, Austin, Texas 78712, U.S.A.
Somorjai, R.	(Canadian)	National Research Council of Canada, 100 Sussex DR, Ottawa, Ontario K1A0R6, Canada
Spirig, F.	(Swiss)	Sem. f. Angew. Math., ETH-Zentrum, CH-8092, Zürich, Switzerland

Szebehely, V.	(U.S.A.)	Dept. of Aerospace Engineering, University of Texas, Austin, Texas 78712, U.S.A.
Taylor, D.	(U.K.)	Royal Greenwich Observatory, Halisham, Herstmonceaux Castle, East Sussex, U.K.
Ülküdas, M.	(Turkish)	Ege Universitesi Ins., P.O. Box 61, Izmir, Turkey
Valsecchi, G.	(Italian)	Institute for Astronomical Sciences, Planetologia, Viale Universita 11, 00185 Rome, Italy
van de Kamp, P	(Dutch)	University of Amsterdam, Amstel 244, 1017 AK Amsterdam, Netherlands
Vasquez, L.	(Spanish)	Dept. de Fisica Teorica, Universidad Complutense, Madrid, Spain
Vicente, R.	(Portuguese)	Dept. of Applied Mathematics, Univ. of Lisbon, Lisbon, Portugal
Vulpetti, G.	(Italian)	Dept. of Space Technology, Telespazio S.P.A. per le Telecommunicazioni Spaziale, Corso d'Italia 43, 00198 Rome, Italy
Waldvogel, J.	(Swiss)	Sem. f. Angew. Math., ETH-Zentrum, CH-8092, Zürich, Switzerland
Wisdom, J.	(U.S.A.)	California Institute of Technology, Pasadena, California 91125, U.S.A.
Yi-Sui, S.	(Chinese)	Dept. of Astronomy, University of Nanking, Nanking, China
Yoshida, H.	(Japanese)	Dept. of Astronomy, University of Tokyo, Tokyo 113, Japan

Group Photograph of Participants of NATO Advanced Study Institute held in Cortina d'Ampezzo, August 2-14, 1981

View of Cortina d'Ampezzo with Circles denoting
the Antonelli Institute

THE DYNAMICS OF CLOSE PLANETARY SATELLITES AND RINGS

K. Aksnes

Mathematics Section, Norwegian Defense Research Establishment, Box 25, 2007 Kjeller and University of Tromsø, Norway

ABSTRACT. The observational history and the evolving dynamical theories are reviewed for the rings of Saturn, Uranus, and Jupiter with particular emphasis on very recent results from the Voyager space missions. An account is then given of three Jovian satellites and eight Saturnian ones discovered from ground or space in 1979 and 1980. The coupled motions of some of these satellites and their possible interactions with the rings are explained in some detail. The paper concludes that the ring and satellite dynamics is as yet poorly understood and poses challenging problems to observers and theoreticians alike.

1. INTRODUCTION.

At the time when this is written the American spacecraft Voyager 2 is only a few weeks away from encounter with Saturn on 24 August 1981; it follows in the tracks of its sistership Voyager 1, which last November televised to the Earth an amazing series of close-up photographs of Saturn and several satellites - and especially of the rings. In March and July 1979 the two spacecraft had uncovered an incredible amount of detail in the Jovian system, highlighted by the discoveries of Jupiter's ring and Io's volcanism.

The Voyager and earlier Pioneer findings have started an entirely new chapter in the history of exploration of the outer planets and their satellite and ring systems: a history that began in 1610 when Galileo Galilei discovered the four big moons of Jupiter with his first crude object glass - of which much perfected models are now carried on board the Voyager spacecraft.

V. Szebehely (ed.), Applications of Modern Dynamics to Celestial Mechanics and Astrodynamics, 1–20.

Galileo's contribution was not so much the satellite discovery itself (whose priority has even been contested by Simon Marius) as his immediate realization of its great importance to cosmology and philosophy. He had for a long time been a firm believer in Copernicus' beautifully simple but unproved picture of the world; now the closed Jovicentric orbits that he charted for his newly found satellites proved that not all bodies are revolving about the Earth.

Galileo can be said to have started modern dynamics as based on both theory and observation. In the following, we shall attempt to trace the further development of dynamics as it applies to planetary rings and close satellites, ending with a description of the Voyager observations, many of which are as yet poorly understood.

2. THE RINGS OF SATURN.

A few months after discovering the four big satellites of Jupiter, Galileo saw the first traces of Saturn's rings, which he at first thought were two close satellites on opposite sides of the planet. He was therefore very surprised to find on subsequent days, that the two "satellites" never left Saturn's side but stuck to it like two ears. His surprise rose to perplexity two years later, at the end of 1612, when he found that the "ears" had disappeared altogether. This prompted Galileo to ponder in a letter of 4 December 1612 if Saturn had perhaps devoured his own children! Today we know that the rings became edgewise to the Sun on 28 December that year and were therefore invisible in Galileo's telescope.

In his Systema Saturnium, Huygens in 1659 offered the first reasonably correct interpretation. He perceived the observed appendages of the planet as a fixed, compact ring whose changing aspects, as viewed from the Earth in reflected sunlight, are consistent with Saturn's 29-year orbital period and an inclination of ~26° between the ring plane and Saturn's orbit.

In 1675 J. D. Cassini discovered the gap named after him between what are now termed the A-ring and the B-ring. His son, Jacques, in 1715 announced that the rings are likely to consist of swarms of small particles; but he was too far ahead of his time and his idea did not gain general acceptance until Maxwell in 1857 proved that Cassini was, indeed, right.

The philosopher Kant suggested, in his work on Nature and Theory of the Universe (1755), a nebular theory of the origin of the solar system. He applied the theory to the Saturnian system as well, regarding the rings as leftover material that did not succeed in condensing into satellites.

In his Exposition du systeme du Monde (1796), Laplace extended Kant's qualitative ring theory and redressed it in mathematics. Already in 1785, Laplace deduced that a compact ring would break up due to tidal forces from Saturn. However, he did not reject the idea of a solid ring, he merely subdivided it into several narrower rings, thinking that this would relieve tidal tensions to a sufficient degree.

Then another French mathematician, Roche, published in 1848 (in Mem. de L'Acad de Sciences -- de Montpellier) his celebrated paper in which he showed that a fluid satellite would break up within 2.44 Saturn radii (R). He suggested that the rings originated from a fluid satellite stretched by tidal tension between the limits 1.15R and 2.26R.

It is easy to establish, through elementary gravity considerations, Roche type of instability criteria for (a) the formation of a satellite, and (b) retention of loose material on its surface (Figure 1):

(a) The forces on the two spheres balance if

$$\frac{Gm}{(2r)^2} = GM[(a-r)^{-2} - (a+r)^{-2}] \approx \frac{GM}{a^2}\frac{4r}{a} \ .$$

Substituting $\frac{M}{m} = \frac{\rho}{\rho'}\frac{R^3}{r^3}$,

we conclude that a satellite will be able to form if

$$\frac{a}{r} > (16\frac{\rho}{\rho'})^{1/3} = 2.52(\frac{\rho}{\rho'})^{1/3} \ .$$

(b) Assume that the satellite rotates synchronously with an angular velocity equal to its mean motion n . The forces on the particle balance if

$$\frac{Gm}{r^2} + n^2(a-r) = \frac{GM}{(a-r)^2} \approx \frac{GM}{a^2}(1+2\frac{r}{a}) \ .$$

Substituting for M/m from above and for n^2 from Kepler's equation,

$$n^2a^3 = GM \ ,$$

we find that the satellite will retain surface material if

$$\frac{a}{r} > (3\frac{\rho}{\rho'})^{1/3} = 1.44(\frac{\rho}{\rho'})^{1/3} \quad .$$

Maxwell (1859) in his treatise on the stability of the motions of Saturn's rings, which had won him the Adams Prize for the year 1856, proved mathematically that "the only system of rings which can exist is one composed of an indefinite number of unconnected particles, revolving round the planet with different velocities according to their respective distances." According to his theory, a set of p particles, each of mass m moving in a circle would be stable only if

$$mp^3 < 2.3\ M \ .$$

With spacings equal to k times the diameter $2r$ of the particles, which are assumed to have the same mean density as the planet, we have

$$\frac{m}{M}p^3 = (\frac{r}{R})^3 \ (\frac{2\pi a}{2kr})^3 = (\frac{\pi}{k}\frac{a}{R})^3 < 2.3 \quad .$$

This means that stability is possible only if $k > 1.4\ a/R$, a result which is independent of the particle size. Since a/R is of the order of 2 for Saturn's rings, stability requires that the spacings between the particles is at least 3 times their diameters.

Kirkwood (1867) noted a similarity between the gaps in the distribution of the asteroids and in Saturn's rings (Figure 2). He assumed that near-commensurabilities in the motions of the asteroids and ring particles with respectively Jupiter and the inner Saturnian satellites, were somehow giving rise to resonance effects which cleaned out the gaps. There are obvious problems with his theory; e.g., why do the asteroids tend to avoid the 1:4, 1:3, 2:5, 3:7, and 1:2 resonances while there is clustering around the 1:1 (Trojans), 3:4, and 2:3 resonances? For Saturn's rings Kirkwood considered only the Cassini (1:2) and Encke (3:5) divisions which he thought mainly Mimas responsible for.

Keeler (1895) proved in a very elegant and powerful way, through spectroscopic observations of the Doppler effect that the rings must consist of free particles in near-circular Kepler orbits about Saturn.

Jeffreys (1947) pointed out the effect of cohesion on Roche's limit and showed that the finite rigidity of a solid satellite would keep it together far inside the Roche limit provided the satellite were sufficiently small (<~200 km). His second significant contribution was to draw attention to the importance of inelastic collisions in the evolution of the rings and the

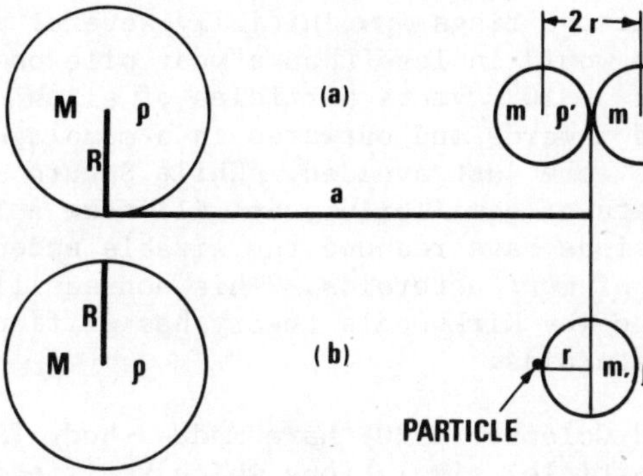

Figure 1. (a) Two small, identical spheres of radius r , mass m , and density ρ' revolve in contact in a circular orbit of radius a about a planet of radius R , mass M , and average density ρ . In (b) one of the small spheres is replaced by a test particle.

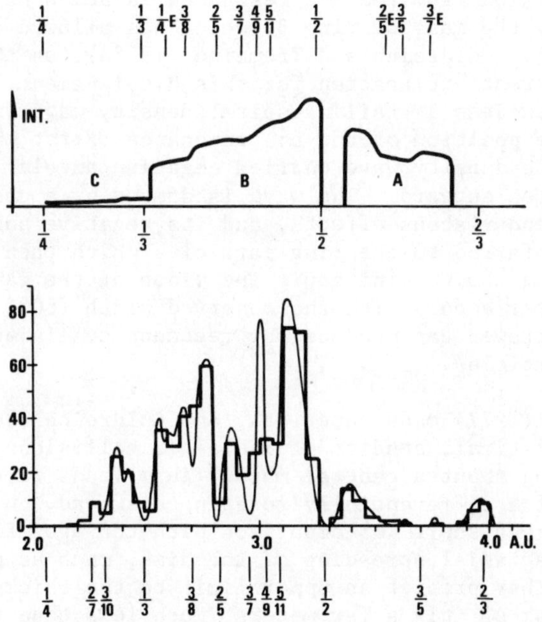

Figure 2. Photometric tracing (top diagram) across Saturn's A, B, and C rings on an arbitrary density scale. The ratios of numbers mark the resonant locations relative to Mimas or Enceladus (E). The corresponding resonant locations of asteroids relative to Jupiter are shown on the same scale below in a histogram of the asteroidal distribution (after Kuiper 1974).

asteroid belt. If the rings were initially several particles thick, collisions would in less than a year pile one particle upon another. In $\leq 10^6$ years particles of sizes ≥ 1 cm would have spread inwards and outwards in a monolayer until internal collisions were just avoided. While Saturn's rings are probably in a state of equilibrium, it will take a long time yet before collisions have reduced the sizable eccentricities and inclinations of most asteroids. This non-equilibrium may be part of the reason why Kirkwood's theory has difficulty especially with the asteroids.

Franklin and Colombo (1970) have made 3-body (Saturn-satellite-ring particle) simulations which verified the role of collisions envisioned by Jeffreys and reproduced the main features in Dollfus' (1968) photometric tracings of the rings. With Mimas in a circular orbit ($e_M = 0$) only the ring gaps around the 1:2 and 2:3 commensurabilities are produced. For $e_M = 0.02$, additional, less prominant dips appear at the locations of the 1:3 and 3:5 resonances. The slight inward displacement of the Cassini division from the 1:2 resonance location is thought to be caused by the mass of ring B, which has allowed an estimate of that mass. Goldreich and Tremaine (1978a), on the other hand, have a different explanation for this displacement. They show that Mimas excites a trailing spiral density wave in Saturn's rings at the position of the 1:2 resonance (first Lindblad resonance). The density wave carries negative angular momentum and propagates outward. The wave is damped by a combination of non-linear and viscous effects, and its negative angular momentum is transferred to the ring particles which then spiral inward, opening the Cassini gap. The width of the gap produced is also in better accord with the observed width (6000 km) than is the much narrower gap produced by resonant collisions among individual particles.

Brahic (1977) made numerical, and Goldreich and Tremaine (1978b) analytical, studies of inelastic collisions between particles moving about a central mass. This leads to the rapid formation of a differentially rotating disk and, on a much longer time scale, yet short in comparison with the age of the solar system, to a radial spreading of the disk, much as predicted by Jeffreys. They predict an upper limit to the thickness of Saturn's disk of only a few meters, much less than the observed ~1 km which the authors believe is in error. Goldreich finds a maximum spreading time of 5×10^9 years, which is made much shorter by erosion and accretion if the particles consist of water ice. Modern ground-based optical and radio observations of Saturn's rings suggest particle sizes ≥ 6 cm.

Burns et al. (1979) have shown that periodic perturbations of the ring particles normal to the ring plane may amount to 0.1 m, 0.3 m, and 2 m due to Mimas, Tethys, and Titan, respectively. These perturbations may aid the erosion process.

Voyager 1's flyby of Saturn in November 1980 has added a wealth of new information about the rings. The highest ring resolution achieved was about 1000 km, which is comparable with the very best resolution obtained in telescopes on Earth during short glimpses of exceptionally good seeing. Of equal or greater importance than the high resolution was the fact that the Voyager cameras viewed the rings from many different angles (aspects) as the spacecraft moved from above the sunlit northern side of the rings to below the dark side, and back up again, while the solar phase angle (sun-ring-camera angle) ranged over nearly 180°. These angles are, of course, severely restricted for a Saturn observer on the Earth.

The photograph on Figure 3 shows a perplexingly large number, perhaps a thousand or more, of individual ringlets that run together without any clear-cut gaps, apart from the dark, broad Cassini division and the much narrower Encke division near the outer edge of ring A. But even these divisions are far from void of material, and it is now quite clear that they are not centered on the 1:2 and 3:5 resonances with Mimas, in accordance with the Kirkwood theory. The very narrow outermost F-ring appears to be held in place by two small shepherding satellites photographed by Voyager 1 but not visible in this picture. It is tempting to speculate that the sharp outer cutoff of the A-ring is caused by yet another satellite discovered by Voyager 1 just outside the ring. As for the ringlets, it appears unlikely that the known satellites outside the ring system are the controllers, although such roles could conceivably be played by a large number of small, imbedded, and as yet undiscovered satellites. Such objects might be resolvable with Voyager 2's photopolarimeter (that instrument failed on Voyager 1).

The strangely twisted appearance of the F-ring when observed at close range and a semi-permanent spoke pattern in the B-ring seem to defy gravitational explanations, strongly suggesting that electromagnetic forces are also at play. Thus, the spokes are most prominent at the distance from Saturn where a particle would be corotating with the planet's strong magnetic field. This could generate electrostatic forces lifting very light particles out of the ring plane.

For a more detailed description of the Voyager ring pictures, reference is made to Collins et al (1980).

Figure 3. Saturn's rings photographed by Voyager 1 on 13 November 1980 at a distance of 1.5 million km.

3. THE RINGS OF URANUS AND JUPITER.

The rings of Saturn held a unique distinction until the rings of Uranus were discovered accidentally on 10 March, 1977. An occultation of an 8th magnitude star had been predicted to occur on that date. Everybody was taken by complete surprise when Elliott et al. (1977) and Millis et al. (1977), observing respectively from the Kuiper Airborne Observatory southwest of Perth and from Perth, announced that the light from the star flickered on and off for brief periods several times before and after the occultation by Uranus itself. This very simple occultation technique not only revealed that Uranus has rings, but also pinpointed their locations to around 50 km - a far greater accuracy than ever achieved for Saturn's rings.

Figure 4 shows the original tracings of the pre-immersion and post-emersion portions of the occultation of the star (SAO 158687) by Uranus on 10 March, 1977, as recorded by Elliot et al. (1977) in the Kuiper Airborne Observatory.

They labeled the most pronounced dips in the tracings α, β, γ, δ, and ε and comment: "The intensity scale has been normalized to the unocculted intensity of the star, and on this scale the amount of starlight occulted ranges from 0.4 for the β and γ events to about 0.9 for the ε event. An approximate distance scale has been computed under the assumption that the bodies causing the occultations lie in the orbit plane of the Uranian satellites. The equal number and symmetry of the occultations suggest that four narrow rings (~10 km wide) and a broader ring of variable thickness (Ring ε, 50-100 km) encircle Uranus. Beginning at 21:52 UT, the commencement of twilight caused a slow rise in the signal level."

Millis et al. (1977) confirmed the existence of the ε, γ, and β ring and added two more rings, labelled 4 and 5, on the inside. Their tracing of the ε ring shows clearly an internal structure in the ring (Figure 5). Careful reexamination of both sets of observations (Elliot et al., 1978) revealed yet two more rings, an η ring between the β and γ rings, and ring 6 closest to Uranus. The geometry of these altogether nine rings and the star's occultation paths for the various participating observation sites are shown on Figure 6. Later ring occultations of fainter stars on 23 December 1977, 10 April 1978, 10 June 1979, 20 March 1979, and 15 August 1980 have confirmed the existence of the nine rings without giving certain evidence of any new ones.

Elliot et al. (1981) have analyzed jointly nearly all the available occultation observations for the above dates. They adopt a kinematic model in which all nine rings are considered to

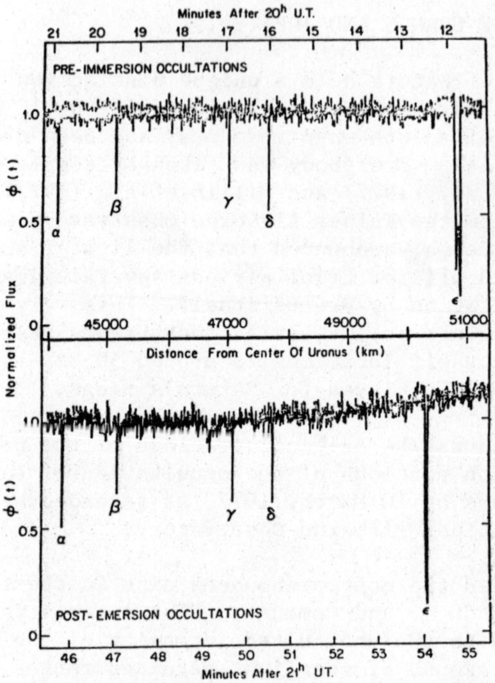

Figure 4. Photometric Tracing across the Uranian rings on 10 March 1977 by Elliot et al. (1977).

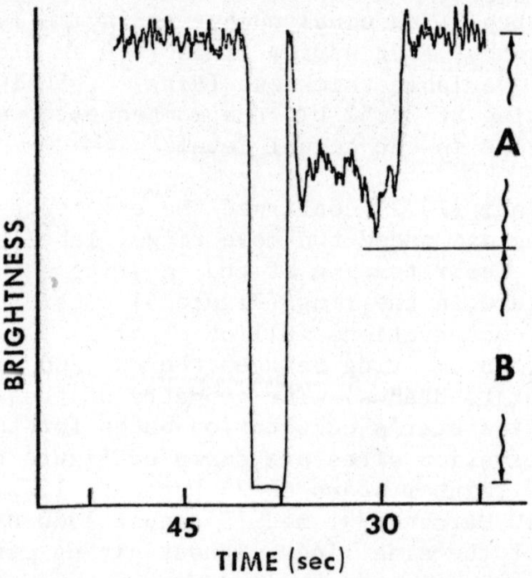

Figure 5. Photometric tracing across the ε ring by Millis et al. (1977). A and B represent the contribution of light from the star and Uranus, respectively. (The very low dip around 40 sec represents the dark current during centering.)

be coplanar ellipses of zero inclination, centered on the center of Uranus and precessing owing to the zonal harmonics in the planet's gravity potential. In all, 105 data points were fit by least squares to a model with 41 free parameters whose resulting values and standard errors are summarized in Table I. The derived ring radii have formal errors of only 33 km. The η, γ, and δ rings show no appreciable departures from circularity, while significant eccentricities are obtained for the other rings among which the ε ring stands out as relatively strongly elliptical ($e = 0.008$) and with a variable width (observed cross-sections between 20 and 100 km). It is still not ruled out, though, that inclined circular models could do as well for rings 4, α, and β.

It was first pointed out by G. Colombo in 1977 that the simplest plausible model for the rings consists of an aggregation of free particles in elliptic orbits with a common apsidal line which precesses in the plane of the satellites, due to Uranus' oblateness, at the rate

$$\dot{\tilde{\omega}} = \left(\frac{GM}{a^3}\right)^{1/2} \left[\frac{3}{2}\frac{J_2}{a^2}\frac{R^2}{(1-e^2)^2} - \frac{15}{4}\frac{J_4}{a^4}\frac{R^4}{(1-e^2)^4}\right],$$

given by the ring's mean elements a and e and the mass M, equatorial radius R, and the second and fourth zonal harmonics J_2 and J_4 of Uranus. The importance of this relation is that it affords a very accurate determination of Uranus' dynamical oblateness from the observed precession rates of the rings:

$$J_2 = (3.352 \pm 0.006) \times 10^{-3}, \quad J_4 = (-2.9 \pm 1.3) \times 10^{-5}.$$

We note that the circular, coplanar orbits of the satellites give no information about these harmonics.

Nicholson et al. (1978, 1981) have demonstrated that the width of the ε ring at a given longitude is very nearly proportional to the mean ring radius r at that longitude. To the first order in e

$$W = 2\Delta a + \frac{2\Delta e}{e}(r - a),$$

where $\Delta a = 30$ km and $\Delta e = 3.7 \times 10^{-4}$. The ring can be considered bounded by two ellipses with elements $a \pm \Delta a$ and $e \pm \Delta e$.

A first attempt at a dynamical theory for the Uranian rings was made by Dermott and Gold (1977). They explained the locations

Table I. Fitted model parameters.[a]

(a) Orbital elements

Ring	Semimajor axis, a (km)	Eccentricity ($e \times 10^3$)	Azimuth of periapse, ω_0 (deg)[b]	Precession rate from fitted J_2 and J_4 (deg/day)	Precession rate fitted individually (deg/day)
6	41 863.8 ± 32.6	1.36 ± 0.07	235.9 ± 2.9	2.7600	2.7706 ± 0.0034
5	42 270.3 ± 32.6	1.77 ± 0.06	181.8 ± 2.5	2.6678	2.6614 ± 0.0030
4	42 598.3 ± 32.7	1.24 ± 0.09	120.1 ± 2.7	2.5963	2.5957 ± 0.0031
α	44 750.5 ± 32.8	0.72 ± 0.03	331.4 ± 2.8	2.1832	2.1785 ± 0.0043
β	45 693.8 ± 32.8	0.45 ± 0.03	231.3 ± 4.0	2.0288	2.0272 ± 0.0064
η	47 207.1 ± 32.9	(0.03 ± 0.04)	(291.7 ± 88.3)	1.8094	—
γ	47 655.4 ± 32.9	(0.04 ± 0.04)	(301.6 ± 49.1)	1.7503	—
δ	48 332.0 ± 33.0	0.054 ± 0.035	139.0 ± 30.2	1.6657	—
ϵ	51 179.7 ± 33.8	7.92 ± 0.04	215.6 ± 0.5	1.3625	1.3625 ± 0.0004

(b) Harmonic coefficients of the gravity potential[c]

$J_2 = (3.352 \pm 0.006) \times 10^{-3}$

$J_4 = (-2.9 \pm 1.3) \times 10^{-5}$

(c) Pole of the ring plane

$\alpha_{1950} = 5^h06^m26^s.1 \pm 10^s.7$

$\delta_{1950} = +15°13'15'' \pm 3'13''$

(d) Relative coordinate corrections

Occultation date	Correction to relative right ascension, $D(\Delta\alpha \cos\delta)$ (km)	Correction to relative declination, $D(\Delta\delta)$ (km)
10 March 1977	56.8 ± 14.3	−17 467.4 ± 28.7
10 April 1978	−11 002.8 ± 16.7	−2 024.6 ± 30.8
10 June 1979	−7 135.3 ± 20.2	3 116.5 ± 95.0
20 March 1980	−4 042.2 ± 15.7	−5 376.0 ± 33.7
15 August 1980	−4 686.2 ± 15.7	−1 932.4 ± 64.3

[a] For $M_u = 8.669 \times 10^{28}$ g and $G = 6.670 \times 10^{-8}$ dyn cm^2 g^{-2}.
[b] At 20:00 UT on 10 March 1977.
[c] For a reference radius R = 26 200 km.

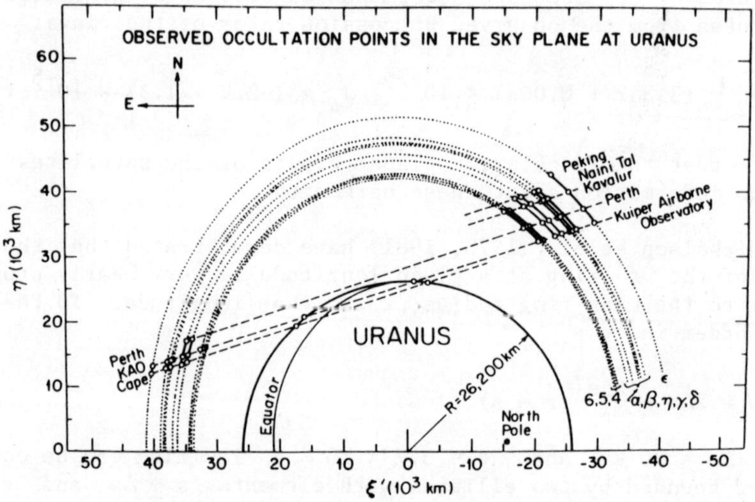

Figure 6. Geometry of Uranian ring crossings as observed from different sites (after Elliot et al. 1978).

of Uranus' rings in terms of three-body resonances between ring particles and pairs of satellites, such that each particle librates about consecutive conjunctions of the satellites. The libration argument is given by

$$\Theta = \lambda_1 - (p + 1)\lambda_2 + p\lambda_3 ,$$

where $\lambda_i = n_i t + \text{constant}$, $i = 1, 2, 3$, are the mean longitudes of the ring particle and the satellites, and $p \equiv (n_1 - n_2)/(n_2 - n_3)$ is an integer. This equation predicts a striking agreement with the locations of rings 5 , δ , γ , and ε for several satellite pairs. The authors concluded that the observed ring pattern was primarily due to the Ariel-Titania and Ariel-Oberon pairs on the basis of the sizes of their mass products, $m_2 m_3$ (the ring particle's mass m_1 is assumed negligible). However, Aksnes (1977) showed that the strength c of the resonance as it enters into the pendulum equation,

$$\frac{d^2\Theta}{dt^2} = c \sin \Theta ,$$

is given approximately as

$$c \propto m_2 m_3 \left(\frac{a_2}{a_3}\right)^p ,$$

and the factor $(a_2/a_3)^p$ makes the Miranda-Ariel pair, with $p = 7, 8, 9, 10$ dominate over all others despite Miranda's tiny mass. Goldreich and Nicholson (1977) reached the same conclusion and also showed that the 4:1 two-body resonance between Miranda and a ring particle is even slightly stronger. In both cases, the resonances seem far too weak to play significant roles. The Dermott-Gold theory is further weakened by the fact that it cannot explain the locations of the last discovered four rings.

In an article entitled "Towards a Theory for the Uranian Rings", Goldreich and Tremaine (1979) show that interparticle collisions, radiation drag (Poyinting - Robertson effect) and differential precession all tend to disrupt the rings of Uranus. The first two effects lead to radial spreading which would disrupt a free ring in $\leq 10^8$ years.

They propose that the rings are confined in radius by gravitational torques from a series of small satellites that orbit within the ring system. Differential precession tends to destroy the apse alignment of the elliptical ring, but selfgravity may restore the alignment. To confine the ε ring, whose mass is $\sim 5 \times 10^{18}$ g, a pair of satellites of roughly comparable mass, in circular orbits around 500 km inside and outside the boundaries of the ring, are required.

In a modification of the confining satellite theory for the Uranian rings, Dermott, Gold, and Sinclair (1979) propose that each ring contains a small satellite which maintains particles in horseshoe orbits about the Lagrangian equilibrium points L_3, L_4, and L_5 (Figure 7). They consider it significant that of all the outer planets, only for Uranus is the radius of the synchronous orbit

$$a_{syn} = \left(\frac{T^2 GM}{4\pi^2}\right)^{1/3},$$

where T is the satellite's rotation period, considerably greater than the earlier derived Roche limits for satellite formation and retention of surface material,

$$a_L(\text{hydrost.}) = 2.52\left(\frac{\rho}{\rho'}\right)^{1/3}, \quad a_L(\text{sphere}) = 1.44\left(\frac{\rho}{\rho'}\right)^{1/3}.$$

For satellite densities ρ' in the range 1-2 g cm^{-3} the Uranian rings fall in the "Roche-zone" between these two limits. If originally near the hydrostatic limit, the satellite would begin to spiral inward due to tidal drag and at the same time shed surface particles which are believed to go into the horseshoe orbits and form sharply delineated rings.

On 5 March, 1979, the spacecraft Voyager 1 discovered Jupiter's ring - again by chance, since if Voyager's TV camera had not happened to be operating and pointed to the right place at the very instant when Voyager 1 was crossing the ring plane (Jupiter's equator), Jupiter's ring should probably still remain undetected.

Jupiter's ring is located inside about 1.8 R_J with a width of ~6000 km and thickness of ~30 km. The outer edge is quite sharp while the inner edge is fuzzy and may in fact extend all the way to Jupiter. The ring is surrounded by a faint halo of very small particles (0.5 μm).

Burns et al. (1980) believe that the particles making up the Jovian ring may be debris which has been excavated by micrometeoroids from the surfaces of many unseen ($R \leq 1$ km) moons residing in the ring. A distribution of particle sizes exists; large objects are sources for small visible ring particles and also account for the absorption of charged particles noted by Pioneer 10.

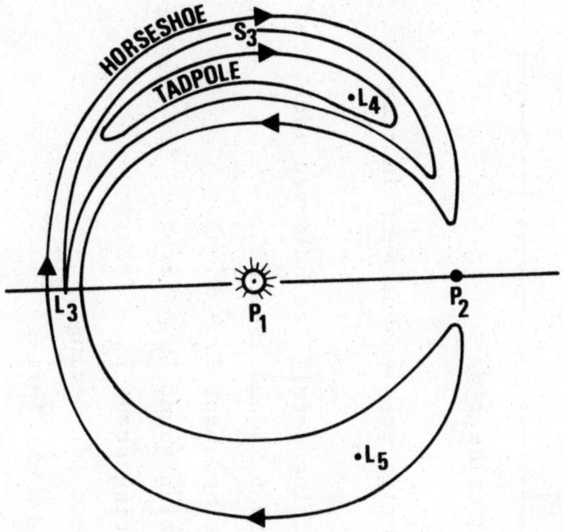

Figure 7. Horseshoe and tadpole orbits of the massless third body in the restricted three-body problem (in a frame rotating with the second body P_2 which revolves in a circular orbit about P_1). After Garfinkel (1977).

It seems obvious that the sharp outer edge of the ring is controlled by the two small satellites 1979 J1 and 1979 J3, discovered on Voyager images at mean distances of 1.80 R_J and 1.79 R_J , the latter being perhaps also responsible for an observed brightness enhancement at that location in the ring. Voyager 1 also discovered a third satellite 1979 J2 at $a = 3.12\ R_J$.

Grün et al. (1980) think that Io's volcanic activity may be a source of the ring particles.

4. NEW SATELLITE DISCOVERIES.

Table 2 includes three satellites of Jupiter and eight of Saturn, which have been discovered (or rediscovered) in 1979 and 1980, either by the Voyager spacecraft or from ground. A temporary satellite nomenclature is used consisting of the year of discovery, the first letter in the name of the primary, and a number for the order of the discovery within the year. Most of the intervening numbers not given in the table belong to the "parent" satellites listed there.

Table 2 Satellite Discoveries in 1979 and 1980*

Name	Diam. (km)	a(km)	a/R	e°	i°	Comments
1979 J1	~ 25	129,000	1.80	0	0	Just outside ring
1979 J2	~ 75	223,000	3.12	0	0	
1979 J3	~ 40	128,000	1.79	0	0	Near brightness enhancement in ring
1980 S1	100×90	151,472	2.51	0.007	0.14	Coorbiting satellites believed identical
1980 S3	90×40	151,422	2.51	0.009	0.34	with 1966 S1 and 1966 S2
1980 S6	~ 160	378,000	6.26	0.005	0.15	Librating about L_4 - Dione
1980 S13	~ 35	294,000	4.88	0	0	Librating about L_4 - Tethys
1980 S25	~ 35	294,000	4.88	0	0	Librating about L_5 - Tethys
1980 S26	~ 200	141,700	2.35	0.004	0.05	Just outside F-ring
1980 S27	~ 220	139,400	2.31	0.003	0.00	Just inside F-ring
1980 S28	40×20	137,700	2.28	0.002	0.30	Just outside A-ring

*The quoted values for the eccentricities (e) and inclinations (i) and some of the diameters are uncertain.

We have already briefly described the three new Jovian satellites (Jewitt et al. 1979, Synnott 1980, 1981). This brings the total number of satellites of Jupiter up to 16, while there are at least 17 satellites of Saturn. Where to draw the distinction between a ring particle and a satellite, and especially how to keep track of the individual chunks that are likely to be seen by Voyager 2, are difficult questions that now face us.

The history of the two coorbiting satellites 1980 S1 and 1980 S3 goes back to 1966 - the previous time when Saturn's rings were seen edge-on from the Earth. One of these two satellites in near-identical orbits are likely to be identical with the satellite Janus discovered by Dollfus (1967) on 10 photographic plates taken on October 29 and December 15, 16, and 17, 1966. Ten years later these plates and some additional ones were reexamined by Fountain and Larson (1977). They verified Dollfus' calculations for Janus but also found an additional satellite about 8000 km inside Janus' orbit - very near the mean distances from Saturn which 1980 S1 and 1980 S3 were subsequently found to have. Aksnes and Franklin (1978) demonstrated that, because of poor spatial and temporal distribution of the observations, both of the two derived orbits lacked uniqueness and that the region just outside ring A might well contain one, two, or even many satellites.

It had been hoped that the Voyager photographs would remove the uncertainty surrounding the two satellites. While Voyager 1 has proved beyond doubt the existence of the pair of coorbiting satellites, their orbits are still not known with sufficient accuracy to permit linking the 1980 observations, both from Voyager 1 and from Earth, uniquely to the observations in 1966 or even to enable differentiating between the two objects on future photographs. In addition to the problem of poor observational distribution, the situation is further complicated by the exceedingly complex motions that the satellite pair exhibits. Numerical integrations by Harrington and Seidelmann (1981) have shown that the two satellites appear to move in quasistable orbits resembling the family of tadpole/horseshoe periodic orbits (Figure 7) seen in the restricted three-body problem and investigated in detail by Garfinkel (1977, 1978, 1980). This mechanism may prevent the satellites, whose mean orbital radii differ by only 50 km, from ever colliding; the difference in longitude between the satellites appears to librate between approximately 6° and 354° in 3000 days, but these results depend on the masses adopted for the satellites.

The satellites 1980 S6, 1980 S13, and 1980 S25 were discovered in groundbased images when the obstructing light from the nearly edge-on rings was at a minimum. As reported by Lecachaux et al. (1980) and Reitsema et al. (1980), the orbit of 1980 S6 librates about the L_4 Lagrangian point of Dione with a period of about 800 days. The satellites 1980 S13 and 1980 S25 share a common orbit with Tethys, librating respectively about the L_4 and L_5 points (Marsden 1981, Seidelmann et al. 1981).

The three aforementioned Voyager 1 satellite discoveries 1980 S26–28 represent respectively the two shepherding satellites just outside and inside the F-ring and the satellite just outside the A-ring.

5. LOOKING AHEAD.

The last few years have produced an almost unbelievably rich harvest of radically new observations, both from ground and space, of planetary rings and satellites. The interpretation and follow-up of these observations are presenting real challenges to observers and theoreticians alike.

On the observational side, better techniques and instruments need to be devised for use both on Earth and on space platforms to enable picking up finer details in the rings and very faint satellites close to or even within the rings. The use of CCD cameras with big telescopes on Earth has already produced very good results and the Large Space Telescope scheduled for launch in the mid-1980's, should enable very high resolution photography of the ring and satellite systems over extended intervals.

As for the dynamics of the rings and closest satellites of Jupiter, Saturn, and Uranus and the roles played by gravitational interactions, one must admit that the understanding is as yet very incomplete. One suspects that it may be necessary to bring hydrodynamics, electrodynamics, and plasma physics to bear on the ring problem in addition to gravitational theory. We have seen several examples of what may be the first known manifestations in nature of horseshoe type of orbits - despite their proven instability in the idealized "restricted" case. This ought to spur on continued work on the real-life problem of perturbed horseshoe/tadpole orbits.

REFERENCES.

Aksnes, K.: 1977, Nature, 269, 783.

Aksnes, K., and Franklin, F. A.: 1978, Icarus, 36, 107.

Brahic, A.: 1977, Astron. & Astrophys., 54, 895.

Burns, J. A., Hamill, P., Cuzzi, J. N. and Durisen, R. H.: 1979, Astron. J., 84, 1783.

Burns, J. A., Showalter, M. R., Cuzzi, J. N. and Pollack, J. B.: 1980, Icarus, 44, 339.

Collins, S. A., Cook II, A. F., Cuzzi, J. N., Danielson, G. E., Hunt, G. E., Johnson, T. V., Morrison, D., Owen, T., Pollack, J. B., Smith, B. A. and Terrile, R. J.: 1980, Nature, 288, 439.

Dermott, S. F. and Gold, T.: 1977, Nature, 267, 590.

Dermott, S. F., Gold, T. and Sinclair, A. T.: 1979, Astron. J., 84, 1225.

Dollfus, A.: 1967, Comptes-Rendus Acad. Sc. Paris, Series B, 264, 822.

Dollfus, A.: 1968, l'Astronomie, 82, 253.

Elliot, J. L., Dunham, E. and Mink, D.: 1977, Nature, 267, 328.

Elliot, J. L., Dunham, E., Wasserman, L. H., Millis, R. L. and Churms, J.: 1978, Astron. J., 83, 980.

Elliot, J. L., French, R. G., Frogel, J. A., Elias, J. H., Mink, D. J. and Liller, W.: Astron. J., 86, 444.

Fountain, J. W. and Larson, S. M.: 1977, Science, 197, 915.

Franklin, F. A. and Colombo, G.: 1970, Icarus, 12, 338.

Garfinkel, B.: 1977, Astron. J., 82, 368.

Garfinkel, B.: 1978, Celes. Mech., 18, 259.

Garfinkel, B.: 1980, Celes. Mech., 22, 267.

Goldreich, P. and Nicholson, P.: 1977, Nature, 269, 783.

Goldreich, P. and Tremaine, S.: 1978a, Icarus, 34, 240.

Goldreich, P. and Tremaine, S.: 1978b, Icarus, 34, 227.

Goldreich, P. and Tremaine, S.: 1979, Nature, 277, 97.

Grün, E., Morfill, G., Schwehm, G. and Johnson, T. V.: 1980, Icarus, 44, 326.

Harrington, R. S. and Seidelmann, P. K.: 1981, Bull. Am. Astron. Soc., 13, 573.

Jeffreys, H.: 1947, Mon. Not. Roy. Astron. Soc., 107, 260.

Jewitt, D. C., Danielson, G. E. and Synnott, S. P.: 1979, Science, 206, 951.

Keeler, E.: 1859, Astrophys. J., 1, 416.

Kirkwood, D.: 1867, Meteoric Astronomy, J. B. Lippencott, Philadelphia.

Kuiper, G. P.: 1974, Celes. Mech., 9, 321.

Lecachaux, J., Laques, P., Vapillon, L., Auge, A. and Despiau, R.: 1980, Icarus, 43, 111.

Marsden, B. G.: 1981, IAU Circular No. 3605.

Maxwell, J. C.: 1859, On the Stability of the Motions of Saturn's Rings (Macmillan and Company, Cambridge and London).

Millis, R. L., Wasserman, L. H. and Birch, P. V.: 1977, Nature, 267, 330.

Nicholson, P. D., Persson, S. E., Matthews, K., Goldreich, P. and Neugebauer, G.: 1978, Astron. J., 83, 1240.

Nicholson, P. D., Matthews, K. and Goldreich, P.: 1981, Astron. J., 86, 596.

Reitsema, H. J., Smith, B. A. and Larson, S. M.: 1980, Icarus, 43, 116.

Seidelmann, P. K., Harrington, R. S. and Pascu, D.: 1981, Astron. J., in press.

Synnott, S. P.: 1980, IAU Circular No. 3507.

Synnott, S. P.: 1981, Science, 212, 1392.

THE MOTION OF SATURN'S CO-ORBITING SATELLITES 1980S1 AND 1980S3

G. Colombo

University of Padova, Italy, and the Center for Astrophysics, Cambridge, Massachusetts

ABSTRACT. The Trojan asteroids located at Jupiter's L_4 and L_5 Lagrangian points, 60° ahead and behind Jupiter, respectively, are members of a large family of objects which are locked in 1:1 orbital resonance with the planet. These objects, in the average, move about the Sun at the same rate as Jupiter. Until recently the Trojans were representing the only case of orbital locking in such a resonance. The Voyager 1 and 2 flybys of Saturn revealed five more small objects which belong to this type of orbital resonance; one which librates about Dione's L_4 point, one each about Tethis' L_4 and L_5 points, and finally, the two co-orbiting pair 1980S1 and 1980S3.

The last two are the most unusual Trojan-like objects, with mean orbital radius of 151×10^3 km and angular motion of 518°.3/day. They have, unlike other 1:1 librators, comparable masses ($M_{S3} = 1/4 \sim 1/5\ M_{S1}$, Smith, et al., 1981). Both masses are, however, very small compared with Saturn's mass M, the ratio being $M_{S1}/M \cong 5 \times 10^{-9}$. The most unusual feature of this motion is the large amplitude of the librations of both satellites with respect to the mean orbit.

An available code (Icarus) for studying the dynamics of three-body systems, Saturn, S1 and S3, was used. We began with the simple circular and planar case, neglecting the observed eccentricity and inclination of S1 and S3, but starting from the observed semi-major axis and phase difference. The equations of motion were integrated with initial conditions corresponding to the observed orbital elements (Synnot, et al., 1981). The two satellites can approach to within 15,000 km but are prevented from passing each other by their mutual gravitational interaction.

V. Szebehely (ed.), Applications of Modern Dynamics to Celestial Mechanics and Astrodynamics, 21–23.

With respect to a frame rotating with the mean-mean motion while S1 is executing an almost harmonic oscillation of 60° amplitude, S3 is executing a horseshoe orbit of 285° amplitude. The maximum relative velocity occurs when the two satellites are 60° apart.

In both cases, the variations of orbital elements were so small and the results so similar that it was expected that a simple analytical theory would give good results. In fact, simple analytical approximations to the motion of all Trojan-like asteroids may be derived as long as their orbits are nearly circular and coplanar (C.F. Yoder, 1981). The theory is a direct extension of the textbook solution of Brown and Shook (1933), avoiding most of the formal mathematical treatment. Only the equation of the relative tangential motion is used, while the changing in orbital radii is controlled by Kepler's law for a massless body about Saturn. This simplified theory gives an estimate of the time variation of the orbital radii and angular separation of the two satellites which agrees very well with the numerical integration results. The minimum angular separation depends on the masses of the two satellites and, for a reasonable density value (close to 1), it becomes 6°.7. The amplitude of the librations of the two satellites in a frame rotating with the mean-mean motion is of the order of 60° for S1 and 285° for S3, as obtained by the numerical integrations. The variations of semi-major axes and the amplitude of librations are roughly inversely proportional to the masses.

The minimum angular separation as computed numerically for the actual eccentric and inclined orbits is smaller, but still of the order of 5°, more than sufficient for the analytical approximation to hold, and sufficient at least for the short term stability of the configuration.

Perturbations of other satellites and gravity field harmonics of Saturn should not have any large secular effect since the mean motion of the two satellites are the same.

The apparent actual stability of the configuration does not shed light on the origin and evolution of the system, although a remark may be appropriate in this respect. The mean-mean motion of the two satellites (518°.3/day) is slightly smaller than $2n_E$ (525°/day) and slightly larger than $4/3n_M$ (509°/day). Considering that the orbits of Enceladus and Mimas have been evolving by tidal effect toward the orbital lock with Dione and Tethis, respectively, one may speculate the origin of the system to be related to the evolution of the two well-known resonant satellite pairs: Mimas-Tethis, Enceladus-Dione.

This work has been supported by the Italian National Research Council.

When this lecture was delivered at the NATO Advanced Study Institute in August, 1981, only the analysis of the numerical results by Silvia Fernandez, Gian Andrea Bianchini, Flavio Pellegrini and F. A. Franklin were known. The analytical approach has been worked out by C. F. Yoder at J.P.L. during the fall of 1981.

REFERENCES.

Brown, E. W. and C. A. Shook (1969). Planetary Theory, Dover, New York, p. 302.

Smith, B. A. et al. (1981). "Encounter with Saturn Voyager 1 Imaging Science Results," Science, 212, pp. 163-198.

Synnott, S. P. et al. (1981). "Orbits of the Small Satellites of Saturn," Science, 212, pp. 191-192.

Yoder, C. F. (1981). "Theory of Motion of Saturn's Co-Orbiting Satellites" (submitted for publication, November 1981).

A QUALITATIVE STUDY OF STABILIZING AND DESTABILIZING FACTORS IN PLANETARY AND ASTEROIDAL ORBITS

John D. Hadjidemetriou

University of Thessaloniki, Thessaloniki, Greece

ABSTRACT. The stability of planetary systems with two planets is studied. The method is based on the mechanism by which instabilities develop in a planetary system with two massless planets when a Hamiltonian perturbation is applied. The results are compared with the KAM theorem. It is shown that instabilities develop at the resonances 1/3, 3/5, 5/7,... and to a lesser extend at 1/2, 2/3, 3/4... At higher order resonances the instabilities are in most cases negligible. At all the above resonances there exists a Hamiltonian perturbation which generates instability. The numerical computations have shown that the increase of the masses always generates instability at the resonances 1/3, 3/5,... for nearly circular orbits, but stable systems exist at the resonances 1/2, 2/3,... These latter systems may have large eccentricities. The effect of a random Hamiltonian perturbation, for example the close encounter with a comet, on the generation of instability may be important for the resonances 1/3, 3/5,... . On the other hand, particular Hamiltonian perturbations exist which stabilize an unstable resonant system.

1. INTRODUCTION.

The purpose of this paper is to discuss the stability of a Hamiltonian system which we consider as a perturbed system of an integrable Hamiltonian system. It is assumed that the motion of the unperturbed system is known and we wish to study the qualitative behaviour of the orbits after the perturbation is applied. In particular, we wish to study whether an unperturbed motion at, or near, a resonance will remain near that resonance after the perturbation is applied. What follows is applicable to a wide class of systems, but in order to be more definite and make the

V. Szebehely (ed.), Applications of Modern Dynamics to Celestial Mechanics and Astrodynamics, 25–44.

main ideas clearer we shall study planetary systems with two planets, where one or both planets have non-zero masses.

Consider for example, the case of two planets, P_1 and P_2, which revolve around the Sun, P_0, in the same plane. If the masses of the planets are negligible, their orbits are two uncoupled Keplerian orbits. In what follows, we shall assume that these orbits are bounded, i.e. elliptic. We have four degrees of freedom and four single valued, analytic, independent integrals of motion, namely the energy integral and the angular momentum integral of each planet. Introduce now the coupling between the two planets by increasing their masses. The system is no longer integrable because we have only two integrals, the energy integral and the angular momentum integral of the whole system. Will the perturbed system remain bounded ? Will the semi-major axes and eccentricities of the osculating orbits of the two planets deviate from their unperturbed values ? How are the above problems related to the ratio ω_1/ω_2 of the frequencies of the unperturbed orbits of the two planets ?

Questions of this type will be discussed in the following. We shall use for this purpose a linear analysis based on the method of parametric resonance and the results will be compared with the theorem of Kolmogorov, Arnold and Moser hereafter called KAM theorem. However, no theory gives a complete answer to problems related to the qualitative aspects of the stability of planetary systems. Numerical results are always useful in giving an insight and providing answers to particular situations but they do not provide a proof. What is more, in some cases where the perturbation is small, the numerical computations do not seem to coincide with what theory predicts, although nature seems to go along with theory. This is the case with the Kirkwood gaps in the distribution of the asteroids. We do have gaps where we should expect from theory, but the numerical results do not show them. Perhaps the computing time is not long enough to show what appears in nature or some other factors should also be taken into consideration.

We shall study first systems with two degrees of freedom and in particular we shall study the stability of an asteroid (of negligible mass) under the attraction of Jupiter which is assumed to describe a circular orbit around the Sun. The results will be extended to systems with four degrees of freedom by considering non-zero masses for both planets.

The main object of this study is to give a global view of the qualitative aspects of the stability of asteroidal and planetary orbits. In this way, we shall find the regions of phase space which are stable and the regions where instability is expected. Having this as a guide one can proceed further and study in detail, quantitatively, the particular regions of interest, for example, the resonant configurations which in the qualitative analysis prove to be unstable. The qualitative analysis is also useful because it gives a general view of the factors which affect the stability, and thus one can select for further, quantitative study those factors which are meaningful from the physical point of view, for example, in the study of the Kirkwood gaps of the asteroids.

The reader is expected to be familiar with the theory of Kolmogorov, Arnold and Moser on nearly integrable Hamiltonian systems. The books of Arnold and Avez (1968) and Arnold (1978) and the review papers by Berry (1978), Helleman (1982), Moser (1978), Ford (1978) and Treve (1978) are very useful as an introduction to the subject. Also, the reader should have knowledge of the theory of parametric resonance. An extensive treatment is given in the book of Yakubovich and Starzhinskii (1975).

2. STABILITY OF THE ASTEROIDS.

A) Invariant tori in the unperturbed case.

We consider a massless body, which we shall identify with an asteroid, moving under the gravitational attraction of the Sun and Jupiter in the plane of motion of these bodies. The orbit of Jupiter around the Sun is considered circular. Thus we have the well known restricted circular 3-body problem (e.g. Szebehely 1967) and we shall study the motion of the asteroid in the usual rotating frame xOy with origin at the center of mass of Sun and Jupiter. The unit of mass is taken such that $1-\mu$ and $\mu \ll 1$ are the masses of Sun and Jupiter, respectively, the unit of distance is taken as the radius of the orbit of Jupiter around the Sun and the unit of time is determined from $G = 1$, where G is the gravitational constant. We have two degrees of freedom and we select as generalized coordinates the polar coordinates (r,ϕ) of P_2 (asteroid) in the rotating frame. The Hamiltonian, for small μ, can be expressed as

$$H = H_o + \mu H_1 \tag{1}$$

where

$$H_o = \frac{1}{2}\left(P_r^2 + \frac{P_\phi^2}{r^2}\right) - \frac{1}{r} - \omega P_\phi \tag{2}$$

and $P_r = \dot{r}$, $P_\phi = r^2(\dot{\phi} + \omega)$ are the momenta per unit mass of the asteroid and ω is the angular velocity of rotation of the rotating frame xOy. In the normalization mentioned above we have $\omega = 1$. The Hamiltonian H_o describes the unperturbed (Keplerian) motion of the asteroid, in the rotating frame $x0y$, the Sun being at the origin 0. The term μH_1 is the perturbation due to the non-zero mass of Jupiter. In what follows we shall consider bounded motion of the asteroid when $\mu = 0$.

In the unperturbed problem we have two degrees of freedom and two integrals of motion,

$$H_o = h, \quad P_\phi = c, \tag{3}$$

where h and c are the energy and angular momentum constants, respectively. These integrals are in involution, i.e., Poisson's bracket is equal to zero, $[H_o, P_\phi] = 0$, and consequently the unperturbed system is integrable. We shall introduce now action-angle variables. Working in the standard way (e.g. Goldstein, 1970) we find for the actions (See also Berry, 1978),

$$I_r = -I_\phi + \frac{1}{\sqrt{-2(h+\omega I_\phi)}} \,, \qquad I_\phi = P_\phi \,, \tag{4}$$

and the Hamiltonian takes the form

$$H(I) = -\frac{1}{2(I_r+I_\phi)^2} - \omega I_\phi \,. \tag{5}$$

The natural frequencies of the system are

$$\omega_{or} = \frac{\partial H}{\partial I_r} = \frac{1}{(I_r+I_\phi)^3}, \qquad \omega_{o\phi} = \frac{\partial H}{\partial I_\phi} = -\omega + \frac{1}{(I_r+I_\phi)^3} \tag{6}$$

We shall give now a more familiar form to the frequencies (6). We note first from (2) that E, given by

$$E = h + I_\phi \omega, \tag{7}$$

is the energy of the Keplerian motion of the asteroid in the inertial frame. Using the well-known formula of elliptic motion

$$a = -\frac{1}{2E} \,, \tag{8}$$

where a is the semi-major axis, we find from (4),

$$I_r + I_\phi = a^{1/2} \tag{9}$$

and finally

$$\omega_{or} = \omega_k, \qquad \omega_{o\phi} = \omega_k - \omega, \tag{10}$$

where

$$\omega_k = a^{-3/2} \tag{11}$$

is the frequency of the Keplerian motion using the normalization mentioned before.

The phase space for the motion of the asteroid is the 4-dimensional space θ_r, θ_ϕ, I_r, I_ϕ, where θ_r, θ_ϕ are the angle variables. Since we consider bounded motion only, the motion of the asteroid in phase space lies on a 2-torus where I_r, I_ϕ are the "radii" and θ_r, θ_ϕ the angular coordinates on the torus (Fig. 1). This torus is completely determined if the values of I_r and I_ϕ for a particular motion are given. Note that for a fixed orbit of the asteroid the energy is constant and also $I_\phi = P_\phi$ = constant and consequently I_r = constant. It can be easily seen that there exists an infinite number of orbits with the same energy constant,

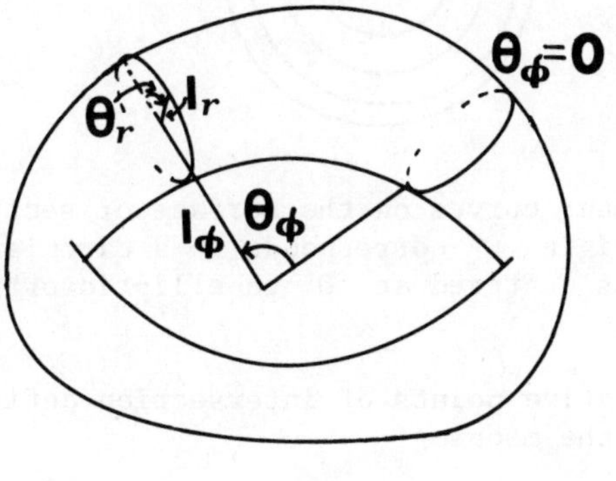

Figure 1: The 2-torus of the unperturbed elliptic motion and the surface of section at $\theta_\phi = 0$.

h, which differ in the angular momentum constant P_ϕ. This means that for a fixed value, h, of the energy we have a set of invariant tori for different values of $P_\phi = I_\phi$. We define now the surface of section

$$H_o = h, \quad \theta_\phi = 0, \tag{12}$$

which is a 2-dimensional submanifold. We can use as coordinates on this surface of section the action-angle coordinates (θ_r, I_r).

Note that a particular set of values θ_r, I_r on the surface of section completely specifies the orbit, because $\theta_\phi=0$ and I_ϕ can be obtained from the equation (5) for $H_o = h$. The invariant curves are defined as the intersections of the 2-torus on which lies the orbit in phase space with the surface of section (12). These are circles with radius I_r (Fig. 2), i.e. 1-torus where (θ_r, I_r) are used as polar coordinates. Evidently, the intersections of an orbit of energy h with the surface of section at $\theta_\phi=0$ lie all on the same circle with a fixed radius I_r which is the value of the action variable I_r specifying the torus on which lies this orbit.

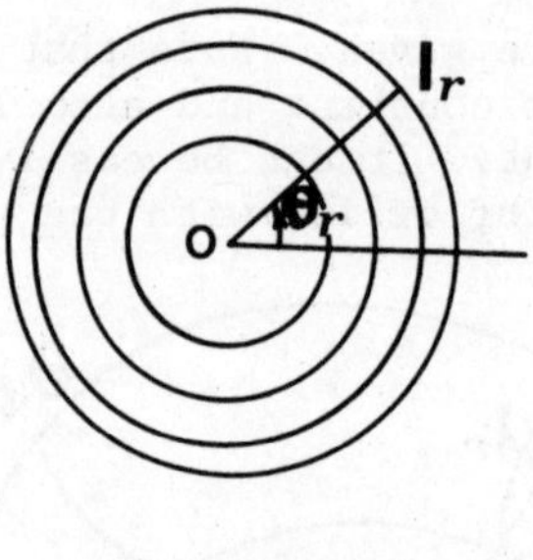

Figure 2: Invariant curves on the surface of section for $\mu = 0$. The origin 0 corresponds to a circular orbit and the circles centered at 0 to elliptic orbits.

Thus, the consecutive points of intersection define on the surface of section the mapping

$$I_r \rightarrow I_r, \quad \theta_r \rightarrow \theta_r + a(I_r) ,$$

where

$$a(I_r) = \omega_{or}\tau$$

is the rotation angle, i.e. the angle between two consecutive points of intersection, τ being the time interval between these two intersections, given by

$$\tau = \frac{2\pi}{\omega_k - \omega} .$$

Thus, the above mapping takes the form

$$I_r \to I_r, \quad \theta_r \to \theta_r + 2\pi \frac{\omega_k}{\omega_k - \omega} . \tag{13}$$

Note that the rotation angle $2\pi\omega_k/(\omega_k-\omega)$ depends on the ratio ω/ω_k of the frequencies of the perturbing body (Jupiter) and the asteroid, which is a function of I_r. This means that the mapping (13) is a twist mapping.

From the above it is clear that in the set of invariant curves (Fig. 2) we have both rational and irrational values for $a(I_r)/2\pi$. The rational values represent periodic orbits of the asteroid in the rotating frame and the corresponding invariant curves are densely distributed, though of measure zero. If $a/2\pi = \omega_k/(\omega_k-\omega) = s/r$, where $r < s$ are integers, the orbit closes after r revolutions in the rotating frame xOy. In this case any point on the corresponding invariant curve I_r = constant is a fixed point of the mapping A^r, where A is the mapping (13). (Note that to each periodic orbit corresponding to s/r there are r fixed points on the invariant curve).

It is useful to express the above results in terms of the elements of the orbit of the asteroid. We shall consider direct orbits only ($P_\phi > 0$) so that

$$I_\phi = P_\phi = a^{1/2}(1-e^2)^{1/2} , \tag{14}$$

where e is th eccentricity. Using (7), (8) and (9) we obtain from (4) and (5),

$$I_r = a^{1/2}[1-(1-e^2)^{1/2}] , \tag{15}$$

$$-\frac{1}{2a} - \omega a^{1/2}(1-e^2)^{1/2} = h. \tag{16}$$

Equation (16) gives the totality of motions of the asteroid in the rotating frame, for fixed energy h (direct motion only). It can be proved (see Szebehely 1967, p. 407) that for $e = 0$ there exists only one solution for a, i.e., for a fixed value h of the energy we have only one direct circular orbit. For this particular orbit, the value of I_r is obtained from (15) to be equal to zero.

Note that for $e \neq 0$, $I_r \neq 0$. Thus we conclude that in the set of

invariant curves in Fig. 2, for a given h, the radii of the invariant curves range from $I_r = 0$ to a value of I_r corresponding,

from (16), to the degenerate ellipse e=1. Consequently, the origin 0 in Fig. 2 corresponds to a circular, direct, orbit and all the circles I_r = constant correspond to elliptic orbits of the asteroid. It is easy to verify that a circular orbit in the inertial frame is a symmetric periodic orbit in the rotating frame xOy. The period of this circular orbit in the rotating frame is

$$T = \frac{2\pi}{\omega_k - \omega} \qquad (17)$$

where $\omega_k = a^{-3/2}$ and a is the radius (semi-major axis) of the circular orbit.

From the above we see that for each value of h we have on the surface of section only one single invariant point 0 and a set of invariant curves which are circles centered at 0 (for direct motion) as shown in Fig. 2. As h varies, the period (17) of the central fixed point varies. This has important consequences on the evolution of stability, as we shall discuss later.

B) Continuation of periodic orbits when $\mu > 0$.

The structure of phase space (for $\mu > 0$) depends critically on the distribution of periodic orbits. Equivalently, the structure on the surface of section depends on the fixed points of the corresponding mapping. In this section we shall study which fixed points (periodic orbits) that existed for $\mu = 0$ survive the perturbation, i.e., still exist at $\mu > 0$. In the next section the stability of these orbits for $\mu > 0$ will be studied and thus we shall have a picture of the phase space on the surface of section.

Let us consider the totality of orbits of the asteroid with fixed energy, as represented on the surface of section in Fig. 2. The central fixed point 0 corresponds to a circular orbit of the asteroid whose period, in the rotating frame, is given by (17). If $T \neq 2\pi n$, n=1,2,3,... (in the normalization used), this orbit is continued when $\mu > 0$ to a periodic orbit of the asteroid. with nearly circular orbit. These are the periodic orbits of the first kind (Griffin, 1920; Hadjidemetriou, 1975,76). From (17), using the fact that $\omega=1$, we obtain that the continuation of a periodic orbit to $\mu > 0$ is possible, for a sufficiently small value of μ, when $\omega/\omega_k \neq n/(n+1)$. The continued periodic orbit is represented by a single fixed point 0 near the origin, on the surface of section (12).

The continuation theorem is not applicable when a resonance condition $\omega/\omega_k=n/(n+1)$ exists, i.e., when $\omega/\omega_k=1/2, 2/3, 3/4, \ldots$ Numerical integrations have shown that for a fixed μ two different branches of periodic orbits of the asteroid in the rotating frame appear. Along each branch the resonance $\omega/\omega_k \simeq 1/2, 2/3, \ldots$ still holds for all the periodic orbits, but in this case the orbit of the asteroid is no longer circular but nearly elliptic and the eccentricity increases as we go along the branch, The above two resonant branches differ in phase only. (Guillaume, 1969; Hadjidemetriou, 1976). All these orbits are periodic orbits of the second kind and are simple periodic orbits, i.e., the asteroid starts perpendicularly from the x axis and crosses it again perpendicularly (in the same sense) at the first intersection. Such a periodic orbit is represented by a single fixed point on the surface of section (12) which, however, is not near the origin $I_r = 0$.

Next, let us study the continuation of the elliptic periodic orbits of the asteroid, i.e., the elliptic orbits which are in resonance: $\omega_k/(\omega_k - \omega) = s/r$, or $\omega/\omega_k = (s-r)/s$, $s > r$.

(Note that a particular case is the one mentioned above for $\omega/\omega_k = 1/2, 2/3, \ldots$, for $s=r+1$). The periodic orbits at the resonance $\omega/\omega_k = (s-r)/r$, for $\mu = 0$, correspond to elliptic orbits of the asteroid (not necessarily symmetric) and are represented by an invariant curve I_r = const. For a fixed energy h, and a fixed resonance $(s-r)/r$ we have a unique value for I_r from (15) because the semi-major axis a is obtained in terms of r/s (note that $\omega = 1$) and from (16) we have for this resonance a fixed eccentricity e. Consequently, we conclude that the invariant curve I_r = constant for $\mu = 0$ corresponds to all the elliptic orbits of the asteroid, with a fixed energy h and a fixed eccentricity. An infinite number of such periodic orbits exists, corresponding to the fact that the line of apsides of the orbit of the asteroid forms an angle with the radius

Sun-Jupiter at $t = 0$, which can take any value. The question is what happens when $\mu > 0$. It can be proved, by making use of the Poincaré-Birkhoff fixed point theorem, that out of this infinite number of periodic orbits for $\mu = 0$ only a finite number survives the perturbation $\mu > 0$. This is due to the fact that the mapping (13) is an area preserving twist mapping, i.e., the rotation angle $2\pi\omega_k/(\omega_k-\omega)$ depends on I_r, and for $\mu > 0$ becomes an area preserving perturbed twist mapping (see Arnold and Avez, 1968, p. 88). The periodic orbits which survived the perturbation $\mu > 0$ are represented on the surface of section (12) by a set of r fixed points lying on a perturbed circle near the original circular invariant curve I_r = const. These are also periodic orbits of the second kind and in general are multiple periodic orbits.

C) The stability of periodic orbits for $\mu > 0$.

We shall discuss first the evolution of the stability of a circular periodic orbit such that $\omega/\omega_k \neq 1/2, 2/3, 3/4, \ldots$. This is evidently an orbitally stable periodic orbit when $\mu = 0$ and from Keplerian theory one obtains (e.g. Hadjidemetriou 1979) that apart from a pair of unit eigenvalues $\lambda_1=\lambda_2=1$, there exists the pair

$$\lambda_{3,4} = e^{\pm i\omega_k T} \tag{18}$$

where by (11) and (17),

$$\omega_k T = \frac{2\pi}{1-\frac{\omega}{\omega_k}} . \tag{19}$$

It is clear that $\omega_k T$ depends on the value of the energy, h. The eigenvalues λ_3, λ_4, given by (18), are the eigenvalues of the linearized mapping at the vicinity of the fixed point 0 which represents the above periodic orbit on the surface of section (12). All the eigenvalues $\lambda_1, .. \lambda_4$ lie on the unit circle as shown in Fig. 3.

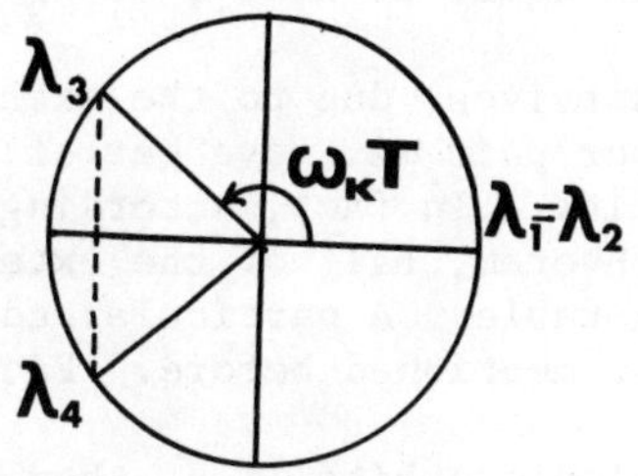

Figure 3: The eigenvalues of the circular orbit ($\mu = 0$).

Let us see now what is the evolution of the eigenvalues of the new periodic orbit, for $\mu>0$, obtained from the continuation of the $\mu=0$ case. In particular, we are interested to see whether the eigenvalues can evolve out of the unit circle and thus generate instability. First of all we note that the pair of unit eigenvalues $\lambda_1=\lambda_2=1$ survives because the energy integral exists also when $\mu>0$. As for the pair λ_3,λ_4 it can be proved (Hadjidemetriou, 1981) that they can move out of the unit circle only when they meet. This happens when $\lambda_3=\lambda_4=\pm1$, for particular values of h, at the following resonances, as obtained from (19):

$$\lambda_3=\lambda_4=\begin{cases} 1 & \text{when} \quad \omega/\omega_k=n/(n+1), \\ -1 & \text{when} \quad \omega/\omega_k=(2n-1)/(2n+1). \end{cases} \tag{20}$$

Note, however, that the resonant orbits $\omega/\omega_k=n/(n+1)$ are not continued as circular orbits when $\mu>0$, as mentioned in the previous section. Consequently, we come to the conclusion that the only circular orbits of the asteroid which can become unstable periodic orbits of the first kind are the resonant cases (2n-1)/(2n+1), i.e. 1/3, 3/5, 5/7, ... The important point at these resonances is that there always exists a Hamiltonian perturbation which generates instability. Numerical computations have shown that the non-zero value of μ is such a perturbation, i.e. the periodic orbits of the first kind for small values $\mu>0$, at the resonances 1/3, 3/5, 5/7, ... are linearly unstable.

Next, we consider the stability characteristics of the periodic orbits of the second kind, i.e. nearly elliptic orbits of the asteroid. The unperturbed orbits out of which these orbits for $\mu>0$ have evolved are resonant r-multiple periodic orbits of the form $\omega/\omega_k=(s-r)/s$ and from (18), (19) we obtain that all the eigenvalues for $\mu=0$ are equal to unity, $\lambda_1=\lambda_2=1$, $\lambda_3=\lambda_4=1$. When $\mu>0$ the pair $\lambda_1=\lambda_2=1$ survives, due to the existence of the energy integral, but the other pair may move out of the unit circle and thus generate instability. In fact, according to the Poincaré-Birkhoff fixed point theorem, half of the extended orbits are stable and half are unstable. A particular case is the resonance $\omega/\omega_k=1/2$, $2/3$, $3/4$, ... mentioned before. Each of these orbits for $\mu>0$ are simple periodic orbits, i.e. they are represented by a single fixed point on the surface of section, whereas in general the periodic orbits of the second kind are multiple periodic orbits ($s-r>1$) and are represented on the surface of section by a set of r invariant points.

Finally, we note that linear instability of a periodic orbit (or a fixed point) implies instability when the non-linear effects are taken into account. Consequently, the structure of phase space on the surface of section can be determined, qualitatively, when the linear stability characteristics are known.

D) The structure of phase space when $\mu>0$.

After the above remarks we are in a postion to determine the structure of phase space, on the surface of section (12), when $\mu>0$. The unperturbed structure is shown in Fig. (2). This can be achieved by the application of the Kolmogorov, Arnold, Moser (KAM) theorem. The Hamiltonian of the perturbed motion, $\mu \ll 1$, is given by (1). This is a nearly integrable system and we wish to study what happens to the invariant curves which we had for $\mu=0$, shown in Fig. 2. We note that

$$\det \begin{vmatrix} \partial\omega_j/\partial I_k & \omega_j \\ \omega_k & 0 \end{vmatrix} \neq 0$$

and consequently the KAM theorem is applicable (Moser 1978). This condition guarantees isoenergetic non-degeneracy, which implies that the ratio ω/ω_k varies, for fixed h, as I_r varies (Arnold 1978, p. 403). Indeed, it can be proved by making use of (5), (6), (7) that ω/ω_k increases as we go outwards, starting from the central fixed point 0 in Fig. 2.

The perturbed phase diagram may be more complicated than that described, for example, in Arnold and Avez (1968, p.91) because in this case the central fixed point 0 may become unstable, at particular resonances, namely 1/3, 3/5, ... as stated before.

Let us consider now a phase diagram on the surface of section (12), for a particular fixed point 0 corresponding to the resonance

ω/ω_k. We distinguish between the following cases:

Case (a): $\omega/\omega_k \neq n/(n+1)$, $(2n-1)/(2n+1)$.

The eigenvalues $\lambda_{3,4}$ are not at +1 or -1 on the unit circle (Fig.3), as can be seen from (19). The continuation to a simple symmetric periodic orbit when $\mu>0$ is always possible. This is a nearly circular orbit for the asteroid. The eigenvalues $\lambda_{3,4}$ cannot move out of the unit circle and the stability of the fixed point 0 is preserved.

Let us see now what happens near the fixed point 0. As mentioned before, the intersections of the orbits with the surface of section, for $\mu=0$, lie on invariant curves which are circles, I_r= const. centered at 0. There is a continuous set of such invariant curves for fixed h and the rational invariant curves $(\omega_k-\omega)/\omega_k=r/s<1$ form a dense subset. Each point on a rational invariant curve for $\mu=0$ is a fixed point of the mapping A^r and the eigenvalues of such an invariant point are equal to unity, $\lambda_3=\lambda_4=1$. By the Poincaré-Birkhoff fixed point theorem, each rational invariant curve generates, when $\mu>0$, an even number of fixed points half of them being of the elliptic type and half of the hyperbolic. Thus homoclinic points appear, due to the intersections of the invariant curves which start from the hyperbolic fixed points along the eingenvectors of the unstable eigenvalues (Arnold and Avez, 1968, p. 88). As a consequence, a narrow zone around the corresponding rational invariant curve which existed for $\mu=0$ is dissolved and this corresponds to the generation of instablility. The measure of phase space where the invariant curves have dissolved is very small for $\mu<<1$. Moreover, the area of instability depends on the order of the resonance r/s (i.e. on the value of s) and consequently it is practically negligible for all resonances except those with low order, for example 2/5, 3/7, ...

The invariant curves which correspond to irrational values of $(\omega_k-\omega)/\omega_k$ sufficiently far from a rational invariant curve survive the perturbation ($\mu>0$) and thus an orbit which started on a rational invariant curve for $\mu=0$ is trapped between irrational invariant tori and thus cannot escape far from the original position. This is a result obtained by the application of the KAM theorem and is valid for two degrees of freedom only. However, the maximum value μ of the perturbation up to which this theorem is applicable is not known and it seems that it is rather small, perhaps much smaller than the actual value of μ for the case of the asteroids. The numerical computations have shown (Contopoulos 1971, Froeschlé 1970, Hénon and Heiles 1964, Walker and Ford 1969) that the dissolution of invariant curves is negligible and not observable, within the computer accuracy, for small values of the perturbation, but as the perturbation increases the dissolution becomes evident. This dissolution starts first at the neighborhood of the unstable points on the surface of section and extends as the perturbation increases. Irrational invariant curves dissolve as well because the KAM theorem is no longer applicable. This dissolution is due to the interaction of resonances from rational invariant curves adjacent to the irrational invariant curve (see Ford 1978). Concluding, we state that the phase space on the surface of section near an irrational fixed point 0 looks like Fig. 4.

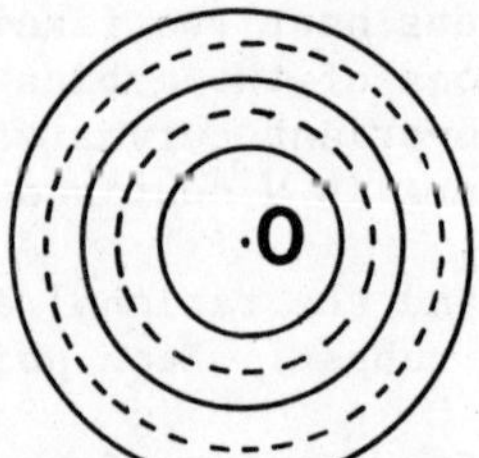

Figure 4: Invariant curves around an irrational fixed point 0. The dotted lines are the rational invariant curves where the dissolution starts first.

The dotted lines represent the invariant curves with $(\omega_k-\omega)/\omega_k=r/s$ where the dissolution starts. The resonance however is in most cases of high order and consequently that area of dissolution is so small that the configuration practically remains stable because there is no interaction of resonances.

Case (b): $\omega/\omega_k=n/(n+1)$.

In this case, the corresponding circular unperturbed orbit of the asteroid cannot be continued to a periodic orbit of the first kind (nearly circular orbit) when $\mu>0$. Instead, a periodic orbit of the second kind is generated (nearly elliptic orbit) which is a simple periodic orbit represented on the surface of section by a single fixed point which is not, in general, near the origin 0 (as the eccentricity may be quite large). Numerical computations have shown that for particular values of h these periodic orbits for $\mu>0$, at the resonances 1/2, 2/3, 3/4, ... are stable. Thus, on the surface of section we have the corresponding single fixed point representing one of these resonances surrounded by closed invariant curves. These closed invariant curves may cover a substantial part of phase space around the resonant fixed point because there are no other fixed points at the same resonance near it. This is equivalent to say that a part of the invariant unperturbed torus which contained the resonant orbit 1/2, 2/3, ... survives the perturbation as a smaller invariant torus. Thus, the same reasoning as in case (a) applies and we may have trapping of asteroids at these resonant orbits.

Case (c) $\omega/\omega_k=(2n-1)/(2n+1)$, $[(\omega_k-\omega)/\omega_k=2/(2n+1)]$.

In this case, instability is always generated at the central fixed point 0. At the corresponding periodic orbit the eigenvalues for $\mu=0$ are $\lambda_3=\lambda_4=-1$. The continuation to a new, nearly circular, periodic orbit for $\mu>0$ is always possible. It can be proven (Hadjidemetriou, 1981) that there always exists a Hamiltonian perturbation which makes the eigenvalues to move out of the unit circle, and the numerical computations have shown that the increase of the mass of Jupiter ($\mu>0$) generates instability. The perturbed orbit is a simple symmetric periodic orbit which is represented on the surface of section by a single fixed point 0, which is hyperbolic with reflection. Thus the phase space around this fixed point is as shown in Figure 5.

Figure 5: The surface of section at an unstable fixed point. Homoclinic points appear near 0.

The mapping around 0 is a hyperbolic twist mapping. (Such mapping has been studied by Easton (1979)). The separatrices which start from 0 intersect at an infinity of homoclinic points and consequently, the smooth invariant curves which we had at $\mu=0$ are dissolved. For low order resonances and small values of μ the dissolved part of phase space is confined by smooth irrational invariant curves, but as μ increases the dissolution extends. What is more, as the energy varies, we encounter more and more resonant orbits of the form (2n-1)/(2n+1) which have an accumulation point at the orbit of Jupiter. Also the numerical computations showed that the unstable eigenvalues at these resonances grow rapidly as n increases, i.e. as we approach the orbit of Jupiter. Thus, the picture of invariant curves shown in Fig. 4 disappears and we have the complicated picture shown in Fig. 5. The instability generated at the resonances (2n-1)/(2n+1) is quite clear on the numerical computations of perturbed orbits, when $n \geq 2$, i.e. at the resonances 3/5, 5/7, ... At n=1 (resonance 1/3), the separatrices seem to behave as in an integrable system, as we found by numerical computations, but this is due to the limited computing time and the low resolution at the scale used in the drawing.

E) Conclusions.

From the above we arrive at the following conclusions concerning the stability of the asteroids:

1. Asteroids of nearly circular orbits can exist at all irrational ratios of the frequencies ω/ω_k because these orbits are trapped around the corresponding fixed point 0 which is surrounded by irrational invariant curves (and rational invariant curves of high order whose dissolution is negligible) as shown in Fig. 4.

2. Asteroids of nearly circular orbits can exist at rational frequencies ω/ω_k different from (n-1)/n or (2n-1)/(2n+1), n=1,2,3,..., for the same reason as mentioned above. Elliptic orbits, however, at these frequencies, represented by closed invariant curves for $\mu=0$, should not exist because the corresponding invariant curves dissolve and thus instability is generated. This instability however is negligible at high order resonances and only in the low order resonances the effects of instability should be observable, for example, at the resonances 2/5, 3/7, etc.

3. Nearly elliptic orbits at the resonances $\omega/\omega_k=(n-1)/n$ should be expected in actual asteroid orbits because there exist simple symmetric periodic orbits at these resonances (single fixed points on the surface of section) which are stable. The explanation is the same as in (1) above (phase picture as in Fig. 4). Note that these cases include the low order resonances 1/2, 2/3, 3/4,... where asteroids with elliptic orbits exist (Hilda group). The same holds for elliptic orbits at the 1/3, 3/5, ... resonances.

4. Nearly circular orbits at the resonances $\omega/\omega_k=(2n-1)/(2n+1)$, i.e. at 1/3, 3/5, 5/7, ... should not be expected in asteroid orbits, because the unperturbed fixed point 0 corresponding to these resonances is always unstable for $\mu>0$ (Fig. 5). The unstable region however, where chaotic behavior is observed, is small and is surrounded by smooth invariant curves which ensure the non-escape far from this resonance. This is especially true at the frequencies 1/3, 3/5, and for this reason this instability is not observable by numerical integrations (Froeschlé and Scholl, 1979, Lecar and Franklin, 1973) because it is a very slow process.

The chaotic region around the unstable fixed point 0 at the resonances 1/3, 3/5, ... (Fig. 5) depends on the magnitude of the perturbation, i.e. on the mass of Jupiter. If it is true that in the beginning of the Solar System the mass of Jupiter was larger than it is now, the resulting instability at the above resonances would be more effective and the gaps may have developed at that time.

5. The importance of the unstable resonant orbits 1/3, 3/5, 5/7, ... lies in the fact that there always exists a Hamiltonian perturbation which generates instability. Consequently, a random Hamiltonian perturbation is expected to generate instability and thus depopulate these resonant orbits much faster than the perturbation due to the non-zero mass of Jupiter ($\mu>0$). (Such a perturbation could be provided by the close approach of a comet). This could explain the gap in the 1/3 resonance. Numerical computations should be made to test at what extent this happens.

3. PLANETARY SYSTEMS WITH NON-ZERO MASSES OF BOTH PLANETS.

The results of the previous section can be readily extended to the case where also the mass of P_2 (called asteroid in section 2) is non negligible. In this case we have four degrees of freedom and the unperturbed system, consisting of the Sun, P_0, and the planets P_1 and P_2, is integrable for zero masses of the planets. In a suitably selected rotating frame whose origin is at the center of mass of P_0, P_1 and the x-axis is the line P_0P_1 we have three degrees of freedom and families of periodic orbits exist in this rotating frame, for fixed masses (Hadjidemetriou 1975, 1976). It can be shown that instabilities appear for the same resonances ω_1/ω_2 as in the asteroid case (Hadjidemetriou 1981) and consequently the same conclusions hold for resonant planetary orbits. In this case there is another parameter which affects the stability of the perturbed orbit, apart from the masses m_1 and m_2. This is the ratio of the masses m_1/m_2 which influences the stability of the elliptic resonant orbits 1/2, 2/3, 3/4, ... Numerical computations have shown that the system is stable at these resonances if the inner planet has the smaller mass (Delibaltas 1976).

Thus, we state, based on the results mentioned in section 2, that no planetary system can exist at the resonances 1/3, 3/5, 5/7, ... In particular, we could not have a planet at the above resonances with Jupiter, i.e. at a radius equal to 2.5, 3.7, 4.2, 4.4, ...A.U., respectively. Note that all these values are within the asteroid belt. Note also that the inner planets Mars, Earth, Venus and Mercury are far away from these unstable areas.

4. STABILIZING AND DESTABILIZING FACTORS.

The increase of the masses of the planets generates instability. For very small values of the masses the instability starts at the resonant orbits $\omega_1/\omega_2=1/3$, 3/5, ... This means that a very small part of the extended family of periodic orbits (for fixed masses) at the above resonances is unstable. But as the masses increase, this unstable part extends and covers a larger set of ratios ω_1/ω_2. On the other hand, the magnitude of the unstable eigenvalue $\lambda>1$ is increasing as the masses increase and the effects of the instability are more evident. For example, when $m_1 = m_2 = .1$, $m_0 = .8$ the unstable region covers the resonant orbit 2/5, i.e. the Sun-Jupiter-Saturn system could not exist with the above masses.

A factor which could stabilize a perturbed unstable periodic orbit of the planetary type, when the masses are increased, is the addition of another Hamiltonian perturbation, for example, the oblateness of the Sun. As an example, consider the unstable system at the resonance 1/3, where the unperturbed eigenvalues are $\lambda_1=\lambda_2=1$, $\lambda_3=\lambda_4=\lambda_5=\lambda_6=-1$. The increase of the masses has as a consequence the evolution of at least one pair $\lambda_3=\lambda_4=-1$, $\lambda_5=\lambda_6=-1$ out of the unit circle. Let us add now another Hamiltonian perturbation. It can be proved (Yakubovich and Starzhinskii, 1975, p. 208) that if the new perturbation adds to the system of variational equations corresponding to the unperturbed orbit 1/3, a term $\varepsilon Q(t)x$, where $Q(t)$ is a T-periodic positive definite matrix $[(Qc,c)>0$ for all $c]$ and x is the displacement vector, then for $|\varepsilon|<\delta<<1$ the eigenvalues at -1 move <u>on</u> the unit circle when $\varepsilon\neq0$ and $m_1=m_2=0$. This means that the eigenvalues move away from the critical point -1 and consequently the system is stabilized. This is so because the effect of the increase of the masses and of the perturbation εQx is additive, in the linear approximation, and the value of ε may be such that the masses m_1,m_2 cannot bring the eigenvalues back to the point -1 and thereafter out of the unit circle. It seems that the oblateness of the Sun may play such a stabilizing role (Ichtiaroglou, 1981).

As a final stabilizing effect let us consider a nearly elliptic orbit of an asteroid, for example at the resonance 2/3. The unperturbed orbit is critical with eigenvalues $\lambda_3=\lambda_4=1$ and a Hamiltonian perturbation always exists which generates instability. However, the increase of the mass of Jupiter may generate stability, i.e. the pair $\lambda_3=\lambda_4=1$ moves <u>on</u> the unit circle away from +1. Thus, if μ is large enough, the displacement away from the critical point is quite large so that it cannot be overcome by the destabilizing effect of a random Hamiltonian perturbation, which alone would generate instability.

All the above apply for the case where the perturbation is Hamiltonian. What would happen if nonconservative effects were included ? This would be the case if a parameter of the problem, for example, the mass of the Sun varies. If the variation is slow it can be shown (Landau and Lifshitz 1976) that the actions are adiabatic invariants. This means that the structure of the phase space around the fixed points is still like that shown in

Figures 5,6, but a particular orbit is not locked in a particular phase diagram for a fixed energy but drifts from one surface of section to the other as its energy varies due to non-conservative forces. It seems that the asteroid moves faster when it enters an unstable zone where the invariant curves have dissolved, but no proof exists of this conjecture.

REFERENCES.

Arnold, V. I. (1978), "Mathematical Methods in Classical Mechanics", Springer-Verlag.

Arnold, V. I. and Avez, A (1968), "Ergodic Problems in Classical Mechanics", Benjamin.

Berry, M. V. (1978), in S. Jorna (ed.) "Topics in Non-linear Mechanics, A Tribute to Sir Edward Bullard", American Institute of Physics, New York, p. 16.

Contopoulos, G (1971), Astron. J., 76, p. 147.

Delibaltas, P. (1976), Astrophys. Space Science, 45,p. 207.

Easton, R. (1979), in V. Szebehely (ed.) "Instabilities in Dynamical Systems", D. Reidel Publ. Co., p. 41.

Ford, J. (1978), in S. Jorna, (ed.) "Topics in Non-linear Mechanics, A Tribute to Sir Edward Bullard", American Institute of Physics, New York, p. 121.

Froeschlé, C (1970), Astron. Astrophys., 9, p. 15.

Froeschlé, C and Scholl, H. (1979), in V. Szebehely (ed.) "Instabilities in Dynamical Systems", D. Reidel Publ. Co., p. 115.

Goldstein, H. (1970), "Classical Mechanics", Addison-Wesley Publ. Co.

Griffin, F. L. (1920), in F. R. Moulton (ed.) "Periodic Orbits", Ch. 14, Carnegie Inst.

Guillaume, P. (1969), Astron. Astrophys., 3, p. 57.

Hadjidemetriou, J. D. (1975), Celes. Mech., 12, p. 155

Hadjidemetriou, J. D. (1976), Astrophys, Space Science, 40, p. 201.

Hadjidemetriou, J. D. (1979), in V. Szebehely (ed.) " Instabilities in Dynamical Systems", D. Reidel Publ. Co., p. 135.

Hadjidemetriou, J. D. (1980), Celes. Mech., 21, p. 63.

Hadjidemetriou, J. D. (1981), Celes. Mech.,(to appear).

Hénon, M. and Heiles, C (1964), Astron. J., 69, p. 73.

Ichtiaroglou, S. (1981), Doctor's thesis, Univ. of Thessaloniki.

Landau, L. and Lifschitz, E. (1976), "Mechanics"(3rd ed.), Pergamon Press.

Lecar, M. and Franklin, F. A. (1974), I.A.U. Symposium No. 62, (Y. Kozai, ed.), D. Reidel Publ. Co.

Moser, J. (1978), in S. Jorna (ed.) "Topics in Non-linear Mechanics, A Tribute to Sir Edward Bullard", American Institute of Physics, New York, p. 1.

Szebehely, V. (1967), "Theory of Orbits", Academic Press, N.Y.

Treve, Y. M. (1978),in S. Jorna (ed.) "Topics in Non-linear Mechanics, A Tribute to Sir Edward Bullard", American Institute of Physics, New York, p. 147.

Walker, G. H. and Ford, J. (1964), Astron. J., 69, p. 73.
Yakubovich, V. A. and Starzhinskii, V. M. (1975), "Linear Differential Equations with Periodic Coefficients", Vol. 1, Halsted Press, Ch. III.

PERTURBATIONS IN STELLAR PATHS*

P. van de Kamp

Swarthmore College Observatory, Pennsylvania

INTRODUCTION. It is a pleasure for me to partake in the activities of this Institute. I am primarily an observing astronomer with his two feet on the ground, so to speak. In my chosen field of (photographic) astrometry, mathematics and celestial mechanics play a role on a very elementary level, compared with that characteristic of the lectures of this Institute. It is, therefore, a comfort and a pleasure to recall my personal contacts with persons whose names have frequently been mentioned here: Ernest W. Brown, the "man of the moon" of the Yale Observatory, an outstanding figure at meetings of the Astronomical Society half a century ago and Carl Siegel, long-time member of the Institute for Advanced Studies in Princeton, whose lecture on the Dreierstoss (triple collision) I remember for its outstanding clarity and lucidity. I must add that ultimate-ultimate celestial mechanician, Albert Einstein, with whom I never entered into scientific discussion, but whom I joined in a chambermusic evening in Swarthmore, Pennsylvania, on June 5, 1938. There we were: Arnold Dresden, piano; Einstein, violin; my brother Jaap, violoncello; and myself, viola (see my articles in the Astronomische Nachrichten (1979), 300, 273; and in l'Astronomie (1981), 95, 285).

1. HISTORICAL REMARKS.

It all goes back to the beginnings of astrometry, the oldest branch of one of the oldest professions: astronomy. A first important name is Hipparchus, the discoverer of precession

* This presentation is based on two lectures given at the Advanced Study Institute in Cortina d'Ampezzo on August 12 and 13, 1981. For further details and references, the reader may consult the author's "Stellar Paths," published by the Reidel Publ. Co., in 1981.

V. Szebehely (ed.), Applications of Modern Dynamics to Celestial Mechanics and Astrodynamics, 45–57.

(125 B.C.), and it is interesting and fitting that his name should be attached to current scientific astrometric space projects.

As to the subject of this paper, it began with Edmund Halley, who destroyed the illusion of fixed stars with his discovery (1718) of stellar proper motions, which appears to follow the law of inertia. Exceptions were found by Friedrich Wilhelm Bessel, who in 1844 announced the perturbation in the proper motions of Sirius and Procyon and attributed these to unseen companions. Bessel, who was far ahead of his time, clearly understood and evaluated the impact of his discovery in these statements: "But light is no real property of mass. The existence of numberless visible stars can prove nothing against the existence of numberless invisible ones." Bessel realized the implications of his discovery: "If a change of motion can be proved in only two cases up to the present time, this still will make all other cases be liable to suspicion. It will be equally difficult to free by observations other proper motions from the suspicion of change, and to get such knowledge of the change as to admit of its amount being calculated."

How true! Another historical perturbation was that in the orbit of Uranus, which led to the visual detection of Neptune in 1846; the planet Pluto was discovered in similar fashion (1930).

The study of orbital motion in binary stars goes back to 1802, when William Herschel noted this phenomenon for the components of Castor and other double stars. Early perturbations in binary orbits were found by Seeliger (1888) for Zeta Cancri, and by Nörlund (1905) for Xi Ursae Majoris, thus making these objects triple systems (by now both are known to be quadruple systems).

2. LONG-FOCUS PHOTOGRAPHIC ASTROMETRY.

Well over half a century ago some of us began to realize that perturbations would be a very suitable subject for the technique and methods of long-focus photographic astrometry used for determining stellar parallaxes. Effectively established and developed at the beginning of the twentieth century by Frank Schlesinger with the long-focus refractor (19.37 m focal length, 102 cm aperture) of the Yerkes Observatory, his example was followed by a number of observatories: Sproul, McCormick, Allegheny and others, all with long-focus telescopes. Early reflecting telescopes were not found particularly suitable, but recent quartz reflectors have proved to be excellent for astrometry, the outstanding instrument being the quartz reflector (15.22 m focal length, 155 cm aperture) of the United States Naval Observatory, located at the USNO station in Flagstaff, Arizona.

The value of the long-focus photographic technique lies in the combination: optical instrument, photographic plate, measuring machine and the fact that measurements are limited to a small angular section of the celestial sphere, usually well below one degree. It is the relative or differential nature of the technique which maintains the high accuracy obtained in long-focus photographic astrometry.

The "parallax star" is referred to a background of faint distant reference stars; a minimum of three is required, more than six or eight is hardly necessary or advisable. High positional accuracy results to a great extent from the long focal length, but also from improved measuring machines. The information extracted from one photographic exposure, central relative to reference stars, yields an accuracy of about 1 micron or 0".02 (probable error). A position based on several exposures of up to four plates taken in any night may yield an error as small as 0.25 micron or 0".005; an average of positions obtained on several nights may reduce the error to as low as 0.1 micron or 0".002. It appears difficult, if not impossible, to obtain any higher accuracy; in any one year there is annual error of 0".002, which prevents further improvement (in accuracy). More about this in Section 6.

3. PARALLAX, ORBITAL MOTION AND QUADRATIC TIME EFFECT.

The measured positions on plates obtained over an extended time interval are reduced by a linear transformation to the same origin, scale and orientation. An important aspect of this reduction is a knowledge of the weights, i.e., the barycentric coordinates of the reference stars with respect to the central ("parallax") star, named dependences by Schlesinger. Knowledge of these is essential in making a proper choice of reference stars.

In parallax determinations traditionally based on observations taken over a few years, the reduced positions in any coordinate (usually RA and Decl.) are given by

$$X = c + \mu t + \pi P \quad , \tag{1}$$

where

- μ is the annual proper motion,
- t is the time in years,
- π is the parallax, relative to the reference stars, and
- P is the parallax factor, i.e., the fractional portion of one astronomical unit in angular measure.

In parallax determination the RA coordinate carries most weight because it is inclined only 23°.5 to the ecliptic. Formula (1) refers to a single star. For the component of a double star, we write

$$X = c + \mu t + \pi P + \alpha Q \ , \tag{2}$$

where α is the semi-axis major of the orbit of the component with respect to the barycenter of the binary, Q is the orbital factor, i.e., the fractional portion of α . Q is calculated from the dynamical elements P , T and e and the orientation elements i , Ω and ω .

In these formulae, c , μ , π and α are ultimately expressed in angular measure, i.e., seconds of arc. For the brighter component of a binary

$$\alpha = a_A = \frac{M_B}{M_A + M_B} a \tag{3}$$

for the fainter component

$$\alpha = a_B = \frac{M_A}{M_A + M_B} a \ , \tag{4}$$

where M_A and M_B are the respective masses of the two components: a_A , a_B and a are the semi-axes major of the orbits of A and B around their center of mass and of the relative orbit of A and B , respectively. If the components are blended into a <u>photocenter</u>, we write

$$\alpha = \left(\frac{M_B}{M_A + M_B} - \frac{\ell_B}{\ell_A + \ell_B} \right) a \ , \tag{5}$$

where α is the semi-axis major of the <u>photocentric orbit</u> and ℓ_A and ℓ_B are the respective luminosities of the two components. This equation is usually written as

$$\alpha = (B - \beta) a \quad \text{or} \quad B = \frac{\alpha}{a} + \beta \ , \tag{6}$$

where the fractional mass and fractional luminosity of the secondary component are given by

$$B = \frac{M_B}{M_A + M_B} \tag{7}$$

$$\beta = \frac{\ell_B}{\ell_A + \ell_B} = (1 + 10^{0.4\Delta m})^{-1} \ . \tag{8}$$

In practice the value of the blending factor β deviates from the ideal relation (8) and decreases more rapidly with increasing magnitude difference, Δm between the components.

Over long time-intervals such as are becoming quite usual by now, another term has to be added to equation (2):

$$X = c + \mu t + qt^2 + \pi P + \alpha Q \ , \tag{9}$$

where q is the quadratic time coefficient (half acceleration). Adequate material is required to separate the quadratic time effect from any orbital effect.

The quadratic time effect is due to (1) a perspective acceleration, and (2) to a spurious acceleration, caused by the proper motions, small though they may be, for the reference stars.

The perspective acceleration amounts to

$$\frac{d\mu}{dt} = -2''\!.05 \times 10^{-6} \ \mu p \ V \ yr^{-2} \tag{10}$$

i.e., directly proportional to the annual parallax (p) and proper motion (μ) expressed in arc seconds and to the radial velocity (V) expressed in km sec^{-1}. The annual spurious quadratic time effect is

$$\left[\Delta D \ \mu_r\right] \tag{11}$$

where ΔD are the annual dependence changes and μ_r the annual proper motions of the reference stars.

4. RESULTS.

a. Parallaxes.

Thousands of parallax determinations have been made. The obtained accuracy depends on the number of plates taken which may range anywhere from 20 to over 1000. The corresponding probable errors range from about 0''.01 to as low as 0''.001. These relative parallaxes have to be reduced to absolute by adding a small correction, a few thousands seconds of arc.

A summary of stars with parallaxes larger than 0''.045 has been given by W. Gliese and most of the parallaxes of intrinsically faint stars have been obtained at the USNO. The absolute magnitudes $M = m + 5 + 5 \log p$ plotted against color, clearly reveal the lower main sequence and the white dwarf or degenerate branch (Fig. 1).

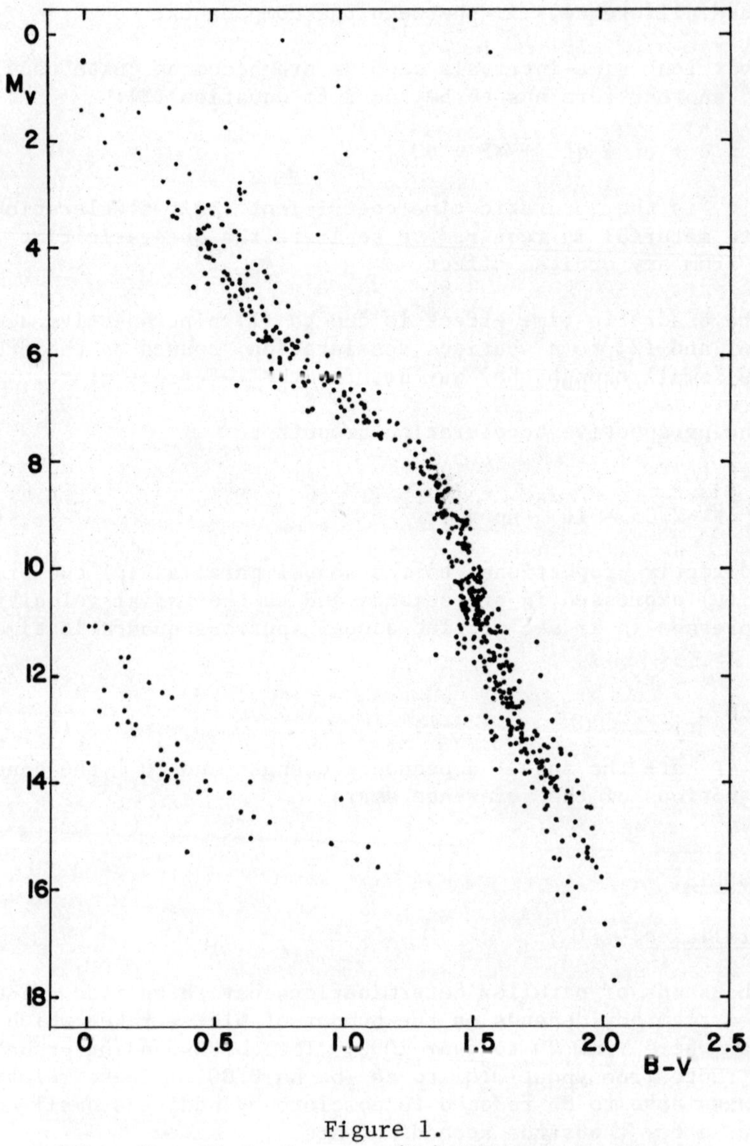

Figure 1.
Color (B − V)-luminosity (M_V) Diagram for Stars Nearer than 22 Parsec (Gliese).

b. Mass-Ratio and Masses.

Accurate mass-ratios have been obtained for a limited number of double stars. The mass-ratio M_B/M_A is given by a_A/a_B (equations 3 and 4) or by $\frac{B}{1 - B}$, where B is obtained from formula (6). The total mass of the binary is given by the harmonic relation

$$M_A + M_B = \frac{a^3}{P^2} \qquad (12)$$

expressed in astronomical units (sun's mass, mean distance Earth-Sun, year) or by

$$M_A + M_B = \frac{a^3}{p^3} \frac{1}{P^2} \qquad (13)$$

where the semi-axis major a and parallax p are expressed in seconds of arc and P is expressed in years. The cube of p puts a severe limit on attainable accuracy even for relatively nearby binaries. Fig. 2 is based on a recent summary by Jahreisz of well-determined masses of the components of visual binaries nearer than 10 parsecs. Note the two white dwarf companions of Sirius and Procyon, and toward the lower right the faint component of Ross 614, the star of smallest known well-determined mass (Section 5).

c. Perspective Acceleration and Radial Velocity.

The largest known perspective acceleration is that of Barnard's star, for which $\mu = 10''.31$, $V = -108 \pm 2.5$ km s^{-1} , $p = 0''.547$; hence we predict

$$\frac{d\mu}{dt} = +0''.00125 \pm 0''.00003 \ .$$

Observations at Sproul Observatory over the interval 1916-1979 corrected for spurious acceleration (equation 10) yield:

$$\frac{d\mu}{dt} = +0''.00130 \pm 0''.00003$$

in good agreement with the predicted value.

An astrometric determination of the perspective acceleration furnishes a possible method for determining radial velocity without recourse to observed Doppler-effect, which may or may not be contaminated by other physical effects, such as gravitational redshift. Examples are given for two nearby stars. We now write equation 10 as follows:

$$V = -4.88 \times 10^{-5} \frac{d\mu}{dt} (\mu t)^{-1} \text{ km s}^{-1} \ . \qquad (14)$$

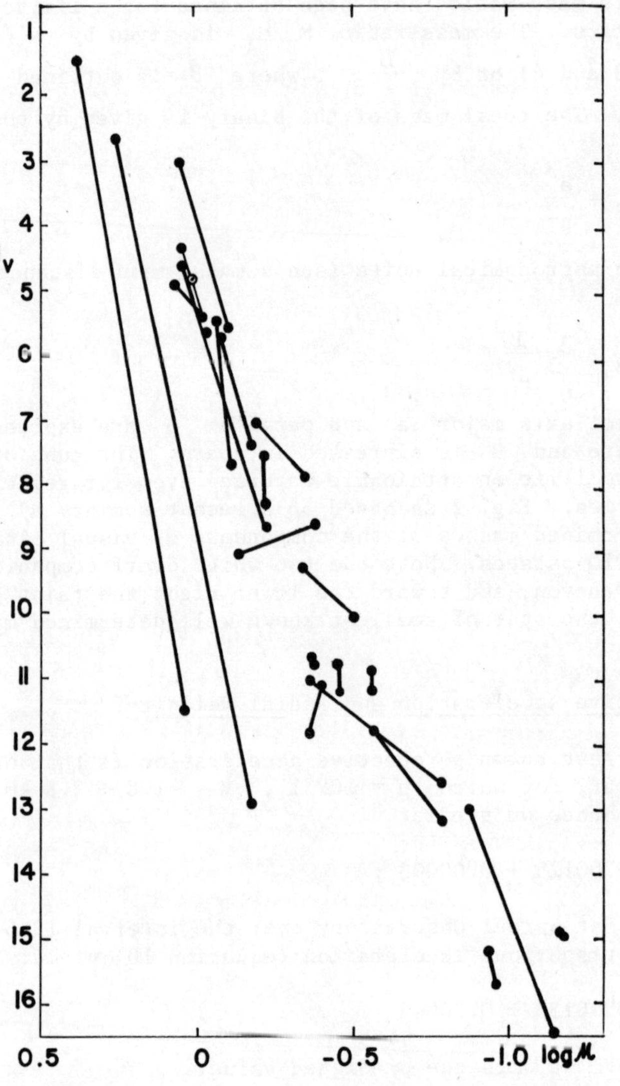

Figure 2.
Mass-Luminosity Relation for Binary Components Nearer then 10 Parsec. The Sun is Indicated by Θ.

The measured acceleration for Barnard's star yields

$$V = -112 \pm 3 \text{ km s}^{-1} .$$

In other words, at present we can determine the radial velocity of Barnard's star equally accurately from perspective acceleration as from Doppler effect!

Another interesting illustration is van Maanen's star, a faint white dwarf, for which, half a century ago, a very large radial velocity of recession $+ 238$ km s^{-1} was found, in conflict with the theory of galactic rotation and therefore possibly attributable to gravitational redshift. The astrometric approach via quadratic time-effect, both at Sproul and Allegheny observatories yields a radial velocity of only about 30 km s^{-1}. Meanwhile, later spectroscopic radial velocity determinations yielded a much lower value, requiring only a small redshift (if any).

d. Eclipsing Binaries.

Wouldn't it be nice if we could make parallax determinations from the orbit of Saturn, the size of which is nearly ten times that of the Earth? However, another method is available, occasionally, measuring the orbit of a double star both astrometrically (in seconds of arc) and spectroscopically (in astronomical units). At present there are two, all too rare, examples.

The eclipsing binary VV Cephei has a period of 20.4 yr and a semi-axis major a of 25 a.u. Long-focus photographic images refer to the photocenter, whose orbit has a semi-axis major α of 8.75 a.u. Observations at Sproul Observatory (1938-1976) yielded annual parallax = 0".0022 ± 0".0011 , reduced to absolute; semi-axis major α = 0".0120 ± 0".0013 = 8.75 a.u. whence orbital parallax = 0".0014 ± 0".0002 with an accuracy obviously much higher than provided by the annual parallax determination.

An even higher accuracy is reached for the eclipsing binary Epsilon Aurigae (period 27.08 yr), where the astrometric measurements refer to the bright component:

annual parallax = 0".003 ± 0".001 , reduced to absolute

semi-axis major a_1 = 0".0227 ± 0".0010 = 13.2 a.u. ,

whence

orbital parallax = 0".00172 ± 0".00008 .

5. PERTURBATIONS.

We have already referred to the classical discoveries of the perturbations in the paths of Sirius and Procyon, as well as in the orbits of Zeta-Cancri and Xi Ursae Majoris. The twentieth century has witnessed renewed progress in this field. The long-focus photographic technique, so eminently suitable for parallax and mass-ratio determinations, is equally applicable to the problem of perturbations.

An orbital analysis may be made from a perturbation in a proper motion if sufficient material is available. Of primary interest are the scale α reduced to linear measure (a.u.) and the period P (years). The harmonic relation is now replaced by the mass-function:

$$\frac{\alpha^3}{P^2} = (M_A + M_B)(B - \beta)^3 , \qquad (15)$$

where A and B refer to the primary and unseen companions, respectively. We may write

$$M_B = \alpha P^{-2/3} (M_A + M_B)^{2/3} + \beta(M_A + M_B) . \qquad (16)$$

The mass M_A of the primary is assumed to be known and is definitely not negligible. Interpretation of this formula requires adopted values for M_A and for β. The interpretation is very simple for negligible values of β. Experience has shown that $\beta = 0$ for $\Delta m > 3$. If, therefore, there is no visual evidence of the companion, and if $\alpha P^{-2/3}$, called the orbital constant, is small, we may write, to a high degree of approximation:

$$M_B = \alpha P^{-2/3} M_A^{2/3} . \qquad (17)$$

In 1937 a systematic search for perturbation among nearby stars was initiated at the Sproul Observatory. Other observatories, notably Allegheny, McCormick and, increasingly so, USNO, are also active in the field.

Historically, two interesting, outstanding photographic discoveries are those of Ross 614 by Reuyl in 1936 and of VW Cephei by Hershey in 1975. The perturbation of Ross 614 was not planned but simply (as several later on) was the by-product of a parallax determination. VW Cephei was put on the observing program of Sproul Observatory for the wrong reason, but nevertheless yielded an interesting perturbation!

Relevant data are:

Name	Parallax(")	P(yr)	α(")	$M_B/M_\odot$
Ross 614	0.243	16.6	0.312 ± 0.002	0.062
VW Cephei	0.041	30.45	0.130 ± 0.002	0.6

Both companions are main-sequence stars and both have now been seen, i.e., visually detected afterwards, in 1955 and 1975, respectively, as were the companions of Sirius and Procyon in 1862 and 1896.

A score or more well established perturbations are now known. We list a selection from these:

Name	Parallax(")	α(")	P(yr)	$M_B/M_\odot$
χ^1 Orionis	0.102	0.095	14.25	0.17
Wolf 1062	0.120	0.028	2.4	0.15
BD + 66° 34 A	0.100	0.125	15.95	0.13
G 24-16	0.115	0.028	1.5	0.07 - 0.11
CC 986	0.136	0.049	3.72	0.06 - 0.08

Perturbations have been found not only for "single" stars but also for established binaries and multiple stars. Note that BD + 66° 34 A , and its unseen companion a have a visual component B , orbiting around the center of mass of A and a in approximately 320 years. Note also that the unseen companions of G 24-16 and of CC 986 appear to have masses only slightly above the lower limit 0.06 $M_\odot$, according to theories for a bona-fide star, of which Ross 614 B is the best known visible example.

Most unseen companions inferred from perturbations are faint main sequence stars; the unseen companion of Zeta Cancri C is a white dwarf. However, there are a small number of unseen companions which, to the best of our knowledge and interpretation, are sub-stellar, i.e., have masses well below 0.06 $M_\odot$ but still well above the largest known mass for a visible planet, i.e., Jupiter, which has a mass slightly below 0.001 $M_\odot$. They include

Name	Parallax(")	α(")	P(yr)	$M_B/M_\odot$
BD + 68° 946	0.213	0.033	26.4	0.01
BD + 43° 4305	0.200	0.021	45	0.003 - 0.005
CC 1228	0.082	0.012	6.3	0.02

Furthermore, there appears to be one case of a faint red dwarf with two planetary companions (Section 7). The discovery of unseen companions is thus raising the multiplicity among stars. Zeta Aquarii (Strand, 1942) and BD + 66°34 (Hershey, 1973) are known to be triple systems. Zeta Cancri, Xi Ursae Majoris and G 70-106/7 are now known to be quadruple systems.

6. ULTIMATE ACCURACY.

The success in discovery and analysis of perturbations by long-focus photographic astrometry is determined and limited by the attainable accuracy. As mentioned before, for any individual star this appears to be limited by a year-error of 0."002.

Recent studies by Hershey, et al, from eight long-range series of stars with intensive observational coverage, obtained with the Sproul refractor (10.93 m focal length, 61 cm aperture), show this very nicely. Apart from a short interval (1941-1949), which required a systematic color correction in RA, the Sproul telescope reveals remarkable stability and accuracy. After eliminating proper motion, parallax, orbital motion and quadratic time-effect when needed, normal points for residuals over intervals of about one year, reveal an instrumental profile with an average deviation from zero close to 0."002 (or 1000 Å). These 8 stars were selected for long temporal coverage, and because they revealed no measurable perturbation.

An analysis like this is important because it provides a background, a test for the reality of perturbations. Fortunately, there are numerous perturbations with amplitudes well above 0."02 and their reality is thus assured. As to smaller perturbations with total amplitudes hovering around or not much above 0."02, one obviously has to be careful and rely on intensive observational coverage, on observational confirmation of predicted future positions or on observations made by more than one observatory. It is gratifying that some of the very small perturbations find confirmation from observations made both at the USNO and at the Sproul Observatory. Examples are Wolf 1062 and G 24-16.

7. BARNARD'S STAR.

This red dwarf, apparent visual magnitude 9.5, spectrum M5, is of particular interest since it is the nearest star in the northern hemisphere (RA = $17^h\ 55^m.4$, Decl. = +4° 33' ; 1950). It was discovered in 1916 as the star of the largest known proper motion, 10."31 annually, it has a parallax of 0."547 (distance 6.0 light years). Its radial velocity at present is -108 km/sec. By AD 11800 the star will be at the perihelion of its path, its parallax will then be 0".85 (distance 3.85 light years), its annual proper motion 25". At present the proper motion increases at the rate of 0".00125 annually, resulting in a substantial quadratic time effect in the position of Barnard's star.(Section 4c).

Barnard's star has been observed at the Sproul Observatory since 1938, at the annual rate of some 100 plates with up to five exposures each, on thirty nights. Material covering more than a thousand nights over the interval 1938-1981 has been measured using three reference stars. An analysis for proper motion parallax and quadratic time-effect leaves annual residuals which clearly indicate a perturbation. The pattern of these yearly mean residuals may be explained as the result of two components: circular perturbations with periods of 13.5 and 19.0 years, and perturbations with radii of about 0".001 each. Their reality can hardly be doubted, in view of the high instrumental precision profile (Section 6) of the telescope and, for several years now, the close agreement of current observation with (earlier) predictions. The greatest elongation of the inferred companions reaches extreme values of about one second of arc, assuming a value of 0.14 $M_\odot$ for the mass of the primary, but no trace of companions has been detected on the well over ten thousand exposures.

A simple calculation yields masses of about 2/3 times that of Jupiter for each of the companions. The orbits of the two companions are not too far from being coplanar and could be co-rotational, the angle between the two orbits being about 25°. The limited accuracy of the observation does not exclude the possibility of much closer co-planarity.

This then would seem to be evidence for the existence of planetary companions for a nearby star. Had Barnard's star been twice as far away, the discovery of the companions would have been much more difficult. A similar situation for Proxima Centauri would have been more easily discovered, but unfortunately that star, as well as other nearby stars in the Southern equatorial hemisphere, await much needed astrometric observations.

It would seem that astrometric searches of this kind should be continued, as long as no other effective techniques are established for the discovery of planets beyond our solar system. Remember that existing techniques, other than the perturbation approach, could not yet reveal a Jupiter-like planet orbiting one of the components of the nearest stellar system Alpha Centauri!

MODERN LUNAR THEORY

J. KOVALEVSKY

CERGA, Grasse, France

ABSTRACT. After a few general remarks on the various perturbations acting on the motion of the Moon, the main problem of the lunar theory is presented.

Formal solutions of the problem are presented and the effects of small divisors are described. There exists a converging iterative procedure that permits the construction of a formal solution to any order despite the difficulty introduced by small divisors and particularly the critical argument $2D - F - \ell + 2\ell'$. Semi numerical solutions are also described. Iterative methods do not converge but there are techniques that permit to deal with small divisors.

Finally, both types of theory are compared from three points of view : series manipulation, convergence of the computation and convergence of the series.

1. INTRODUCTION

The goal of modern lunar theory is to produce the geocentric coordinates of the center of mass of the Moon as function of time with an accuracy such that they match the observations of lunar retroreflectors by lunar lasers. These are presently made with an internal precision of better than 10 cm and it is contemplated that, in the future, they might reach the 1 or 2 cm accuracy limit. This means that the relative exactness of the expressions representing the coordinates of the Moon should be ensured to at least 10^{-10} or 10^{-11}.

V. Szebehely (ed.), Applications of Modern Dynamics to Celestial Mechanics and Astrodynamics, 59–76.

This objective can definitely be met by numerical integration, provided that the physical model representing the Earth-Moon system and its dynamical environment is known sufficiently well so that the necessary accuracy of the results can be achieved for a sufficient interval of time (10 to 20 years presently). Table 1 presents the sizes of the various perturbations of the lunar motion. It can be seen that, compared with the requirements, all of them should be taken into account, sometimes with high relative precision.

Table 1

effects	order of magnitude
2 body problem	1
Main problem	4.10^{-2}
Direct planetary perturbations	7.10^{-5}
Indirect planetary perturbations	5.10^{-6}
Earth oblateness (J_2)	3.10^{-6}
Earth potential (J_3)	5.10^{-8}
Earth potential (J_4)	1.10^{-10}
Nutation	2.10^{-8}
General relativity	2.10^{-9}
Tidal interaction	5.10^{-9}

We are not going to describe here the many methods that were used to compute any of these effects. Let us just say that all have been recently investigated and a complete semi-numerical theory exists with an accuracy of about 20 meters (Chapront and Chapront-Touzé, 1982). Some methods are straightforward. Others are rather involved as in the case of relativistic effects (Lestrade, 1980) and planetary perturbations (Chapront-Touzé and Chapront, 1980). The latter are at present the less precisely determined despite the fact that they have been computed to the third order of planetary masses and that their expression requires several times more terms than a much more accurate solution of the main problem.

Although numerical integration may suffice for practical needs of to-day, it remains essential to obtain analytical solutions in order to understand fully the dynamical behaviour of the Moon and to detect the terms sensitive to various physical parameters. An accurate lunar theory remains one of the major objectives of Celestial Mechanics. With respect to the present achievements, one has still to gain one or two orders of magnitude. Therefore, it is not out of order to discuss now some of the difficulties encountered in constructing a lunar theory and, particularly, the solution of of the main problem.

These difficulties are essentially due to :

- the very large number of terms in the series,
- the convergence of the computing procedure,
- the mathematical convergence of the series.

2. THE MAIN PROBLEM OF THE LUNAR THEORY

Let us select a geocentric reference system with fixed directions in space. The TXY plane is a plane parallel to the ellipse representing the apparent motion of the Sun around the barycenter G of the Earth-Moon system. The following notations will be adopted in this paper :

M : mass of the Moon (L)
E : mass of the Earth (T)
m' : mass of the Sun (S)

Jacobian coordinates (see figure) are adopted :

$\vec{TL}$: (x,y,z;r)

$\vec{GS}$: (x',y',z';r')

$\theta = \cos(\vec{GS}, \vec{GL})$

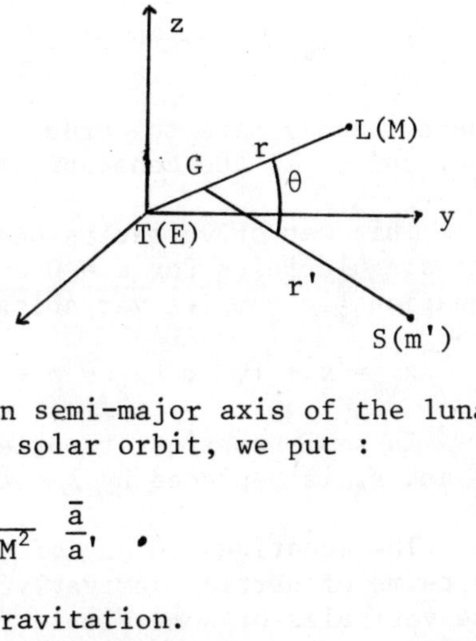

If $\bar{a}$ and a' are respectively the mean semi-major axis of the lunar orbit and the semi-major axis of the solar orbit, we put :

$$\alpha = \frac{E - M}{E + M}\frac{\bar{a}}{a'} \quad , \quad \mu = \frac{EM}{E^2 - M^2}\frac{\bar{a}}{a'} \, .$$

Finally, k is the constant of gravitation.

The equations of motion can be expressed in terms of a disturbing function

$$R = km' \left[\frac{E + M}{E}\frac{1}{|LS|} + \frac{E + M}{M}\frac{1}{|TS|}\right] \tag{1}$$

Assuming that r/r' is a small quantity, R may be developed as:

$$R = n'^2 a^2 \left[\left(\frac{r}{a}\right)^2\left(\frac{a'}{r'}\right)^3 P_2(\theta) + \left(\frac{r}{a}\right)^3\left(\frac{a'}{r'}\right)^4 \alpha P_3(\theta) + \left(\frac{r}{a}\right)^4\left(\frac{a'}{r'}\right)^5 (\alpha^2 + \alpha\mu) P_4(\theta) + \ldots\right] \tag{2}$$

where we have replaced km'/a'^3 by n'^2 , n' being the mean motion of the solar orbit.

Many different sets of variables can be used :

- The elliptic elements : a, e, I, $h = \Omega$, $g = \omega$, $\ell = n(t-\tau)$ with the Lagrange equations ;

- Delaunay variables and the corresponding system of canonical equations ;

- rectangular coordinates.

In recent theories, another system was used. Introduced by Chapront and Mangeney (1969), it is the following :

$$\left.\begin{aligned} & m = \frac{n'}{n} , \quad x = e\cos\ell , \quad y = e\sin\ell \\ & p = \sin\frac{I}{2}\cos(v+g), \quad q = \sin\frac{I}{2}\sin(v+g) \\ & r_0 = m_0\cos(v+g+h-v'), \quad s_0 = m_0\sin(v+g+h-v') \end{aligned}\right\} \quad (3)$$

where v and v' are the true anomalies relative to the Moon and the Sun, and m_0 is the constant of integration of the equation in m.

This set of variables has the double advantage of removing the singularities for $e = 0$ and $i = 0$ and of leading to simple equations in complex variables defined as follows :

$$z_1 = x + iy \; ; \; z_2 = p + iq \; ; \; z_3 = r_0 + is_0$$

In semi-numerical theories, the redundant couple of variables r_0 and s_0 is replaced by $\lambda = \ell + g + h$.

The equations in any of these sets of variables are expressed in terms of partial derivatives of R or parts of R with respect to the variables or some simple function of them. In fact, they are all equivalent in the sense that the Jacobian of the transformation from any set of variables $(x_1, x_2 \ldots x_6)$ into another $(y_1, y_2 \ldots y_6)$ is non-vanishing and not singular for all the values of the variables within the domain of validity of the lunar theory. Hence, the conclusions reached with one of these sets can be applied to any other.

To solve these equations, two completely different approaches may be chosen and have both been recently used :

1. Analytical or formal theory,

2. semi-numerical theory.

3. FORMAL SOLUTIONS

The principles and the signification of formal solutions of a system of differential equations are found in Poincaré's "Méthodes nouvelles de la Mécanique Céleste", chapter VIII (1893). In these solutions, one expresses the coordinates or any other variables as series of time in such a way that all the parameters and all the constants of integration appear in a litteral form.

In the lunar theory, a number of parameters and of constants of integration are small quantities. Advantage is taken of this property in order to develop the solution in power series of these small quantities. The following quantities are chosen :

$$\begin{aligned} m_0 &= 0.0748\ldots \\ e_0 &= 0.0549\ldots && \text{(mean eccentricity of the lunar orbit)} \\ e' &= 0.0167\ldots && \text{(eccentricity of the solar orbit)} \\ \gamma_0 &= 0.0445\ldots && \text{(mean sin I/2)} \\ \alpha &= 0.00251\ldots \\ \mu &= 0.0000316\ldots \end{aligned} \tag{4}$$

Any function f(x) where x is one of the parameters and x_0 the corresponding small quantity may be written :

$$f(x) = f(x_0 \cdot \frac{x}{x_0})$$

where $y = x/x_0$ is close to 1 and is a new variable.

One may develop f(x) in the form :

$$f(x) = \sum_{i=0}^{\infty} x_0^i A_i y^i \tag{5}$$

The expression (5) is the formal development of f(x). However, in order to simplify the formulation, we shall write (5) as :

$$f(x) = \sum_{i=0}^{\infty} A_i x^i \tag{6}$$

and keep formally x as a variable. All the computations can be correctly made using the form (6).

In particular, one may develop the disturbing function R (2) and express it in power series of the parameters (4) or in terms of m, e, γ instead of m_0, e_0 and γ_0 and in trigonometric series of the angular variables ℓ, g, h and ℓ' (see for instance, Brouwer and Clemence, 1961). Traditionally, however, one uses :

$$F = \ell + g$$
$$D = \ell + g + h - \ell'$$

rather than g and h.

One obtains finally :

$$R = \sum_{k_1} \sum_{k_2} \sum_{k_3} \sum_{k_4} A_{(k)}(m,e,e',\gamma,a,\mu)\cos(k_1D+k_2F+k_3\ell+k_4\ell') \qquad (7)$$

where (k) stands for k_1, k_2, k_3, k_4

The factor n'^2a^2 present in (2) can be written as $m_0^2a^2n^2$. Hence, R is at least of order two in the small quantities.

Let us also remember that (7) being obtained through some algebraic manipulations of the developments of the two body problem, the coefficients A reflect the parity and order properties of Bessel functions. This is called the D'Alembert characteristics : the coefficient $A_{(k)}$ of $\cos(k_1D+k_2F+k_3\ell+k_4\ell')$ is of the order of :

$$e^{|k_3|} \,.\, e'^{|k_2|} \,.\, \gamma^{|k_4|} \qquad (10)$$

Furthermore, k_2 is even.

In deriving the solution of the equations of motion written using (7), one must make an a priori choice of the relative orders of the small quantities. This choice is rather arbitrary. Delaunay has set to 1 the order of m_0, e_0, e', γ_0 and to 2 the order of α. This was also the choice of Bec-Borsenberger (1978) who added μ to the equations with an order 4. On the other hand, Deprit, Henrard and Rom (1970) owing to the very slow convergence of the series in m , have used the following convention :

m_0 : order 1

e_0, e', γ_0 : order 2

α : order 4

Different methods have been used in deriving the solution. Delaunay has applied its well known method. Deprit and Henrard used Lee series and Bec-Borsenberger used an iterative procedure (see Bec et al., 1973). But in all the cases, the solutions built by increasing orders of small quantities are equivalent in the sense that the same variables have the same expression when they are expressed in terms of the same small quantities (if this was not the case, one of the expressions would not satisfy formally the equations to the given order). The general form for the solution is :

$$X = \sum_{(k)} A_{(k)}(m_0, e_0, e', \gamma_0, \alpha, \mu)\,\mathrm{soc}(k_1\bar{D} + k_2\bar{F} + k_3\bar{\ell} + k_4\ell') \tag{11}$$

$$\text{with } A_{(k)} = \sum B_{(\lambda,k)}\, m_0^{\lambda_1} e_0^{\lambda_2} e'^{\lambda_3} \gamma_0^{\lambda_4} \alpha^{\lambda_5} \mu^{\lambda_6} \tag{12}$$

where we use the notation soc for sin or cos.

In these expressions, $B_{(\alpha,k)}$ are numbers and, by definition, $\ell' = n'(t - \tau')$.

Furthermore $\bar{D} = n_D(t - \tau_D)$,

$\bar{F} = n_F(t - \tau_F)$,

$\bar{\ell} = n\,(t - \tau_\ell)$,

n_D, n_F, n_ℓ having the form (12) and all τ are constants of integration.

4. MAIN DIFFICULTIES OF FORMAL SOLUTIONS

Whatever is the method used to solve the equations, the intermediate expressions have the form (11). At some stage, it is necessary to perform integrations on such terms and compute

$I = \int X\,dt$

Each term of I has the form :

$$E_{(k)} = \frac{\pm A_{(k)}\,\mathrm{soc}(k_1\bar{D} + k_2\bar{F} + k_3\bar{\ell} + k_4\ell')}{k_1 n_D + k_2 n_F + k_3 n_\ell + k_4 n'} \tag{13}$$

Noting that the first terms of the mean motions as developed to the third order are :

$$n_D = n_0\left[1 - m_0 - m_0^2 + O(4)\right],$$

$$n_F = n_0\left[1 - \frac{1}{4}m_0^2 - \frac{9}{32}m_0^3 + O(4)\right],$$

$$n_\ell = n_0\left[1 - \frac{7}{4}m_0^2 - \frac{225}{32}m_0^3 + O(4)\right],$$

$$n' = n_0\left[m_0\right],$$

we deduce that the divisor of $E_{(k)}$, developed in m_0, is

$$D_{(k)} = n_0\Big[(k_1+k_2+k_3) + m_0(k_4-k_1) - m_0^2\ (k_1+\frac{k_2}{4}+\frac{7}{4}k_3)$$
$$- (\frac{9}{32}k_2 + \frac{225}{32}k_3)\ m_0^3 + O(4)\Big]$$

Depending on relations between $k_1 \ldots k_4$, one may get divisors of the order of m_0, m_0^2 and even m_0^3, but never of a higher order.

The appearance of divisors $D_{(k)}$ might suggest to keep them as such and not to try to transform the expression into a development of the form (12). The difficulty of such a procedure is that during the various series manipulations, the numerators and the denominators are multiplied and the numerical coefficients grow very rapidly. After a few additions or multiplications of series, they become completely unmanageable. This implies that one has to develop the denominators using the binomial expansion assuming that $D_{(k)}$ takes the form :

$$D_{(k)} = A_0 m_0^i\ (1 + \Sigma\ A_{(\lambda)} m_0^{\lambda_1}\ e_0^{\lambda_2}\ e'^{\lambda_3}\ \gamma_0^{\lambda_4}\ \alpha^{\lambda_5}\ \mu^{\lambda_6})$$

$$\text{or}\quad D_{(k)} = A_0\ m_0^i\ (1 + \varepsilon) \tag{14}$$

The development of $1/(1 + \varepsilon)$ is convergent only if $|\varepsilon| < 1$. This is verified for the divisors encountered in the first iterations of the solution. But this may not always be true and, as soon as this condition is not met, the construction of a formal solution is no more possible for higher order terms. This case has not been reached in the formal solutions constructed up to now, but there must be a definite limiting order beyond which the formal solution cannot be constructed and the series become formally divergent.

Another difficulty is due to the division by m_0^i, independently of its binomial multiplier. Since m^2 appears as a factor of R, for i = 0, 1 and 2, it is indeed possible to make the division, without altering the form (12) of the coefficients. But in the case when i = 3, the situation becomes different. This happens for critical arguments such as :

$$\phi_c = 2k\ \ (- 2D + F + \ell - 2\ell') \tag{15}$$

The corresponding divisor is :

$$n_{\phi_c} = - \frac{117}{16}\ k_3\ m_0^3 + O(m_0^4)$$

and therefore, since the numerator may have only m_0^2 as a factor, a divisor m_0 will remain in the results. This case was studied by Poincaré (1908) and it appears that as soon as such critical arguments are considered, the singularity of m = 0 explicitely occurs. However, the order is positive since the characteristic is $(e_0^2 e'^4 \gamma_0^2)^{k_3}$. A. Bec-Borsenberger has verified that such a term does not appear in the disturbing function with $k_3 = 1$. However, it may occur with higher order terms during the computation and therefore, one must admit that the form of the solution as given by (11) and (12) must be modified, allowing for a negative exponent λ_1 of m_0. But formally, in terms of global orders, such terms do not produce inacceptable results.

Another consequence of the existence of small divisors factored by m_0^i is that, in order to construct all terms to an order Ω, it is necessary to compute the right hand members of the equations to the order $\Omega + i$. One may wonder whether the construction of the equations to the order $\Omega + i$ can be made without having a complete solution of order Ω . If this was right, the procedure of construction of a formal solution would not be convergent and therefore, it would not be possible to compute such a solution. Let us consider this problem.

5. CONVERGENCE OF THE ITERATIVE PROCEDURE

A detailed study of the behaviour of all types of terms during the construction of a litteral solution of the main lunar problem has recently been made by Borsenberger-Bec (1979) and Borsenberger-Bec and Kovalevsky (1979) in order to investigate whether there exists a procedure that permits to obtain all terms of a given order in a finite number of operations.

In practice, this was considered using variables defined in (3). The method is an iterative one : the result of the Nth iteration is substituted in the right-hand members of the equations and the (N + 1)th solution is then obtained through an integration. Let us assume that a solution complete to the order ω_0 has been obtained. This has been ascertained for $\omega_0 = 5$ through a number of supplementary iterations during which no new term has appeared, and also by comparing with Delaunay series. Then, if one shows that a finite number of operations suffices to construct a complete solution for $\omega_0 + 1$ whatever is ω_0, the convergence of the procedure is proved.

In this proof, one considers a term $\delta x_j = a^i_{jk} \sin \phi_i$ (16)

obtained during a certain iteration for the variable x_j (j=1,2...7) with

$$\phi_i = i_1 D + i_2 F + i_3 \ell + i_4 \ell'$$

$$a^i_{jk} = c^i_j \, m_0^{k_1} \, e_0^{k_2} \, e'^{k_3} \, \gamma_0^{k_4} \, \alpha^{k_5} \, \mu^{k_6},$$

the order being :

$$\Omega_{jk} = k_1 + k_2 + k_3 + k_4 + 2k_5 + 4k_6$$

The equations of motion are written explicitly in terms of increments of the variables.

Putting $x_j = x_j^0 + \delta x_j$, the equations $dx_j/dt = X_j$ are written in the form of

$$\frac{dx_j}{dt} = X_j \, (\vec{x}_0) + \Sigma \, \frac{\partial X_j(\vec{x}_0)}{\partial x_i} \, \delta x_i + \ldots \qquad (17)$$

where $\vec{x}_0$ represents the solution of order ω_0. It is easy to show that, since the order of δx_j is larger than $\omega_0 \geqslant 5$ and since the maximum change of order during an integration is 3, it is sufficient to have $\partial X_i(\vec{x}_0)/\partial x_i$ only to the order 4. These quantities are explicitly computed from the basic solution of order 5 and are identical for any ω_0. Furthermore, when $\omega_0 \geqslant 5$, higher order terms in (16) cannot build up contributions to the solution to the order $\omega_0 + 1$.

To trace all the contributions of any type of x_j in the solutions of (17) is a very tedious task that has been completely described by Borsenberger-Bec (1979).

In a first step, the formation of terms of order $\omega_0 + 1$ during the first iteration is studied. Since divisors may have orders 0, 1, 2 or 3, the right-hand members of (17) are computed to the order $\omega_0 + 4$. Some terms that are formed are of the order $\omega_0 + 1$ and are part of the solution in construction. Other are of order $\omega_0 + 2$ to $\omega_0 + 4$.

During the second iteration, we substitute in the δx_i of equation (17) all the terms formed during the first iteration. The right-hand members of this equation are therefore at least of order $\omega_0 + 2$. Consequently, the new terms of the solution to the order $\omega_0 + 1$ that will be formed at this point, will have been divided by some small divisor.

Each of these new terms is again considered and its new effects are analysed, and so on. In practice, the number of cases to be considered is limited since only those terms that give rise to small divisors are considered. Among those terms, only those whose combinations with the arguments present in the partial derivatives give again rise to small divisors, may contribute to the solution of order $\omega_0 + 1$.

It was shown by Borsenberger-Bec (1979) that for all arguments with small divisors of order 1 or 2, no more contributions are created after the fourth iteration.

However, in the case of a combination of the critical argument ϕ_c as given by (15) and $\pm \ell$, a term of the same order and the same argument is produced at every iteration in the variables x and y. However, we have shown (Borsenberger-Bec and Kovalevsky, 1979) that if c_i is the coefficient of such argument in the i-th iteration, the coefficient becomes

$$c_{i+1} = \frac{225}{234\,k_3} c_i = \lambda c_i \quad \text{in the next iteration.}$$

The successive contributions are terms of a convergent geometric series, so that if c_1 is the corresponding term of the first iteration, the total effect is :

$$\Delta x = \sum_{i=1}^{\infty} \delta x_i = c_1 \left(1 + \lambda + \lambda^2 + \ldots\right) = \frac{c_1}{1 - \lambda} \tag{18}$$

In conclusion, we have shown that the iterative method of solving the equations is a convergent procedure to obtain the formal solution of the equations of motion. A maximum of 4 iterations is necessary to gain one order in the small parameters, provided that the algorithm (18) is used whenever the critical argument is formed in the equations in x and y.

This does not prove that any method would also converge. But this proves that there exists convergent algorithms that permit the construction of a formal solution of the main problem of the lunar theory, even when critical terms appear.

6. SEMI-NUMERICAL METHOD

A major disadvantage of literal theories is that the series representing the solution are very cumbersome and that very heavy and long calculations are involved in deriving them. One way to simplify the theory is to replace some or all literal parameters by numbers.

The form of series (11) is hence :

$$X = \Sigma A_{(k)} \text{soc} \, (k_1\bar{D} + k_2\bar{F} + k_3\bar{\ell} + k_4\ell') \tag{19}$$

where A(k) is a number. Similarly, the coefficients of t in $\bar{D}$, $\bar{F}$, $\bar{\ell}$ and ℓ' are also numbers.

The original work that is closest to this concept is Brown's lunar theory. It is however the advent of high speed computers capable of treating large series of the form (19) that brought this technique to its present form (Kovalevsky, 1959). The application of this method to the main problem of the lunar theory was made by Chapront-Touzé, 1976 and 1980, using the system of variables (3) described above.

In principle, one is led to integrate equations of the form

$$\frac{d\vec{X}}{dt} = \vec{F}\ (\vec{X},t) \tag{20}$$

where $\vec{F}$ is composed of series of type (19).

The integration gives :

$$\vec{X} = \vec{X}_0 + \int \vec{F}(\vec{X},t)dt = \vec{X}_0 + \vec{Y} \tag{21}$$

At this point, it is necessary to determine the constants of integration $\vec{X}_0$. This is done by forcing some coefficient to have a given value. For instance, the coefficient of sin $\bar{\ell}$ in the expression of the longitude :

$$L = \bar{\lambda} + 2e_0 \sin \bar{\ell} + \ldots$$

is one of such adjustment constants. It is therefore necessary to compute L from the series (20), keeping the components of $\vec{X}_0$ constant. Then, the coefficient of sin $\bar{\ell}$ is identified with $2e_0$, imposing a condition on $\vec{X}_0$.

Once the constants of integration are defined in this manner, the new solution is substituted in the right hand members of (19) and the procedure is reiterated until the difference between two successive solutions $\vec{X}_{i+1} - \vec{X}_i$ is negligible.

However, this is not generally true, at least for some terms, especially those which undergo a division by a small divisor during the integration process. After ten iterations, Chapront-Touzé detected a number of such divergent terms.

Let us consider the system used by Chapront-Touzé :

$$\begin{aligned} \frac{dh_i}{dt} &= \varepsilon H_i(h_j) && 1 \leqslant i \leqslant 5 \ ; \ 1 \leqslant j \leqslant 6 \\ \frac{dh_6}{dt} &= h_1 + \varepsilon H_6(h_j) && (h_6 = \lambda = \ell + g + h) \end{aligned} \tag{23}$$

The last equation has a first order term $h_1 = m$. All other terms of the equations are multiplied at least by the small coefficient factoring the disturbing force, i.e. $\varepsilon = m^2$. It is a second order small quantity in terms of the literal theory.

Let us substitute a solution which has the form :

$$h_i = \sum_k h_i^k \operatorname{soc} \phi_k \tag{23}$$

and one gets $H_i(h_j) = \sum_k H_i^k(h_j) \operatorname{soc} \phi_k$

And, by integration of (22), one gets

$$h_1 = \varepsilon \sum \frac{h_1^k (h_j) \sin \phi_k}{\nu_k}$$

and

$$h_6 = \varepsilon \sum \frac{H_1^k(h_j) \cos \phi_k}{\nu_k^2} + \frac{H_6^k(h_j) \cos \phi_k}{\nu_k}$$

where ν_k is the frequency corresponding to ϕ_k.

H_1^k is computed using the preceding iteration. If ν_k is of order 0 or 1, all the terms of h_6 have at least the same precision than H_1^k and hence are improved. But if ν_k is of order 2, h_6 can never be improved and the method does not converge.

A completely different method must be applied to such terms. This was done by Eckert and Smith (1966). A simpler method is due to Chapront-Touzé (1976). Let us assume that h_0 is the result of the N-th iteration and δh is an unknown increment to be determined. One has from (22)

$$\frac{d}{dt}(h_i+\delta h_i) = \varepsilon H_i(h_j+\delta h_j) = \varepsilon H_i(h_j) + \sum \frac{\partial H_i}{\partial h_j} \delta h_j$$

Using the developments (23), we integrate, identify term by term and extract the coefficients of soc ϕ_k with the frequency ν_k.

$$\pm(\nu_k+\delta\nu_k)(h_i^k + \delta h_i^k) = H_i^k(h_j) + \sum \left| \frac{\partial H_i}{\partial h_j} \delta h_j \right|_{(k)}$$

Neglecting the second order product $\delta\nu_k \, \delta h_i^k$, one gets a linear equation in δh_i^k and $\delta\nu_k$. Adding the three supplementary conditions expressing that there are no secular terms in the even developments of y and q and that the secular term of λ is m_0, one obtains as many equations as unknowns and one can determine all δh_i^k one needs (provided that their number is not too large in order to avoid correlations).

7. FIRST ORDER SEMI-NUMERICAL THEORY

A major drawback of a solution obtained as described above is that the constants of integration ($\vec{C}_0$) must be predetermined before the solution is constructed. In order to improve the solution, it is necessary to have the expression of all the solutions for a vicinity of $\vec{C}_0$. Calling $\delta\vec{C}(\delta C_1 \ldots \delta C_6)$ any increment to $\vec{C}_0$, one has :

$$h_j(\vec{C}_0 + \delta\vec{C}) = h_j(\vec{C}_0) + \sum_{i=1}^{6} \frac{\partial h_j}{\partial C_i} \delta C_i \tag{24}$$

h_j is a solution of the differential equation :

$$\frac{dh_j}{dt} = H_j(h_i) + \sum_{i=1}^{6} \frac{\partial}{\partial C_i} \left(\frac{dh_i(\vec{C}_0)}{dt} \right) \delta C_i \tag{25}$$

A substitution of (24) in (25) produces a set of 36 differential equations in $\partial h_j/\partial C_i$ that can be treated by a similar method as the original equation. Once the solution is obtained, one gets (24) in a semi-numerical form, except for δC_i. This first order solution is normally sufficient for the Moon since the value of $\vec{C}_0$ is very well known a priori. The same difficulties of convergence linked with the small divisors as in the zero order solution exist. The same methods may be applied to resolve them.

8. MIXED SOLUTION

As we have seen in section 4, one major difficulty of a litteral solution is the necessity to develop the inverse of the denominator in powers of the small parameters, leading to slowly convergent if not divergent expressions. One may try to express the solution not in function of these parameters, but in function of their increments. Any expression (11) would take the form :

$$X = \sum_{(k)} A_{(k)}(m_0+\delta m, e_0+\delta e, \ldots) \operatorname{soc}(k_1\bar{D}+k_2\bar{F}+k_3\bar{\ell}+k_4\ell') \tag{26}$$

where m_0, e_0 ... are numerical constants, so that the literal parameters are much smaller, allowing for much faster convergence of the series. This was recently advocated by Henrard (1979). It has the advantage of a litteral solution, since the derivatives with respect to all parameters can be readily computed.

But the constant part of the solution of (26) is a semi-analytical solution with constants of integration equal to m_0, e_0 etc. So in their determination, one will run into all the difficulties described in the last two paragraphs. These may be a gain in computing the first and eventually further order theories. But the main core of the computation is strictly the construction of a semi-numerical theory.

9. COMPARISON BETWEEN THE TWO TYPES OF THEORY

During our presentation of literal and semi-numerical theories, advantages and disadvantages were quoted. Some others were not dealt with. Let us compare the principal characteristics of these two approaches.

9.1. Computation and series manipulation

A first obvious fact is that the construction of a literal theory implies much larger series : six parameters and four arguments while only four parameters are needed for a semi-analytical theory. A complete 9-th order polynomial (12) has 296 coefficients B_λ and often more than 50 non zero terms are actually present in such terms. This means that together with a largely increased complexity of the series manipulation programs, a factor of about 50 is to be expected for a literal theory equivalent in precision to the Chapront-Touzé theory (1980) which has about 2000 terms per expression with a precision estimated to 40 cm (0".0002). This is to compare with the 30,000 terms computed by Deprit, Rom and Henrard for an analytical solution (Henrard, 1973) which still has some uncertainties of the order of 2 to 10 meters (0".001 to 0".005) as quoted by Henrard (1978).

A difficult point in all theories is the existence and propagation of round-off errors. Of course, this can be reduced by using more significant figures. In the case of literal theories, at least for the lower order terms, the fact that all coefficients are rational numbers has been used for correcting such errors. This is a small advantage over semi-numerical theories.

9.2. Convergence of the computing procedure

We have proved that the iterative method of constructing a formal theory is convergent, and also seen why the same method does not converge in the semi-numerical iterative method. A reason for this discrepancy is the use of the variable $\lambda = \ell + g + h$ that involves a double integration, while the use of redundant variables r_0 and s_0 does not imply such a double integration with divisors that are often of order 4 and eventually more for critical terms. But since it is rather complex to use redundant variables in a semi-numerical theory, this drawback is better removed by direct algebraic computation of the corresponding terms as the one we have presented.

It is however to be noted that even if the same set of variables is adopted for both theories, there is still no other criterion to establish the convergence of the procedure but to compare series obtained in successive approximations. This procedure is not able to distinguish between the causes of apparent non convergence : truncation errors, round-off errors, effects of small divisors, slowness of the procedure adopted, etc... Any of these effects may produce erroneous contributions to some of the computed terms and effect convergence. In particular, it is not possible to know a priori what arguments may rise after a few iterations and which are spurious arguments introduced by round-off errors.

One may compare how the various terms are produced in a litteral theory where the orders of computed terms are strictly controlled, with the occurence of their equivalent in a semi-numerical theory. One can see that many high order terms obtained before the iterations have reached the necessary accuracy for low order solutions are totally wrong. These are automatically eliminated in the formal theories but in semi-numerical methods, they are kept and these spurious quantities may fastly propagate throughout the series.

In conclusion, analytical methods are much more efficient in controlling the construction of terms ; they avoid truncation errors up to a given order, they automatically remove spurious higher order terms whenever they appear and, above all, there exist a convergent computing procedure that is correct for all orders.

9.3. Convergence of series

The convergence of series representing the solution is the fundamental difficulty for any type of theory. The problem being non-integrable, the Kolmogorov-Arnold-Moser theorem leaves us with no hope of having an analytical or semi-numerical expressions representing a flow of trajectories for a domain of possible initial conditions. It is still convenient to use the Poincaré's conception of asymptotic series to describe the situation. The truncated series represent the solution in a limited interval of time Δt with an error ε. This means that, in the domain of the initial conditions of the lunar theory, one is far from any ergodic or quasi-ergodic region of the phase space. A flow of solutions originated in this domain remains in the interval Δt with a maximum error of $\varepsilon/2$ by analytical functions of constants of integration. Furthermore, the truncation errors should be uniformly smaller than $\varepsilon/2$ in the same interval of time if one keeps all the terms to the order N. No evaluation of the relationship between ε, Δt and N exists. It has however been shown, by comparison with numerical integrations, that the present semi-numerical theories are better than $\varepsilon = 40$ cm in $\Delta t = 20$ years (Kinoshita, 1982). This is a lower bound of the possibilities of such theories.

In the case of literal theories an additional problem arises from the slow convergence of the binomial expansion (see section 4). It is not known starting from what order the development (14) does not converge any more. This order will be another limitation of the application of formal theories. This limit depends upon the numerical values of the parameters m_0, e_0, e', etc... In the case of the lunar theory, this problem did not arise in the Deprit-Henrard solution, although some terms have been computed up to the 20th order.

In practice, this problem of strict convergence may not be as critical. The main problem is only an approximation of the actual problem. Various perturbations listed in 1 are to be added and some of them have the effect of changing slowly the values of the parameters. In particular, this is the case of the tidal interaction and dissipation in the Earth-Moon system, which produce a yearly variation of m_0 of the order of 10^{-10}. This single effect, if added to the main problem of the lunar theory, may change the problem of convergence of series and this should be investigated. Other perturbations, like the various planetary effects are also introduced with secular terms which again produce secular changes in the parameters.

So, in conclusion, let us say that, in practice, to the accuracy that has been achieved, no trace of difficulty due to the lack of convergence, has been detected. Even if another two orders of magnitude in the precision are gained - and this would be necessary to meet the present requirements - there is no sign of difficulty due to convergence problems.

REFERENCES.

Bec, A., Kovalevsky, J. and Meyer, C., 1973, The Moon, **8**, p. 434.

Bec-Borsenberger, A., 1978, *Etude d'une théorie analytique du problème principal de la Lune*, Thesis, Université Paris 6.

Bec-Borsenberger, A., 1979, Celestial Mech., **20**, p. 355.

Bec-Borsenberger, A. and Kovalevsky, J., 1979, in *Natural and artificial satellite motion*, P.E. Nacozy and S. Ferraz-Mello Ed., Univ. of Texas Press, p. 83.

Brouwer, D. and Clemence, G.M., 1961, *Methods of Celestial Mechanics*, Academic Press, p. 311.

Chapront, J. and Chapront-Touzé, M., 1981, in IAU colloq. N°63, Grasse, in press, Reidel Publishing Co.

Chapront, J. and Mangeney, L., 1969, Astron. and Astroph., 2, p. 425.

Chapront-Touzé, M., 1976, *Construction itérative d'une solution semi-analytique du problème central de la Lune*, Thesis, Université Paris 6.

Chapront-Touzé, M., 1980, Astron. and Astroph., 83, p. 86.

Chapront-Touzé, M. and Chapront, J., 1980, Astron. and Astroph., 91, p. 237.

Deprit, A., Henrard, J. and Rom, A., 1970, Astron. J., 75, p. 747.

Eckert, W.J. and Smith, H.F. Jr, 1966, IAU symp. N°25, p. 242.

Henrard, J., 1973, Ciel et Terre, 89, p. 1.

Henrard, J., 1978, Celestial Mech., 17, p. 195.

Henrard, J., 1979, Celestial Mech., 19, p. 337.

Kinoshita, H., 1981, in IAU Colloq. N°63, Grasse, in press.

Kovalevsky, J., 1959, Bulletin Astron., série 3, 23, p. 1.

Lestrade, J.F., 1980, Astron. and Astroph., 92, p. 302.

Poincaré, H., 1908, Bulletin Astron., 25, p. 321.

SOME ASPECTS OF MOTION IN THE GENERAL PLANAR PROBLEM OF THREE BODIES; IN PARTICULAR IN THE VICINITY OF PERIODIC SOLUTIONS ASSOCIATED WITH NEAR SMALL-INTEGER COMMENSURABILITIES OF ORBITAL PERIOD.

P. J. Message

Liverpool University, United Kingdom

INTRODUCTION. Inquiry continues into the question as to which features of the motion of a system of mutually perturbing planets or satellites persist in the long term. In the case of three mutually perturbing bodies, one is led to a consideration of the properties of periodic solutions of the equations of motion, both because they represent a class of motions of which the behaviour certainly is known for all time, once it is known for one period, and also because of the relation of periodic solutions to near-commensurabilities of orbital period, from which arise major difficulties hindering the use of asymptotic perturbation series derived from Poisson, Von Zeipel, or Lie series methods, for the rigourous prediction of behaviour of planetary-type motions over indefinitely long time intervals. In what follows will be sought, in the gravitational problem of three bodies in the plane, the main varieties of motion in the vicinity of periodic solutions associated with a small-integer commensurability, extending to the general problem the approach used earlier in the restricted problem of three bodies in the plane (Message, 1966, where is given references to earlier investigations of the restricted case).

1. Formulation of the Problem.

Consider three particles, P_o, P_1, and P_2, of masses m_o, m_1, and m_2, respectively, moving in a plane under their mutual gravitational attractions. Their relative positions are specified by Jacobi's system of relative position vectors, that is, by $\underset{\sim}{\rho}_1 = P_oP_1$ and $\underset{\sim}{\rho}_2 = G_1P_2$, where G_1 is the mass-centre of P_o and P_1.

V. Szebehely (ed.), Applications of Modern Dynamics to Celestial Mechanics and Astrodynamics, 77–101.

Let a_i, e_i, λ_i, and $\tilde{\omega}_i$ denote the major semi-axis, eccentricity, mean longitude and near-apse longitude in the Keplerian elliptic orbit defined by the position vector $\underset{\sim}{\rho}_i$ and the velocity $\dot{\underset{\sim}{\rho}}_i$. Let Kepler's third-law constant be $\mu_i = G(m_o + m_1)$, where G is the constant of gravitation. The canonical set $(\lambda_1, \lambda_2, \tilde{\omega}_1, \tilde{\omega}_2)$ define the relative motion of the three bodies when taken together with their conjugate momenta $(\Lambda_1, \Lambda_2, \Pi_1, \Pi_2)$ respectively, where

$$\left.\begin{aligned} \Lambda_i &= m_i^+ \sqrt{(\mu_i a_i)} \\ \text{and} \qquad & \\ \Pi_i &= \Lambda_i\{\sqrt{(1-e_i^2)} - 1\}. \end{aligned}\right\} \qquad (1)$$

Here $m_1^+ = \dfrac{m_o m_1}{m_o + m_1}$, $m_2^+ = \dfrac{(m_o + m_1)m_2}{m_o + m_1 + m_2}$

(see, e.g., Message (1980 or 1981)). The Hamiltonian function is equal to

$$\frac{1}{2}\sum_{i=1}^{2} m_i^+ \dot{\underset{\sim}{\rho}}_i^2 - G\sum_{i=0}^{1}\sum_{j=i+1}^{2} \frac{m_i m_j}{P_i P_j} \quad ,$$

and takes the form

$$H = -\sum_{i=1}^{2} \frac{\mu_i^2 (m_i^+)^3}{2\Lambda_i^2} - R, \qquad (2)$$

where

$$R = \frac{G m_1^+ m_2^+}{a_2} \sum_{j \in \mathbb{Z}^3} K_j \cos N_j \qquad (3)$$

and

$$N_j = j_1 \ell_1 + j_2 \ell_2 + j_3(\lambda_1 - \lambda_2). \qquad (4)$$

The sum in R is to be taken over all sets of three integers $j = (j_1, j_2, j_3)$ with $j_1 \geq 0$. Here $\ell_i = \lambda_i - \tilde{\omega}_i$ is the mean anomaly in the orbit defined by $\underset{\sim}{\rho}_i$ and $\dot{\underset{\sim}{\rho}}_i$, and K_j is a function of a_1/a_2, e_1, and e_2, which, because of the d'Alembert property, has the form

$$K_j = \sum_{k \in \mathbb{Z}_+^2} K_{jk}\, e_1^{|j_1|+2k_1}\, e_2^{|j_2|+2k_2} \tag{5}$$

Here K_{jk} is a function of the ratio a_1/a_2 only, and the summation is over all pairs $k = (k_1, k_2)$ of non-negative integers.

The Hamiltonian function has no explicit dependence on the time, so there is the energy integral

$$H = \text{constant}. \tag{6}$$

We shall be concerned with the near-commensurability

$$(p+q)\, n_2 \approx p\, n_1\ , \tag{7}$$

where p and q are small, mutually prime integers, with $q > 0$, and $n_i = \mu_i^{\frac{1}{2}} a_i^{-3/2}$ is the mean motion in the orbit defined by $\underset{\sim}{\rho}_i$ and $\dot{\underset{\sim}{\rho}}_i$. To emphasize the dominant effects arising near this commensurability, let us use the set of coordinates

$$\left.\begin{aligned} \phi &= \lambda_2 - \lambda_1, \\ \theta_1 &= \{(p+q)\lambda_2 - p\lambda_1\}/q - \tilde{\omega}_1 = (p+q)\phi/q + \ell_1\ , \\ \theta_2 &= \{(p+q)\lambda_2 - p\lambda_1\}/q - \tilde{\omega}_2 = p\phi/q + \ell_2\ , \end{aligned}\right\} \tag{8}$$

and $\chi = \frac{1}{2}(\lambda_1 + \lambda_2)$.

The conjugate momenta are

$$\left.\begin{aligned} \Phi &= \tfrac{1}{2}(\Lambda_2 - \Lambda_1) + \{(2p+q)/(2q)\}\, (\Pi_1 + \Pi_2)\ , \\ \Theta_1 &= -\Pi_1\ , \\ \Theta_2 &= -\Pi_2\ , \end{aligned}\right\} \tag{9}$$

and $X = \Lambda_1 + \Lambda_2 + \Pi_1 + \Pi_2\ ,$

so that $\Lambda_1 = -\Phi + \frac{1}{2}X - (p/q)(\Theta_1+\Theta_2)$

and

$$\Lambda_2 = \Phi + \frac{1}{2}X + \{(p+q)/q\}(\Theta_1+\Theta_2).$$

Note that we may write

$$N_j = j_1\theta_1 + j_2\theta_2 + \{j_3 - j_1(p+q)/q - j_2 p/q\}\phi \ , \tag{10}$$

so that χ is an ignorable coordinate in this coordinate system and therefore,

$$X = \text{constant} \tag{11}$$

is an integral of the motion (being the integral of conservation of angular momentum).

Now suppose that a transformation

$$(\phi,\theta_1,\theta_2,\chi;\Phi,\Theta_1,\Theta_2,X) \longmapsto (\phi^*,\theta_1^*,\theta_2^*,\chi^*;\Phi^*,\Theta_1^*,\Theta_2^*,X^*)$$

of the Lie series type (see, e.g., Hori 1966) is chosen, so that the new Hamiltonian

$$H^* = -\sum_{i=1}^{2} \frac{\mu_i^2 m_i^{+3}}{2\Lambda_i^{*2}} - R^* \tag{12}$$

(where Λ_i^* is the same function of the new variables that Λ_i is of the old) is independent of ϕ^* . The relations between the new and old variables will be of the type

$$\theta_i^* = \theta_i^* + \frac{\partial S}{\partial \Theta_i^*} + \text{higher-order terms,}$$

where

$$S = \frac{Gm_1^+ m_2^+}{a_2} \sum_j{}' \frac{K_j^* \sin N_j^*}{j_1 n_1^* + j_2 n_2^*} + \text{higher-order terms.} \tag{13}$$

(giving the terms of lowest order in m_1/m_o and m_2/m_o). The summation is over sets j with $(j_1,j_2)\neq(0,0)$. Here K_j^* , N_j^* , and n_1^* are the same functions of the new variables that K_j, N_j, and n_i, are of the old ones. We see from (10) that R^* can only contain terms with

$$p(j_1+j_2) = q(j_3-j_1) \ , \tag{14}$$

i.e. in which j_1+j_2 is a multiple of q . Since ϕ^* is an ignorable coordinate in this "long-period" problem, we have the additional integral

$$\Phi^* = \text{constant}. \tag{15}$$

This is an "adelphic" integral in Whittaker's (1927) sense, since Φ^* is related to the original coordinates and momenta of the problem by infinite series which are asymptotic, but not uniformly convergent, due to the presence of the denominators of the type $j_1n_1^* + j_2n_2^*$.

2. Conditions for a Periodic Solution.

A periodic motion of the relative configuration (periodic in a suitably chosen uniformly rotating frame) is provided if θ_1^*, θ_2^*, Θ_1^*, and Θ_2^* are constant (as well as ϕ^* and X^*). This follows because the relations between the new and the old variables, and hence the quantities defining the mutual configuration of the three particles, involve only constants, apart from ϕ^* which then has the constant rate of change: $n_2^\# - n_1^\#$, where

$$n_i^\# = \frac{d\lambda_i^*}{dt} = n_i^* + \frac{2}{m_i^+ n_i^* a_i^*}\frac{\partial R^*}{\partial a_i^*} + \frac{1-e_i^{*2} - \sqrt{(1-e_i^{*2})}}{m_i^+ n_i^* a_i^{*2} e_i^*}\frac{\partial R^*}{\partial e_i^*} \ . \tag{16}$$

Here λ_i^*, a_i^*, and e_i^* are the same functions of the starred canonical variables that λ_i, a_i, and e_i are of the unstarred ones. The period of the motion is $T = 2\pi q/|n_2^\# - n_1^\#|$, during which ϕ increases by $2\pi q$, ℓ_1 by $2\pi(p+q)$, and ℓ_2 by $2\pi p$. For a formal proof of the existence of such solutions for sufficiently small values of m_1/m_0 and m_2/m_0, see Message (1980).

Now Θ_i^* is constant if $\dfrac{\partial H^*}{\partial \theta_i^*} = 0$, i.e. $\sum_j{}' \; j_i K_j^* \sin N_j^* = 0$ (17)

which is certainly so if $\sin(j_1\theta_1^* + j_2\theta_2^*) = 0$ for all (j_1,j_2) for which j_1+j_2 is an integer multiple of q. This requires each of θ_1^* and θ_2^* to be an integer multiple of π/q, and for $\theta_2^* - \theta_1^*$ to be zero or a multiple of π. (Note that R^* remains unchanged if θ_1^* and θ_2^* are increased by the same integer multiple of π/q, though R, with the short-period agument ϕ, does not have the corresponding property.)

For θ_i^* to be constant requires $\frac{\partial H^*}{\partial \Theta_i} = 0$

which may be written as

$$\nu^{\#} = \frac{\sqrt{(1-e_i^{*2})}}{m_i^+ n_i^* a_i^{*2} e_i^*} \frac{\partial R^*}{\partial e_i^*} \quad , (i = 1,2) , \qquad (17)$$

where

$$\nu^{\#} = \{(p+q)n_2^{\#} - pn_1^{\#}\}/q . \qquad (18)$$

The two equations (17) give the relations between e_1^*, e_2^*, $\nu^{\#}$ (which indicates the closeness of the commensurability), m_1, and m_2.

In cases where the eccentricities may be small it is better to use the coordinate $\eta_i = \sqrt{(2\Theta_i^*)} \sin \theta_i^*$ and its conjugate momentum, $\xi_i = \sqrt{(2\Theta_i^*)} \cos \theta_i^*$ in place of θ_i^* and Θ_i^*. If we define ε_i by $2\Theta_i^* = \Lambda_i^* \varepsilon_i^2$, then $\varepsilon_i^2 = 2\{1-\sqrt{(1-e_i^2)}\}$,

so that $e_i = \varepsilon_i - \frac{1}{8}\varepsilon_i^3 + O(\varepsilon_i^5)$.

Then we obtain

$$\left.\begin{aligned}\eta_i &= \sqrt{(\Lambda_i^*)}\ \varepsilon_i \sin\theta_i^* \\ \text{and}\quad \xi_i &= \sqrt{(\Lambda_i^*)}\ \varepsilon_i \cos\theta_i^* \ .\end{aligned}\right\} \tag{20}$$

In terms of these, the conditions set out above for a periodic solution are that η_i and ξ_i are constants. The equations of motion are

$$\left.\begin{aligned}\dot{\eta}_i &= \nu\xi_i - \frac{\partial R^*}{\partial \xi_i} \\ \text{and}\quad \dot{\xi}_i &= -\nu\eta_i + \frac{\partial R^*}{\partial \eta_i}\end{aligned}\right\} \tag{21}$$

where $\nu = \{(p+q)n_2 - pn_1\}/q$.

2.1 The "Moderately Close" commensurability case.

This is the case in which $|\nu|$ is small, but still significantly larger than $n_2 m_1/m_0$ and $n_2 m_2/m_0$. The term of R^* containing the lowest power of e_1^* is

$$\frac{Gm_1^+ m_2^+}{a_2^*} K_{1,q-1,0,0}\ e_1^*\ (e_2^*)^{q-1} \cos\{\theta_1^* + (q-1)\theta_2^*\}$$

$$= \frac{Gm_1^+ m_2^+}{a_2^*\sqrt{(\Lambda_1^*)}} K_{1,q-1,\ 0,0}\ (e_2^*)^{q-1}\ \{\xi_1\cos(q-1)\theta_2^* - \eta_1\sin(q-1)\theta_2^*\}.$$

The conditions (17) for a periodic solution become

$$\left.\begin{aligned}\nu^{\#} &= \frac{n_1^* a_1^* \beta_2}{a_2^* e_1^*} K_{1,q-1,\ 0,0}(e_2^*)^{q-1}\cos\{\theta_1^* + (q-1)\ \theta_2^*\} \\ &\qquad + \text{higher powers of } e_1^* \\ \text{and}\quad \nu^{*} &= \frac{n_2^*\beta_2}{e_2^*}\ K_{q-1,1,0,0}(e_1^*)^{q-1}\cos\ \{(q-1)\theta_1^* + \theta_2^*\} \\ &\qquad + \text{higher powers of } e_2^* \ ,\end{aligned}\right\} \tag{22}$$

where

$$\beta_1 = \frac{m_1^+}{m_o+m_1} = \frac{m_o m_1}{(m_o+m_1)(m_o+m_2)}$$

and

$$\beta_2 = \frac{m_2^+}{m_o+m_1} = \frac{m_2}{m_o+m_1+m_2} \; .$$

If q=1, we have

$$\left.\begin{aligned} e_1^* &= \frac{n_1 a_1^*}{\nu^{\#} a_2^*}\,\beta_2\,K_{1000}\cos\theta_1^* + \text{higher powers of } \frac{n_1\beta_2}{\nu}\,, \\ &\text{and} \\ e_2^* &= \frac{n_2}{\nu^{\#}}\,\beta_1\,K_{0100}\cos\theta_2^* + \text{higher powers of } \frac{n_2\beta_1}{\nu}\,, \end{aligned}\right\} \quad (23)$$

so, to the first order in the $(n\beta/\nu)$'s, e_1^*/e_2^* is independent of ν. As $\beta_1 \to 0$ and $\beta_2 \to 0$, $e_1^* \to 0$ and $e_2^* \to 0$, so these periodic solutions are of Poincaré's first sort.

If $q \geq 2$, no solution is possible for non-zero e_1^* and e_2^* unless ν is small enough, but the solution $\xi_1 = \eta_1 = \xi_2 = \eta_2 = 0$ (Poincaré's first sort) is always possible, as equations (21) show. (Note that for q=1, R_* has terms linear in ξ_1 and ξ_2, and so equations (21) show that a periodic solution with $\xi_1 = \xi_2 = 0$ will not generally be possible.)

2.2 The "Very Close" commensurability case.

Here $|\nu|$ is as small as $n_2\beta_1$ and $n_1\beta_2$, and the conditions (17) will not in general lead to small values of e_i^*, which will not therefore tend to zero with β_i. Thus these solutions are of Poincaré's second sort, which are therefore linked to the first sort by continuous transition as successively smaller values of $|\nu|$ are taken. For $q \geq 2$, there is still the first sort of solution $\xi_1 = \eta_1 = \xi_2 = \eta_2 = 0$.

2.3 The "Mixed" case.

This is the case, which can arise when one of the two mass ratios β_1 and β_2 is significantly smaller than the other, say $\beta_2 \ll \beta_1$, when the commensurability is "very close" in comparison to one mass ratio, but only "moderately close" in relation to the other. That is,

$$n_2\beta_1 \gg |\nu| \gg n_1\beta_2 \ .$$

Equation (17) with i=2 will not in general require e_2^* to be small (the right-hand side will have the term $qn_2^*\beta_1 K_{0,q,0,0,}(e_2^*)^{q-2}\cos q\theta_2$), and then (17) with i=1 will give a solution for e_1 even if q=1 . From this case, the restricted problem of three bodies can be obtained by letting $m_2 \to 0$. Then the value of e_1 on the periodic solution will also tend to zero, so we reach the periodic solutions of the first sort in the circular restricted problem.

3. The Linear Equations of Variation

Consider now motion in the vicinity of a particular periodic solution. Suppose that a periodic solution corresponds to

$$\left.\begin{aligned} &\eta_i = \eta_{io} , \quad \xi_i = \xi_{io} \quad (i=1,2) \\ &\Phi^* = \Phi_o^* , \quad X^* = X_o^* , \\ &\phi^* = (n_{20}^{\#} - n_{10}^{\#})\, t + \phi_o^* , \quad \chi^* = \tfrac{1}{2}(n_{10}^{\#} + n_{20}^{\#})\, t + \chi_o^* \end{aligned}\right\} \quad (24)$$

where ϕ_o^* and χ_o^* are constants. Let the motion near this be denoted by

$$\left.\begin{aligned} &\eta_i = \eta_{io} + \delta\eta_i , \quad \xi_i = \xi_{io} + \delta\xi_i , \quad (i = 1,2) , \\ &\Phi^* = \Phi_o^* + \delta\Phi , \quad X^* = X_o^* + \delta X, \quad \phi^* = (n_{20}^{\#} - n_{10}^{\#})t + \phi_o^* + \delta\phi, \\ &\text{and} \\ &\chi^* = \frac{1}{2}(n_{10}^{\#} + n_{20}^{\#})t + \chi_o^* + \delta\chi. \end{aligned}\right\} \quad (25)$$

Because of the symmetries of R^*, we may, without loss of generality, suppose that $\eta_{10} = \eta_{20} = 0$. We form the equations of motion neglecting squares and products of the displacements $\delta\eta_i$, $\delta\xi_i$, $\delta\Phi$, and δX. Noting that H^* is even in (η_1, η_2), and independent of χ^* and of ϕ^*, we obtain the equations

$$\frac{d}{dt}\begin{pmatrix} \delta\eta_1 \\ \delta\eta_2 \\ \delta\phi \\ \delta\chi \end{pmatrix} = \begin{pmatrix} A_{11} & A_{12} & C_1 & D_1 \\ A_{21} & A_{22} & C_2 & D_2 \\ C_1 & C_2 & E & F \\ D_1 & D_2 & F & G \end{pmatrix}\begin{pmatrix} \delta\xi_1 \\ \delta\xi_2 \\ \delta\Phi \\ \delta X \end{pmatrix}$$

and

$$\frac{d}{dt}\begin{pmatrix} \delta\xi_1 \\ \delta\xi_2 \\ \delta\Phi \\ \delta X \end{pmatrix} = \begin{pmatrix} -B_{11} & -B_{12} & 0 & 0 \\ -B_{21} & -B_{22} & 0 & 0 \\ 0 & 0 & 0 & 0 \\ 0 & 0 & 0 & 0 \end{pmatrix}\begin{pmatrix} \delta\eta_1 \\ \delta\eta_2 \\ \delta\phi \\ \delta\chi \end{pmatrix} \qquad (26)$$

where the coefficients

$$A_{ij} = \frac{\partial^2 H^*}{\partial\xi_i \partial\xi_j}, \qquad B_{ij} = \frac{\partial^2 H^*}{\partial\eta_i \partial\eta_j}, \qquad C_i = \frac{\partial^2 H^*}{\partial\Phi^* \partial\xi_i},$$

$$D_i = \frac{\partial^2 H^*}{\partial X^* \partial\xi_i}, \qquad E = \frac{\partial^2 H^*}{\partial\Phi^{*2}}, \quad F = \frac{\partial^2 H^*}{\partial\Phi^* \partial X^*}, \quad G = \frac{\partial^2 H^*}{\partial X^{*2}},$$

are all evaluated with the values (24), and are therefore all constant. Then

$$\delta\ddot{\eta}_i = - \sum_{j=1}^{2} \sum_{k=1}^{2} A_{ij} B_{jk} \delta\eta_k . \quad (i = 1,2) \qquad (27)$$

The stable case is that in which the eigenvalues of the matrix AB are both positive, say ω_1^2 and ω_2^2.

Then

$$\delta\eta_i = \sum_{j=1}^{2} \rho_i^{(j)} \sin \tau_j , \quad (i = 1,2) \tag{28}$$

where $\tau_j = \omega_j t + \delta_j$, δ_1 and δ_2 are constants of integration, and $\begin{pmatrix} \rho_1^{(j)} \\ \rho_2^{(j)} \end{pmatrix}$ is the eigenvector of AB (with an arbitrary multiplying factor which is a constant of integration) corresponding to the eigenvalue ω_j^2 .

Clearly $\delta\Phi$ and δX are constants, and

$$\delta\xi_i = \sum_{j=1}^{2} \breve{A}_{ij} \left\{ \sum_{k=1}^{2} \rho_j^{(k)} \omega_k \cos \tau_k - C_j \delta\Phi - D_j \delta X, \right\} \quad (i = 1,2)$$

$$\delta\phi = \sum_{j=1}^{2} \sum_{k=1}^{2} \sum_{\ell=1}^{2} C_j \breve{A}_{jk} \rho_k^{(\ell)} \sin \tau_\ell$$

$$+ \left\{ \left(E - \sum_{j=1}^{2} \sum_{k=1}^{2} C_j \breve{A}_{jk} C_k \right) \delta\Phi \right.$$

$$\left. + \left(F - \sum_{j=1}^{2} \sum_{k=1}^{2} C_j \breve{A}_{jk} D_k \right) \delta X \right\} t + \delta\phi_o ,$$

and $$\delta\chi = \sum_{j=1}^{2} \sum_{k=1}^{2} \sum_{\ell=1}^{2} D_j \breve{A}_{jk} \rho_k^{(\ell)} \sin \tau_\ell$$

$$+ \left\{ \left(F - \sum_{j=1}^{2} \sum_{k=1}^{2} C_j \breve{A}_{jk} D_k \right) \delta\Phi \right.$$

$$\left. + \left(G - \sum_{j=1}^{2} \sum_{k=1}^{2} D_j \breve{A}_{jk} D_k \right) \delta X \right\} t + \delta\chi_o \tag{29}$$

where $\breve{A}$ is the inverse of A, and $\delta\phi_o$ and $\delta\chi_o$ are constants of integration. To interpret this solution in more detail, consider the various cases.

3.1 The "Moderately Close" case.

For q=1 the e^*_{io} are small with $n_1\beta_2/|\nu|$ and $n_2\beta_1/|\nu|$, while for $q \geq 2$, the e^*_{io} are zero. Hence it is useful to approximate by retaining only the leading few terms of expansions in powers of e^*_{io}. We have

$$A_{11}=\nu^{\#}-2n_1^*\,\alpha\beta_2(K^*_{0010}+K^*_{2000}+A_{111}e^*_{10}+A_{112}e^*_{20})+0(\beta_i e^2_{jo}),$$

$$A_{22}=\nu^{\#}-2n_2^*\,\beta_1(K^*_{0001}+K^*_{0200}+A_{221}e^*_{10}+A_{222}e^*_{20})+0(\beta_i e^2_{jo}),$$

$$A_{12}=A_{21}=-\sqrt{\{n_1^*\,n_2^*\,\alpha\beta_1\beta_2\}}(K^*_{1100}+K^*_{1,-1,0,0}$$

$$A_{121}\hat{e}_{10}+A_{122}e^*_{20})+0(\beta_i e^2_{jo}),$$

$$B_{11}=\nu^{\#}-2n_1^*\,\alpha\beta_2(K^*_{0010}-K^*_{2000})+0\,(\beta_i e^2_{jo})$$

$$B_{22}=\nu^{\#}-2n_2^*\,\beta_1(K^*_{0001}-K^*_{0200})+0\,(\beta_i e^2_{jo})$$

$$B_{12}=B_{21}=+\sqrt{\{n_1^*n_2^*\alpha\beta_1\beta_2\}}(K^*_{1100}-K^*_{1,-1,0,0})+0\,(\beta_i e^2_{jo}), \qquad (30)$$

where the A_{ijk} are linear combinations of K^*_{0000}, K^*_{1000}, K^*_{1010}, K^*_{0100}, K^*_{0110}, and their derivations with respect to $\alpha = a_1^*/a_2^*$. Note that, for $q \geq 3$, $K^*_{2000} = K^*_{0200} = K^*_{1100} = 0$, and, for $q \geq 2$, $e^*_{10} = e^*_{20} = 0$.

Then

$$(AB)_{11} = \nu^{\#2} - 2\nu^{\#}n_1^*\alpha\beta_2(2K^*_{0010}+A_{111}e^*_{10}+A_{112}e^*_{20})$$

$$+\,n_1^*n_2^*\alpha\beta_1\beta_2(K^{*2}_{1,-100}-K^{*2}_{1100}+4K^{*2}_{0010}-4K^{*2}_{2000})+0\,(n\nu\beta_i e_{jo},$$

$$n\nu\beta_j^2,\,n^2\beta_j^2 e_{io}),$$

$$(AB)_{12}=\sqrt{\{n_1^*n_2^*\alpha\beta_1\beta_2\}}\{-2K^*_{1,-100}\nu^{\#}-2n_1^*\,\alpha\beta_2(K^*_{1100}-K^*_{1,-100})$$

$$(K^*_{0010}+K^*_{2000}+A_{111}e^*_{10}+A_{112}e^*_{20})$$

$$+ 2n_2^*\beta_1(K^*_{0001} - K^*_{0200})(K^*_{1100} + K^*_{1,-100} + A_{121}e^*_{10} + A_{122}e^*_{20})\}$$

$$+ 0\ (n\nu\beta_i e^2_{jo},\ n\nu\beta^2, n^2\beta^2 e_{jo});\ (AB)_{21} = \sqrt{\{n_1^* n_2^* \alpha\beta_1\beta_2\}\{-2K^*_{1,-100}\nu^{\#}}$$

$$(AB)_{21} = \{n_1^* n_2^* \alpha\beta_1\beta_2\}\{-2K^*_{1,-100}\nu^{\#} + 2n_1^*\alpha\beta_2(K^*_{0010} - K^*_{2000})(K^*_{1100}$$

$$+ K^*_{1,-100} + A_{121}e^*_{10} + A_{122}e^*_{20}) - 2n_2^*\beta_1(K^*_{1100} - K^*_{1,-100})(K^*_{0001} + K^*_{0200}$$

$$+ A_{221}e^*_{10} + A_{222}e^*_{20})\} + 0(n\nu\beta_i e^2_{jo}, n\nu\beta^2, n^2\beta^2 e_{jo})$$

and

$$(AB)_{22} = \nu^{\#2} - 2\nu^{\#}n_2\beta_1(2K^*_{0001} + A_{221}e^*_{10} + A_{222}e^*_{20})$$

$$+ n_1^* n_2^* \alpha\beta_1\beta_2\ (4K^{*2}_{0001} - 4K^{*2}_{0200} + K^{*2}_{1,-100} - K^{*2}_{1100})$$

$$+ 0\ (n\nu\beta e^2_{jo},\ n\nu\beta^2_j,\ n^2\beta^2_i e_{jo})\ , \tag{31}$$

whence the eigenvalues ω_1^2 and ω_2^2 of the matrix AB are

$$\nu^{\#2} - \nu^{\#}n_1\alpha\beta_2(2K^*_{0010} + A_{111}e^*_{10} + A_{112}e^*_{20})$$

$$- \nu^{\#}n_2\beta_1(2K^*_{0001} + A_{211}e^*_{10} + A_{212}e^*_{20})$$

$$+ n_1^* n_2^* \alpha\beta_1\beta_2(K^{*2}_{1,-100} - K^{*2}_{1100} + 2K^{*2}_{0001} - 2K^{*2}_{0010} - 2K^{*2}_{2000} - 2K^{*2}_{0200})$$

$$\pm\nu^{\#}\{4(n_1^*\alpha\beta_2K^*_{0010} - n_2^*\beta_1K^*_{0001})^2 - n_1^* n_2^* \alpha\beta_1\beta_2$$

$$(4K^{*2}_{0001} + 4K^{*2}_{0010} - 4K^{*2}_{2000} - 4K^{*2}_{0200} - K^{*2}_{1,-100})\}^{\frac{1}{2}} + \ldots \tag{32}$$

and we always have linear stability in this case. Also

$$\{(AB)_{11} - \omega_i^2\}\ \rho_1^{(i)} + (AB)_{12}\rho_2^{(i)} = 0 \tag{33}$$

whence $[n_1^*\alpha\beta_2K^*_{0010} - n_2^*\beta_1K^*_{0001} \pm\{(n_1^*\alpha\beta_2K^*_{0010} - n_2^*\beta_1K^*_{0001})^2$

$-n_1^* n_2^* \alpha\beta_1\beta_2(K^{*2}_{0001} + K^{*2}_{0010} - K^{*2}_{2000} - K^{*2}_{0200} - K^{*2}_{1,-100})\}^{\frac{1}{2}}]\ \rho_1^{(i)}$

$$\approx\sqrt{(n_1^* n_2^* \alpha\beta_1\beta_2)}\{K^*_{1,-100} + (n_1^*\alpha\beta_2/\nu^{\#})(K^*_{1100} - K^*_{1,-100})(K^*_{0010} + K^*_{2000})$$
$$-(n_2^*\beta_1/\nu^{\#})(K^*_{0001} - K^*_{0200})(K^*_{1100} + K^*_{1,-100})\}\rho_2^{(i)}$$
$$(i = 1,2) \qquad (33)$$

so that $\rho_1^{(i)}$, $\rho_2^{(i)}$ will be of the same order of magnitude, and the librations in θ_1 and θ_2 are not decoupled.

Now det $|A| = \nu^2 + 0(\nu n\beta)$, so $A^{-1} = \begin{pmatrix} 1/\nu & 0 \\ 0 & 1/\nu \end{pmatrix} + 0\ (n\beta/\nu^2)$, (34)

and hence $\delta\xi_i \approx \sum_{k=1}^{2} \rho_i^{(k)} \cos\tau_k - (C_i/\nu^{\#})\delta\Phi - (D_i/\nu^{\#})\ \delta X$. (35)

Thus in each (η_i, ξ_i)-plane the motion is the superposition of two approximately circular motions, each executed with uniform angular speed. No clear-cut distinction can be drawn between circulation and libration of each critical angle θ_i, since in a given motion, θ_i may execute complete circulations on some excursions, and at other times have stationary values, according to whether the phases of the constituent motions cause the origin of the (η_i, ξ_i)-plane to be encircled or not during each particular excursion.

3.1.2

This should be contrasted with the situation where the primary body (P_o) is sufficiently oblate for the corresponding contributions to K^*_{0020} and K^*_{0002}, to enable these coefficients to dominate over $K^*_{1,-1,0,0}$ (i.e. for the e_1^2 and e_2^2 terms to dominate over the $e_1e_2\cos(\tilde\omega_1-\tilde\omega_2)$ terms) and over K_{2000} and K_{0200} (i.e. over the $e_1^2\cos 2\theta_1$ and $e_2^2\cos 2\theta_2$ terms). Then the matrix AB becomes approximately diagonal, and the equations (27) for $\delta\eta_1$ and $\delta\eta_2$ approximately de-couple, giving

$\delta\eta_1 \approx \rho_{11} \sin\tau_1$, and $\delta\eta_2 \approx \rho_{22} \sin\tau_2$, so that each of the independent free librations is especially associated with one of the bodies P_1 and P_2, and we may speak of the "free eccentricity" of each. This occurs in the system of Saturn's satellites Enceladus and Dione, due to the sufficiently large oblateness of Saturn. A corresponding situation occurs in Jupiter's system of the Galilean satellites, due to the oblateness of Jupiter. The secular variation theory of minor planets effectively decouples from that of the major planets, allowing one to speak of the "free eccentricity" of a minor planet. This occurs because of the small mass of a minor planet causes it to have negligible effect on the secular motions of the major planets.

3.1.3

To return to the three-body problem with no oblateness of the primary P_o, we have approximately, in view of (34),

$$\left.\begin{aligned} \delta\phi &\approx \sum_{j=1}^{2} \sum_{\ell=1}^{2} (C_j\rho_j^{(\ell)}/\nu) \sin \tau_\ell + (E\delta\Phi + F\delta X)\, t + \delta\phi_o \\ \delta\chi &\approx \sum_{j=1}^{2} \sum_{k=1} (D_j\rho_j^{(\ell)}/\nu) \sin \tau_\ell + (F\delta\Phi + G\delta X)\, t + \delta\chi_o . \end{aligned}\right\} \quad (36)$$

3.1.4

The transition from deep to shallow resonance may be followed by considering the quantities

$$\hat{\eta}_i = \sqrt{(-2\Pi_i)} \sin \tilde{\omega}_i \quad \text{and} \quad \hat{\xi}_i = \sqrt{(-2\Pi_i)} \cos \tilde{\omega}_i, \qquad (i = 1,2). \qquad (37)$$

This would be a suitable set of canonical rectangular-type parameters to study the secular variations of the apses and nodes in the absence of resonance. From our solution for η_i and ξ_i we find that

$$\left.\begin{aligned} \hat{\xi}_i &= \xi_{io} \cos L + \sum_{j=1}^{2} \rho_i^{(j)} \cos \chi_j \\ &\text{and} \\ \hat{\eta}_i &= \xi_{io} \sin L + \sum_{j=1}^{2} \rho_i^{(j)} \sin \chi_j \end{aligned}\right\} \quad (i = 1,2) \qquad (38)$$

where $L = \{(p+q)\lambda_2 - p\lambda_1\}/q$, and $\chi_j = L - \tau_j$,

so that $\chi_j = \nu^{\#} - \omega_j$.

As we consider successively larger values of $|\nu^{\#}|$, that is, successively shallower resonance, the $|\xi_{io}|$ become smaller. The first terms in each of the equations (38) become more and more short periodic, and the χ_j become successively closer to the frequencies in the secular variation theory without resonance, as may be readily seen on comparing the relevant equations.

3.1.5

The approach to the restricted problem of three bodies. As we have already seen, this corresponds to taking $m_2 \to 0$, so that $e_{10} \to 0$. Before taking this limit we must first put

$$\xi_2 = \tilde{\xi}_2 \sqrt{\beta_2}\,, \quad \eta_2 = \tilde{\eta}_2 \sqrt{\beta_2} \tag{39}$$

and we obtain the equations

$$\left.\begin{aligned}
\frac{d}{dt}\begin{pmatrix}\delta\eta_1\\ \delta\tilde{\eta}_2\\ \delta\phi^*\\ \delta\chi^*\end{pmatrix} &= \begin{pmatrix}\tilde{A}_{11} & \tilde{A}_{12} & C_1 & D_1\\ \tilde{A}_{21} & \tilde{A}_{22} & \tilde{C}_2 & \tilde{D}_2\\ C_1 & \tilde{C}_2 & E & F\\ D_1 & \tilde{D}_2 & F & G\end{pmatrix}\begin{pmatrix}\delta\xi_1\\ \delta\tilde{\xi}_2\\ \delta\Phi^*\\ \delta X^*\end{pmatrix}\\
\frac{d}{dt}\begin{pmatrix}\delta\xi_1\\ \delta\tilde{\xi}_2\\ \delta\Phi^*\\ \delta X^*\end{pmatrix} &= \begin{pmatrix}-\tilde{B}_{11} & -\tilde{B}_{12} & 0 & 0\\ -\tilde{B}_{21} & -\tilde{B}_{22} & 0 & 0\\ 0 & 0 & 0 & 0\\ 0 & 0 & 0 & 0\end{pmatrix}\begin{pmatrix}\delta\eta_1\\ \delta\tilde{\eta}_2\\ \delta\phi^*\\ \delta\chi^*\end{pmatrix}
\end{aligned}\right\} \tag{40}$$

where $\tilde{A}_{11} = \nu^{\#}$, $\tilde{A}_{22} = \nu^{\#} - 2n_2^*\beta_1\,(K^*_{0001} + K^*_{0200} + A_{222}e^*_{20})$ $+ 0(\beta_1 e^{*2}_{20})$, $\tilde{A}_{12} = 0$, $\tilde{A}_{21} = -\sqrt{(n_1^* n_2^* \alpha\beta_1)}\,(K^*_{1100} + K^*_{1,-100} + A_{122}e^*_{20})$ $+ 0(\beta)$, $\tilde{B}_{11} = \nu^{\#}$, $\tilde{B}_{22} = \nu^{\#} - 2n_2^*\beta_1(K^*_{0001} - K^*_{0200})$,

$$\tilde{B}_{12} = 0, \quad \tilde{B}_{21} = \sqrt{(n_1^* n_2^* \alpha\beta_1)}\,(K^*_{1100} - K^*_{1,-100}) + 0(\beta_1). \tag{41}$$

In this way we have

$$(\tilde{A}\tilde{B})_{11} = \nu^{\#2} + 0(\beta_1^2), \qquad (\tilde{A}\tilde{B})_{12} = 0 ,$$

$$(\tilde{A}\tilde{B})_{21} = -\nu^{\#}\sqrt{(n_1^* n_2^* \alpha\beta_1)}\,(2K^*_{1-100} + A_{122}e^*_{20}) + 0(\beta_1{}^{3/2})$$

and

$$(\tilde{A}\tilde{B})_{22} = \nu^{\#2} - 2\nu^{\#} n_2\beta_1(2K^*_{0001} + A_{222}e^*_{20}) + 0(\beta_1^2). \tag{42}$$

This leads to $\omega_1^2 = \nu^{\#2}$, and

$$\omega_2^2 = \nu^{\#2} - 2\nu^{\#} n_2\beta_1(2K^*_{0001} + A_{222}e^*_{20}). \tag{43}$$

Then $\rho_1^{(2)} = 0$, so $\rho_2^{(2)}$ gives the proper eccentricity of P_2, with $\omega_2 \approx \nu^{\#} - n_2^*\beta_1(2K^*_{0001} + A_{222}e^*_{20})$.

Also $\sqrt{(n_1^* n_2^* \alpha\beta_1)}\,(2K^*_{1-100} + A_{122}e^*_{20})\,\rho_1^{(1)}$

$$\approx 2n_2^*\beta_2(2K^*_{0001} + A_{222}e^*_{20})\,\rho_2^{(1)}$$

and so a free eccentricity $(\rho_1^{(1)})$ of P_1 (giving the elliptic restricted problem) leads to a forced motion in the eccentricity of P_2. Since

$$\left.\begin{aligned} e_2 \cos\theta_2 &\approx e_{20} + \kappa e_1 \cos\theta_1 \\ e_2 \sin\theta_2 &\approx \kappa e_2 \sin\theta_2 \end{aligned}\right\} \tag{44}$$

where $\kappa \approx \dfrac{n_1^*\,\alpha^{3/2}K^*_{1-100}}{2n_2^*\;K^*_{0001}}$,

we have that if $e_{20} \gg \kappa e_1$, then $e_2^2 \approx e_{20}^2 + 2\kappa e_1 e_{20}\cos\theta_1$

(and the forced term has the factor β_1 through e_{20}) and, since θ_2 is always close to θ_{20}, we have $\theta_2 \approx \theta_{20} + (\kappa e_1/e_{20}) \sin\theta_1$.

But $\theta_1 \approx \tilde{\omega}_2 - \tilde{\omega}_1 + \theta_{20}$, so the periodic terms in e_2 and θ_2 have the argument $\tilde{\omega}_2 - \tilde{\omega}_1$.

3.2 The "Very Close" Commensurability case.

In this case the e_{io} are not in general small, so to take only the leading terms in expansions in powers of e_{io} would probably be misleading. It is convenient in this case to use the angular variables.

$$\theta = \theta_1^* = \{(p+q)\lambda_2 - p\,\lambda_1\}/q - \tilde{\omega}_1 ,$$

$$\text{and} \quad \zeta = \tilde{\omega}_2 - \tilde{\omega}_1 = \theta_1^* - \theta_2^* , \tag{45}$$

and to suppose R^* to be expressed in terms of θ,ζ,Λ_1, Λ_2,Π_1 and Π_2.

The equations of motion take the form

$$\left.\begin{aligned} \dot{\theta} &= \nu^{\#} + \frac{\partial R^*}{\partial \Pi_1} , \\ \dot{\zeta} &= \frac{\partial R^*}{\partial \Pi_1} - \frac{\partial R^*}{\partial \Pi_2} , \\ \dot{\Lambda}_i &= s_i \frac{\partial R^*}{\partial \theta} \\ \text{and } \dot{\Pi}_i &= (-1)^i \frac{\partial R^*}{\partial \zeta} - \delta_{i1} \frac{\partial R^*}{\partial \theta} \qquad (i = 1,2) \end{aligned}\right\} \tag{46}$$

where

$$s_1 = -p/q, \quad s_2 = (p+q)/q,$$

and $\delta_{ij} = +1$ if $i = j$ and zero if $i \neq j$.

The conditions discussed above for a periodic solution become

$$\left.\begin{aligned} &\nu + \frac{\partial R^*}{\partial \Pi_1} = 0 \qquad \frac{\partial R^*}{\partial \Pi_1} = \frac{\partial R^*}{\partial \Pi_2} \\ &\text{as well as } \frac{\partial R^*}{\partial \theta} = \frac{\partial R^*}{\partial \zeta} = 0. \end{aligned}\right\} \tag{47}$$

The linear equations of variation are

$$\delta\dot{\lambda}_i = - \frac{3n_i}{\Lambda_i} \delta\Lambda_i - \sum_{J=1}^{2} \left(\frac{\partial^2 R^*}{\partial\Lambda_i \partial\Lambda_j} \delta\Lambda_j + \frac{\partial^2 R^*}{\partial\Lambda_i \partial\Pi_j} \delta\Pi_j \right)$$

$$\delta\dot{\varpi}_i = - \sum_{j=1}^{2} \left(\frac{\partial^2 R^*}{\partial \Pi_i \partial \Lambda_j} \delta\Lambda_j + \frac{\partial^2 R^*}{\partial \Pi_i \partial \Pi_j} \delta\Pi_j \right)$$

$$\delta\dot{\Lambda}_i = s_i \left(\frac{\partial^2 R^*}{\partial \theta^2} \delta\theta + \frac{\partial^2 R^*}{\partial\theta\partial\zeta} \delta\zeta \right)$$

and
$$\delta\dot{\Pi}_i = \{(-1)^i \frac{\partial^2 R^*}{\partial\theta\partial\zeta} - \delta_{i1} \frac{\partial^2 R^*}{\partial\theta^2} \} \delta\theta$$

$$+ \{(-1)^i \frac{\partial^2 R^*}{\partial\zeta^2} - \delta_{i1} \frac{\partial^2 R^*}{\partial\theta\partial\zeta} \} \delta\zeta , \qquad (48)$$

in which the second partial derivatives of R^* are evaluated on the periodic solution, and so are constants. From these

$$\left. \begin{aligned} \delta\dot{\theta} &= \sum_{i=1}^{2} (P_i \delta\Lambda_i + Q_i \delta\Pi_i) \\ &\text{and} \\ \delta\dot{\zeta} &= \sum_{i=1}^{2} (S_i \delta{<}_i + T_i \delta\Pi_i) \end{aligned} \right\} \qquad (49)$$

where

$$\left. \begin{aligned} P_i &= - \frac{3n_i s_i}{\Lambda_i} - \sum_{j=1}^{2} s_j \frac{\partial^2 R^*}{\partial\Lambda_i \partial\Lambda_j} + \frac{\partial^2 R^*}{\partial\Pi_1 \partial\lambda_i} , \\ Q_i &= - \sum_{j=1}^{2} s_j \frac{\partial^2 R^*}{\partial\Lambda_j \partial\Pi_i} + \frac{\partial^2 R^*}{\partial\Pi_1 \partial\Pi_i} , \\ S_i &= \sum_{j=1}^{2} (-1)^{j+1} \frac{\partial^2 R^*}{\partial\Pi_j \partial\Lambda_i} \\ &\text{and} \\ T_i &= \sum_{j=1}^{2} (-1)^{j+1} \frac{\partial^2 R^*}{\partial\Pi_i \partial\Pi_j} . \end{aligned} \right\} \begin{matrix} (i=1,2) \\ (50) \end{matrix}$$

Therefore,

$$\left. \begin{aligned} \delta\ddot{\theta} &= - A\delta\theta - B\delta\zeta , \\ \delta\ddot{\zeta} &= - C\delta\theta - D\delta\zeta \end{aligned} \right\} \qquad (51)$$

$$\text{where}\quad A = -\sum_{i=1}^{2}\left\{P_i s_i \frac{\partial^2 R^*}{\partial\theta^2} + Q_i\left((-1)^i \frac{\partial^2 R^*}{\partial\theta\partial\zeta} - \delta_{i1}\frac{\partial^2 R^*}{\partial\theta^2}\right)\right\},$$

$$B = -\sum_{i=1}^{2}\left\{P_i s_i \frac{\partial^2 R^*}{\partial\theta\partial\zeta} + Q_i\left((-1)^i \frac{\partial^2 R^*}{\partial\zeta^2} - \delta_{i1}\frac{\partial^2 R^*}{\partial\theta\partial\zeta}\right)\right\},$$

$$C = -\sum_{i=1}^{2}\left\{S_i s_i \frac{\partial^2 R^*}{\partial\theta^2} + T_i\left((-1)^i \frac{\partial^2 R^*}{\partial\theta\partial\zeta} - \delta_{i1}\frac{\partial^2 R^*}{\partial\theta^2}\right)\right\},$$

and

$$D = -\sum_{i=1}^{2}\left\{S_i s_i \frac{\partial^2 R^*}{\partial\theta\partial\zeta} + T_i\left((-1)^i \frac{\partial^2 R^*}{\partial\zeta^2} - \delta_{i1}\frac{\partial^2 R^*}{\partial\theta\partial\zeta}\right)\right\}.$$

The solution of equations (51) is

$$\left.\begin{aligned} \delta\theta &= \sum_{i=1}^{2} \rho_i \sin\tau_i \\ \delta\zeta &= \sum_{i=1}^{2} \sigma_i \sin\tau_i\,, \end{aligned}\right\} \qquad (52)$$

with $\tau_i = \omega_i\tau + \delta_i$ (i=1,2), δ_i being a constant of integrations,

$$\left.\begin{aligned} \text{where}\quad (A - \omega_i^2)\,\rho_i + B\sigma_i &= 0 \\ \text{and}\quad C\rho_i + (D - \omega_i^2)\,\sigma_i &= 0. \end{aligned}\right\} \qquad (i = 1,2) \quad (53)$$

The linearly stable case is that in which the eigenvalues ω_1^2 and ω_2^2 of the matrix $\begin{pmatrix} A & B \\ C & D \end{pmatrix}$ are both positive, and in this case there are two librations, which are independent in the linear approximation. Note that P_i is of the order $1/(a_i^2 m^+_i)$ while Q_i, S_i and T_i are of the order $1/(a_i^2 m_o)$, so that A and B are of order $n^2\beta_i$, while C and D are of order $n^2\beta_i^2$. So approximate expressions for the two ω_i^2 are, choosing ω_1 to be the larger of the two frequencies,

$$\left.\begin{aligned}\omega_1^2 &= A + BC/(A-D) + O(n^2\beta^3)\\ \text{and}\qquad \omega_2^2 &= D - BC/(A-D) + O(n^2\beta^3).\end{aligned}\right\} \tag{54}$$

Thus ω_1 is of order $n\sqrt{\beta}$, and $\sigma_1 \approx \frac{C}{(A-D)}\rho_1 = O(\beta\rho_1)$, so in this libration, of shorter period, the amplitude is of an order of magnitude greater in θ than in ζ. Also ω_2 is of order $n\beta$, and $A\rho_2 \approx -B\sigma_2$, so in this libration, of the longer period, the amplitudes are of the same order of magnitude in θ and in ζ. Note that equations (48) admit the integrals

$$\delta\Lambda_1 + \delta\Lambda_2 + \delta\Pi_1 + \delta\Pi_2 = \delta X \quad \text{and} \quad s_1\,\delta\Lambda_2 - s_2\delta\Lambda_1 = \delta Y \tag{55}$$

corresponding to the conservation of angular momentum, and to the adelphic integral, respectively. Using them with (49) gives

$$\left.\begin{aligned}\delta\Pi_i &= F_i\delta X + G_i\delta Y + \sum_{j=1}^{2} V_{ij}\,\omega_j\cos\tau_j\\ \text{and}\quad \delta\Lambda_i &= (s_i - F_1 - F_2)\delta X + \{(-1)^i - G_1 - G_2\}\delta Y\\ &\quad -s_i\sum_{j=1}^{2}(V_{1j} + V_{2j})\,\omega_j\cos\tau_j\ ,\end{aligned}\right\} \tag{56}$$

where $i = 1,2$,

$$F_i = (-1)^{i+1}\{Q_\ell\sum_{k=1}^{2} s_k S_k - T_\ell\sum_{k=1}^{2} s_k P_k\}/\Delta\ ,$$

$$G_i = (-1)^{i+1}\{(s_1-s_2)(S_1P_2-S_2P_1) + T_\ell(P_1-P_2) + Q_\ell(S_2-S_1)\}/\Delta,$$

and

$$V_{ij} = (-1)^{i+1}\{(\sum_{k=1}^{2} s_k P_k - Q_\ell)\sigma_j + (\sum_{k=1}^{2} s_k S_k - T_\ell)\rho_j\}/\Delta,$$

(ℓ being equal to 2 if i=1, and 1 if i=2),

$$\text{and}\quad \Delta = \left(\sum_{k=1}^{2} s_k P_k\right)(T_1-T_2) + \left(\sum_{k=1}^{2} s_k S_k\right)(Q_2-Q_1) + Q_1T_2 - Q_2T_1.$$

Hence

$$\left.\begin{aligned} \delta\lambda_i &= \delta n_i t + \sum_{j=1}^{2} W_{ij} \sin \tau_j + \delta\varepsilon_i \\ \delta\widetilde{\omega}_i &= \delta g_i t + \sum_{j=1}^{2} \widetilde{W}_{ij} \sin \tau_j - s_1\delta\varepsilon_1 - s_2\delta\varepsilon_2, \end{aligned}\right\} \quad (i=1,2) \qquad (57)$$

where $\delta\varepsilon_1$ and $\delta\varepsilon_2$ are constants of integration,

$$\delta n_i = \left[-\frac{3n_i}{\Lambda_i}(s_i - F_1 - F_2) - \sum_{j=1}^{2} \left\{ (s_j - F_1 - F_2) \frac{\partial^2 R^*}{\partial\Lambda_i \partial\Lambda_j} + F_j \frac{\partial^2 R^*}{\partial\Lambda_i \partial\Pi_j} \right\} \right] \delta X + \left[-\frac{3n_i}{\Lambda_i}\left((-1)^i - G_1 - G_2\right) - \sum_{j=1}^{2} \left\{ \left((-1)^i - G_1 - G_2\right) \frac{\partial^2 R^*}{\partial\Lambda_i \partial\Lambda_j} + G_j \frac{\partial^2 R^*}{\partial\Lambda_i \partial\Pi_j} \right\} \right] \delta Y,$$

$$\delta g_i = - \sum_{j=1}^{2} \left\{ (s_j - F_1 - F_2) \frac{\partial^2 R^*}{\partial\Pi_i \partial\Lambda_j} + F_j \frac{\partial^2 R^*}{\partial\Pi_i \partial\Pi_j} \right\} \delta X - \sum_{j=1}^{2} \left\{ \left((-1)^i - G_1 - G_2\right) \frac{\partial^2 R^*}{\partial\Pi_i \partial\Lambda_j} + G_j \frac{\partial^2 R^*}{\partial\Pi_i \partial\Pi_j} \right\} \delta Y ,$$

$$W_{ij} = \frac{3n_i}{\Lambda_i} s_i (V_{1j} + V_{2j}) + \sum_{k=1}^{2} \left\{ s_k \frac{\partial^2 R^*}{\partial\Lambda_i \partial\Lambda_k} (V_{1j} + V_{2j}) - \frac{\partial^2 R^*}{\partial\Lambda_i \partial\Pi_k} V_{\kappa j} \right\},$$

and $$\widetilde{W}_{ij} = \sum_{k=1}^{2} \left\{ s_i \frac{\partial^2 R^*}{\partial\Pi_i \partial\Pi_k} (V_{1j} + V_{2j}) - \frac{\partial^2 R^*}{\partial\Pi_i \partial\Pi_k} V_{\kappa j} \right\}.$$

Note that $s_1\delta n_1 + s_2\delta n_2$ is approximately equal to the part dependent on δX and on δY of

$$-\frac{3n_1 s_1 \delta\Lambda_1}{\Lambda_1} - \frac{3n_2 s_2 \delta\Lambda_2}{\Lambda_2} ,$$

which is in turn approximately equal to $P_1\delta\Lambda_1 + P_2\delta\Lambda_2$.

This, from the first of (49) is equal to

$$\delta\dot{\theta} - \sum_{i=1}^{2} Q_i \delta\Pi_i = \sum_{i=1}^{2} \rho_i \omega_i \cos \tau_i + \text{terms of order } \beta_i \delta e_j .$$

So $s_1\delta n_1 + s_2\delta n_2$ is smaller than δe_1 and δe_2 by an order of magnitude in β_i, corresponding to the fact that, along each family of periodic solutions of the second sort, e_{10} and e_{20} change but $(p+q)n_2-pn_1$ is very nearly constant, in fact being exactly zero along the family in the limit $\beta_1\to 0$ and $\beta_2\to 0$.

3.3 The Mixed Case

This is the case (introduced in section 2.3) in which $n_2\beta_1 >> |\nu| >> n_1\beta_2$, so that $|\xi_{10}|$ is small (with β_2), but $|\xi_{20}|$ is not necessarily small. In fact if we approach the restricted problem from the very close commensurability case by letting $\beta_2\to 0$, we necessarily pass through this case as β_2 becomes less than ν/n_i. Use of the variables η_1, ξ_1, λ_1^*, Λ_1^*, θ_2^*, Π_2^*, λ_2^*, and Λ_2^* leads to the equations

$$\left.\begin{aligned}
\dot{\eta}_1 &= \nu^{\#}\xi_1 - \frac{\partial R^*}{\partial \xi_1}\ , & \dot{\xi}_1 &= -\nu^{\#}\eta_1 + \frac{\partial R^*}{\partial \eta_1}\ ,\\
\dot{\theta}_2^* &= \nu^{\#} + \frac{\partial R^*}{\partial \Pi_2^*}\ , & \dot{\Pi}_2^* &= -\frac{\partial R^*}{\partial \theta_2^*}\ ,\\
\dot{\Lambda}_i^* &= s_i\left\{\frac{\partial R^*}{\partial \theta_2^*} + \xi_1\frac{\partial R^*}{\partial \eta_1} - \eta_1\frac{\partial R^*}{\partial \xi_1}\right\}, & \dot{\lambda}_i^* &= n_i^* - \frac{\partial R^*}{\partial \Lambda_i^*}\ ,
\end{aligned}\right\} \quad (58)$$

where $\nu^{\#} = \sum_{i=1}^{2} s_i\,(n_i - \frac{\partial R^*}{\partial \Lambda_i^*})$.

Consider the periodic solution

$$\eta_1 = 0,\ \xi_1 = \xi_{10} = \frac{1}{\nu_o^{\#}}\frac{\partial R^*}{\partial \xi_1}\ ,\quad \theta_2 = 0 \text{ or } \Pi,\ \frac{\partial R^*}{\partial \Pi_2^*} = -\nu_o^{\#}\ . \quad (59)$$

Formation of the linear equations of variation proceeds in the usual way, and we find

$$\left.\begin{aligned}
\delta\ddot{\eta}_1 &= -A\delta\eta_1 - B\delta\theta_2\ , \quad \text{and}\\
\delta\ddot{\theta}_2 &= -C\delta\eta_1 - D\delta\theta_2\ ,
\end{aligned}\right\} \quad (60)$$

where $A = (\nu_o^{\#})^2 + O(n\nu\beta_2)$,

$$B = \frac{3n_2^*}{\Lambda_2^*} s_2^2 \frac{\partial^2 R^*}{\partial\theta_2^2} \xi_{10} + O(\beta_1^{3/2} \beta_2) ,$$

$$C = \frac{-3n_2^*}{\Lambda_2^*} s_2^2 \frac{\partial^2 R^*}{\partial\xi^2} + O(n\nu\sqrt{\beta_1}) ,$$

and $$D = \frac{3n_2^*}{\Lambda_2^*} s_2^2 \frac{\partial^2 R^*}{\partial\theta_1^{*2}} + O(n^2\beta_2, n^2\beta_1^2) ,$$

so that B is smaller than A, C, or D. The general solutions of equations (60) is

$$\left.\begin{aligned} \delta\eta_1 &= \sum_{j=1}^{2} \sigma_j \sin \tau_j , \\ \delta\theta_2 &= \sum_{j=1}^{2} \rho_j \sin \tau_j , \end{aligned}\right\} \quad (61)$$

where $\tau_j = \omega_j t + \delta_j$, δ_j being a constant of integration, ω_j^2 an eigenvalue of the matrix $\begin{pmatrix} A & B \\ C & D \end{pmatrix}$, and $\begin{pmatrix} \sigma_j \\ \rho_j \end{pmatrix}$ a corresponding eigenvector.

We may take $\omega_1^2 = A - BC/(A-D) + O(B^2)$, with $\rho_1 \approx C\sigma_1/(A-D)$,

and $\omega_2^2 = D + BC/(A-D) + O(B^2)$, with $\sigma_2 \approx B\rho_2/(A-D)$, which is small compared with ρ_2.

Thus $\delta\eta_1 \approx \sigma_1 \sin \tau_1$ and $\delta\xi_1 \approx \sigma_1 \cos \tau_1$. So σ_1 corresponds to the free eccentricity of P_1, which is reflected in the periodic term in θ_2 of amplitude ρ_1. The free libration in θ_2 of amplitude ρ_2 is, however, only reflected in η_1 and ξ_1 with much smaller amplitude. If σ_1 remains non-zero as $\beta_2 \to 0$, we reach the elliptic restricted problem.

REFERENCES.

Hori, G., (1966), Publications of the Astronomical Society of Japan, 18, p. 287.

Message, P. J., (1966), Proceedings of I.A.U. Symposium No. 25, p. 197.

Message, P. J., (1966), Celestial Mechanics, 21, p. 55.

Whittaker, E.T., (1927), "Analytical Dynamics", Cambridge University Press, Chapter XVI.

THE STABILITY OF N-BODY HIERARCHICAL DYNAMICAL SYSTEMS

Archie E. Roy

Department of Astronomy, University of Glasgow,
Glasgow, Scotland

ABSTRACT. The equations of motion of n-body hierarchical dynamical system (HDS) in a generalized Jacobi coordinate system enable empirical stability parameters to be readily defined. The magnitudes of these parameters, together with the ratios of successive radius vectors in the HDS may make it possible to compute the stability of the HDS, so providing a measure of the time interval in which there is an even chance of the status quo of the HDS being altered by mutual perturbations.

1. INTRODUCTION.

In the Solar System and in smaller number stellar systems (triple stars, quadruple stars and so on), a hierarchical arrangement of the orbit sizes is favoured. Thus the planetary orbits, with the exception of the Pluto-Neptune system, are almost circular and do not cross in a time interval substantially longer than the longest orbital period. Close approaches are avoided (even in the Pluto-Neptune case). Again, the major components of the satellite systems exhibit a hierarchical structure. The exceptions in these systems, for example, the outermost satellites of Jupiter, are small in mass. Likewise the small number stellar systems such as the multiple star Castor have arrangements of orbital sizes that may be termed hierarchical.

Such systems are in general examples of the n-body problem in dynamics but in particular are examples of the n-body hierarchical dynamical system (HDS). A system of n bodies ($n \geq 3$) is termed a HDS if, when described by a suitable coordinate system, the orbital radii may be ordered in ascending size and that

V. Szebehely (ed.), Applications of Modern Dynamics to Celestial Mechanics and Astrodynamics, 103–130.

order is maintained for a time interval at least as long as the longest orbital period in the system.

The question of the system's stability is immediately relevant. There are many definitions of stability. The one adopted in the present investigation of hierarchical dynamical systems is what one may term the status quo definition.

A HDS is said to be stable if in a time interval substantially longer than the longest orbital period in the system (i) the ordering in size of the radius vectors of the bodies does not change, (ii) there are no approaches of any two of the bodies to each other close enough to change in a non-reversible way the semimajor axes, eccentricities and inclinations of their orbits, (iii) no body is ejected from the system. Such a definition may shock the purists in celestial mechanics by its seeming lack of rigour but in practice it is found to be a useful tool in an investigation of any n-body HDS.

We now show how any n-body HDS may be classified as a simple or general HDS. A simple HDS can be described by a classical Jacobi system of coordinates while a general HDS requires a generalized Jacobi system of coordinates. It is then shown how the equations of motion of the bodies in any HDS demonstrate that each body performs a disturbed Keplerian orbit. Following the work of Emslie, Roy and Walker (Walker et al, 1980; Walker and Roy, 1981, 1982a, b; Walker, 1982) a study of these equations enables certain parameters to be defined, termed empirical stability parameters. These parameters may then be used to investigate the stability of the HDS.

2. THE EQUATIONS OF MOTION OF THE SIMPLE N-BODY HDS.

Let n point masses P_i, of masses m_i, have radius vectors $\underset{\sim}{R}_i$, (i=1,2,...,n) with respect to an origin O in an inertial system (Figure 1). Then the mutual radius vector joining P_i to P_j is $\underset{\sim}{r}_{ij}$, where $\underset{\sim}{r}_{ij} = \underset{\sim}{R}_j - \underset{\sim}{R}_i$.

Let the vectors $\underset{\sim}{\rho}_i$ be defined such that

$$\underset{\sim}{\rho}_2 = \underset{\sim}{r}_{12}$$

$$\underset{\sim}{\rho}_3 = \text{vector } C_2 P_3 ,$$

where C_2 is the centre of mass P_1 and P_2,

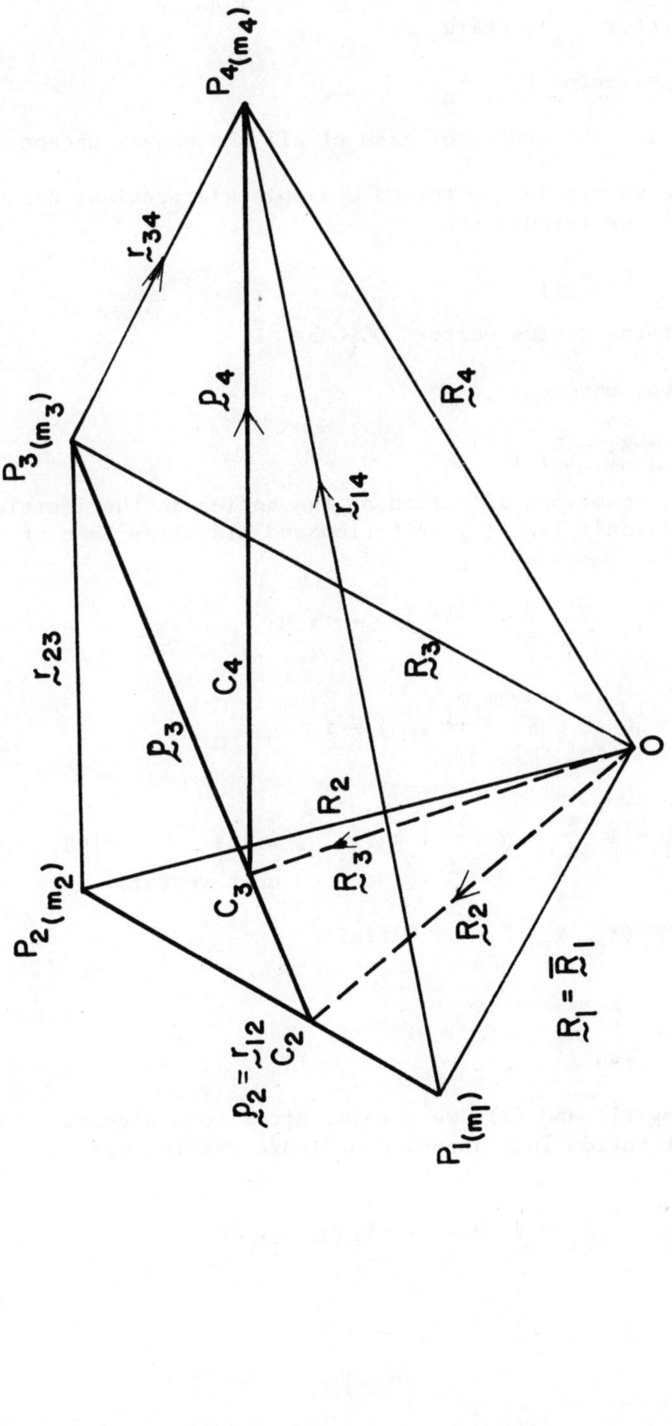

Figure 1. The Simple Hierarchical Dynamical System.

$$\underset{\sim}{\rho}_4 = \text{vector } C_3 P_4 ,$$

where C_3 is the centre of mass of P_1, P_2 and P_3, and so on to vector $\underset{\sim}{\rho}_n$, where

$$\underset{\sim}{\rho}_n = \text{vector } C_{n-1} P_n ,$$

C_{n-1} being the centre of mass of all the masses except P_n.

The system is now termed a simple hierarchical dynamical system if we further take

$$|\underset{\sim}{\rho}_i| < |\underset{\sim}{\rho}_{i+1}| .$$

Let the radius vector OC_i be $\underset{\sim}{\bar{R}}_i$.

Then, obviously,

$$\underset{\sim}{\rho}_i = \underset{\sim}{R}_i - \underset{\sim}{\bar{R}}_{i-1} . \tag{1}$$

The equations of motion of the bodies in the inertial system under Newton's law of gravitation and his three laws of motion are thus:

$$m_i \ddot{\underset{\sim}{R}}_i = \underset{\sim}{\nabla}'_i U , \quad i = 1, \ldots, n \tag{2}$$

where

$$U = \frac{1}{2} G \sum_{i=1}^{n} \sum_{j=1}^{n} \frac{m_i m_j}{r_{ij}} , \quad j \neq i ,$$

$$\underset{\sim}{\nabla}'_i = \underset{\sim}{i} \frac{\partial}{\partial X_i} + \underset{\sim}{j} \frac{\partial}{\partial Y_i} + \underset{\sim}{k} \frac{\partial}{\partial Z_i} ,$$ $\underset{\sim}{i}$, $\underset{\sim}{j}$, $\underset{\sim}{k}$ being unit vectors

and $\underset{\sim}{R}_i = (X_i, Y_i, Z_i)$. Defining

$$M_i = \sum_{j=0}^{i} m_j , \quad (m_o = 0) \tag{3}$$

and using (1) and (2), we obtain, after some algebra, the equations of motion in a Jacobi coordinate system, viz.

$$\frac{m_i M_{i-1}}{M_i} \ddot{\underset{\sim}{\rho}}_i = \underset{\sim}{\nabla}_i U , \quad i = 1, 2, \ldots, n \tag{4}$$

where

$$\underset{\sim}{\nabla}_i = \underset{\sim}{i}\frac{\partial}{\partial x_i} + \underset{\sim}{j}\frac{\partial}{\partial y_i} + \underset{\sim}{k}\frac{\partial}{\partial z_i} ,$$

and $\underset{\sim}{\rho}_i = (x_i, y_i, z_i)$.

From equations (4) the usual integrals of energy and angular momentum may be formed. In essence, we have already used the system's centre of mass integrals in forming the equations of motion in a Jacobi coordinate system.

We now express U as a function of the ρ's .

It may be easily shown that

$$\underset{\sim}{r}_{k\ell} = \underset{\sim}{\rho}_\ell - \underset{\sim}{\rho}_k + \sum_{j=k}^{\ell-1} \frac{m_j}{M_j}\underset{\sim}{\rho}_j .$$

The relationship may be used to obtain U as a function of the $\underset{\sim}{\rho}_i$. An expansion of U in terms of the ratios ρ_i/ρ_j , $(i = 2, 3, \ldots, n-1, j = 3, \ldots, n; j > 1)$ where $\rho_i/\rho_j < 1$, may then be applied, yielding, correct to the second order in ρ_i/ρ_j ,

$$\ddot{\underset{\sim}{\rho}}_i = GM_i\underset{\sim}{\nabla}_i \left[\frac{1}{\rho_i}\{1 + \sum_{k=1}^{i-1} \varepsilon^{ki}P_2(C_{ki}) + \sum_{\ell=1+i}^{n} \varepsilon_{\ell i}P_2(C_{i\ell})\}\right] \tag{5}$$

where

$$\varepsilon^{ki} = \frac{m_k M_{k-1}}{M_k M_{i-1}}\alpha_{ki}^2 \; ; \; \varepsilon_{\ell i} = \frac{m_\ell}{M_i}\alpha_{i\ell}^3 , \quad i = 2, \ldots, n .$$

In these expressions,

$$\alpha_{ij} = \rho_i/\rho_j < 1 \; ; \; C_{ij} = \frac{\underset{\sim}{\rho}_i \cdot \underset{\sim}{\rho}_j}{\rho_i\rho_j} \; ; \; \rho_i = |\underset{\sim}{\rho}_i| ,$$

while $P_2(x)$ is the Legendre polynomial of order 2 in x .

On examination it is seen that the first term of the right hand side of each of equations (5) represents the undisturbed elliptic motion of the i-th mass about the mass-centre of the sub-system of masses $m_1, m_2, \ldots, m_{i-1}$. The other terms, and of course the higher order terms neglected, provide the perturbations of the Keplerian orbit.

3. THE EQUATIONS OF MOTION OF THE GENERAL N-BODY HDS.

Let n point masses be arranged in the system shown in Figure 2, with $n = 2^q$, q being an integer.

Define parameters M_{ij} , a and b by the relations,

$$M_{ij} = \sum_{k=a}^{b} m_k ,$$

where

$$a = (j - 1).2^{q-i} + 1$$

and (6)

$$b = j.2^{q-i}.$$

The parameter M_{ij} denotes the jth subsystem in level i of the whole n-body system.

Consider, for example, the case $q = 3$. Then we have an eight-body system with the following values of M_{ij} , viz:

$$M_{01} = \sum_{k=1}^{8} m_k ,$$

$$M_{11} = \sum_{k=1}^{4} m_k \; ; \; M_{12} = \sum_{k=5}^{8} m_k ,$$

$$M_{21} = m_1 + m_2 \; ; \; M_{22} = m_3 + m_4 \; ; \; M_{23} = m_5 + m_6 \; ;$$

$$M_{24} = m_7 + m_8 ,$$

$$M_{3k} = m_k \; ; \; (k = 1, 2, . . . , 8) .$$

Thus M_{01} is the sum of all the masses in the system. It also represents the zeroth level and that sub-system which is the system itself. It is convenient to take M_{01} in Figure 2 to represent also the position of the mass centre of the system.

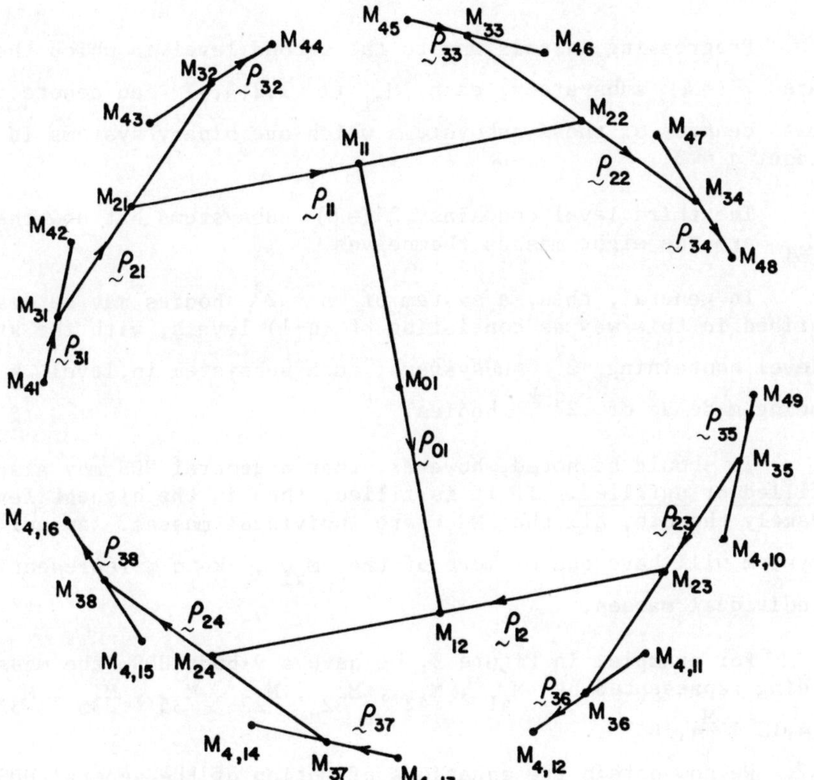

Figure 2. The General Hierarchical Dynamical System with $n = 2^q$; Particular Case $q = 4$.

The first level contains $2^1 (= 2)$ subsystems, the numbering of the masses in M_{11} and M_{12} showing that we are dealing with two separate quadruple systems. Again it is convenient to allow M_{11} and M_{12} to denote the positions of the mass centres of these two quadruple systems.

Progressing in this way to the second level in which there are $2^2 (= 4)$ subsystems, each M_{2i} $(i = 1,2,3,4)$ can denote the mass centres of these subsystems which are binary systems in the case $q = 3$.

The third level contains $2^3 (= 8)$ subsystems but now the M_{3i} are the eight masses themselves.

In general, then, a system of $n = 2^q$ bodies may be described in this way as consisting of $(q+1)$ levels, with the kth level containing 2^k subsystems, each subsystem in level k being made up of 2^{q-k} bodies.

It should be noted, however, that a general HDS may also be filled or unfilled. If it is filled, then in the highest level, namely the qth, all the M_{qi} are individual masses. An unfilled system will have one or more of the M_{ki}, $k < q$, representing individual masses.

For example, in Figure 3, we have a 9-body HDS, the masses being represented by M_{41}, M_{42}, M_{32}, M_{22}, M_{35}, M_{36}, M_{37}, $M_{4,15}$, $M_{4,16}$.

We now obtain the equations of motion of the general HDS in a generalized Jacobi coordinate system.

Let $\bar{\underset{\sim}{R}}_{ij}$ be the position vectors of the M_{ij}, measured from 0 in an inertial system. Thus

$$\bar{\underset{\sim}{R}}_{ij} = \overrightarrow{0\,M_{ij}} .$$

Then,

$$\bar{\underset{\sim}{R}}_{ij} = \frac{1}{M_{ij}} \sum_{h=a}^{b} m_h \underset{\sim}{R}_h \tag{7}$$

where

$$\underset{\sim}{R}_h = \overrightarrow{O\,P_h} \,,$$

P_h being the position of the body of mass m_h and a and b being defined as in (6).

Defining the vector $\underset{\sim}{\rho}_{ij}$ by

$$\underset{\sim}{\rho}_{ij} = \overrightarrow{M_{i+1,2j-1}\,M_{i+1,2j}} = \overrightarrow{O\,M}_{i+1,2j} - \overrightarrow{O\,M}_{i+1,2j-1} \,,$$

we have

$$\underset{\sim}{\rho}_{ij} = \bar{\underset{\sim}{R}}_{i+1,2j} - \bar{\underset{\sim}{R}}_{i+1,2j-1} \,. \tag{8}$$

Using equations (2) and (7) and differentiating equation (8) twice, we find that

$$\ddot{\underset{\sim}{\rho}}_{ij} = \frac{1}{M_{i+1,2j}} \sum_{g=\alpha}^{\beta} \underset{\sim}{\nabla}'_g\, U - \frac{1}{M_{i+1,2j-1}} \sum_{h=\gamma}^{\delta} \underset{\sim}{\nabla}'_h\, U \,, \tag{9}$$

where

$$\alpha = (2j-1).2^{q-i-1} + 1 \,,$$

$$\beta = 2j.2^{q-i-1} \,,$$

$$\gamma = (2j-2).2^{q-i-1} + 1 \,,$$

$$\delta = (2j-1).2^{q-i-1} \,,$$

and $\underset{\sim}{\nabla}'_g$ is the gradient operator associated with $\underset{\sim}{R}_g$.

After a little reduction, equations (9) may be transformed to give the required equations of motion in generalized Jacobi coordinates, viz.

$$\frac{M_{i+1,2j}\,M_{i+1,2j-1}}{M_{ij}}\, \ddot{\underset{\sim}{\rho}}_{ij} = \underset{\sim}{\nabla}_{ij}\, U \,, \tag{10}$$

where $i = 0, \ldots, q-1$, $j = 1, \ldots, 2^i$, and $\underset{\sim}{\nabla}_{ij}$ is the gradient operator associated with $\underset{\sim}{\rho}_{ij}$.

The force function U is now expanded in a manner analogous to the way in which it was expanded in Section 2, the expansion being now carried out in terms of the ratios $\alpha_{k\ell}^{ij}$, defined by

$$\alpha_{k\ell}^{ij} = \rho_{ij}/\rho_{k\ell} ,$$

where $i = 0,1, . . . , q-1$; $j = 1,2, . . . , 2^i$; $k = 0,1, . . . , q-1$; $\ell = 1,2, . . . , 2^k$; $k < i$ and <u>all $\alpha_{k\ell}^{ij}$ are less than unity</u>.

The expansion involves expressing $\underset{\sim}{r}_{k\ell}$ as a function of the $\underset{\sim}{\rho}_{ij}$. After some algebra, details of which may be found in Walker 1982a, the resulting expression is found to be

$$\underset{\sim}{r}_{k\ell} = \sum_{h=0}^{q-1} \{(-1)^{g_h(\ell)} \frac{M_{q-h,f(g_h(\ell))}}{M_{q-h-1,g_{h+1}(\ell)}} \underset{\sim}{\rho}_{q-h-1,g_{h+1}(\ell)}$$

$$- (-1)^{g_h(k)} \frac{M_{q-h,f(g_h(k))}}{M_{q-h-1,g_{h+1}(k)}} \underset{\sim}{\rho}_{q-h-1,g_{h+1}(k)}\} , \qquad (11)$$

where

$$f(j) = j - (-1)^j \; ; \quad g(j) = \text{int}\,(\frac{j+1}{2}) ,$$

int(x) denoting the integer part of x.

Applying expression (11) to the expression for U , viz.

$$U = \frac{1}{2}G \sum_{i=1}^{n} \sum_{j=1}^{n} \frac{m_i m_j}{r_{ij}} , \quad j \neq i ,$$

and expanding, it is found that, correct to the second order in the ratios of the smaller to the larger radius vectors, namely the $\alpha_{k\ell}^{ij}$,

$$U = \frac{1}{2}G \sum_{\ell=1}^{n} \sum_{k=1}^{n} \frac{M_{qk} M_q}{M_{q-a-1,g_{a+1}(\ell)}} \{1 +$$

$$\sum_{h=0}^{a-1} \left[F^2_{qh\ell} (\alpha^{q-h-1,g_{h+1}(\ell)}_{q-a-1,g_{a+1}(\ell)})^2 P_2(C^{q-h-1,g_{h+1}(\ell)}_{q-a-1,g_{a+1}(\ell)}) \right.$$

$$\left. + F^2_{qhk} (\alpha^{q-h-1,g_{h+1}(k)}_{q-a-1,g_{a+1}(k)})^2 P_2(C^{q-h-1,g_{h+1}(k)}_{q-a-1,g_{a+1}(k)}) \right] \} \qquad (12)$$

where

$$F_{qh\ell} = (-1)^{g_h(\ell)} \frac{M_{q-h,f(g_h(\ell))}}{M_{q-h-1,g_{h+1}(\ell)}} ,$$

$$\rho_{i,j} = |\underset{\sim}{\rho}_{i,j}| ,$$

$$\alpha^{i,j}_{k,\ell} = \rho_{i,j}/\rho_{k,\ell} ,$$

$$C^{i,j}_{k,\ell} = \frac{\underset{\sim}{\rho}_{i,j} \cdot \underset{\sim}{\rho}_{k,\ell}}{\rho_{i,j} \ \rho_{k,\ell}}$$

and $P_2(x)$ is the Legendre polynomial of order 2 in x .

Inspection of (10) and (12) shows that the first term on the right hand side of (12) provides the unperturbed Keplerian motion of the $\underset{\sim}{\rho}_{ij}$ radius vectors. The other terms in (12), and, of course, the terms neglected in the expansion, provide the perturbations in the Keplerian orbits.

4. AN UNAMBIGUOUS NOMENCLATURE FOR A GENERAL HDS.

Consider the nine-body system in Figure 3. A short hand description of this unfilled 5-level general HDS is obviously desirable. It is provided unambiguously by the formula 9(5(3,2),4) , arrived at by progressively breaking down the 9-body system until it is composed of a number of simple HDS. Thus the 9-body system is composed of a 5-body system ($M_{4,16}$, $M_{4,15}$, M_{37}, M_{36}, M_{35}) and a 4-body system (M_{41}, M_{42}, M_{32}, M_{22}).

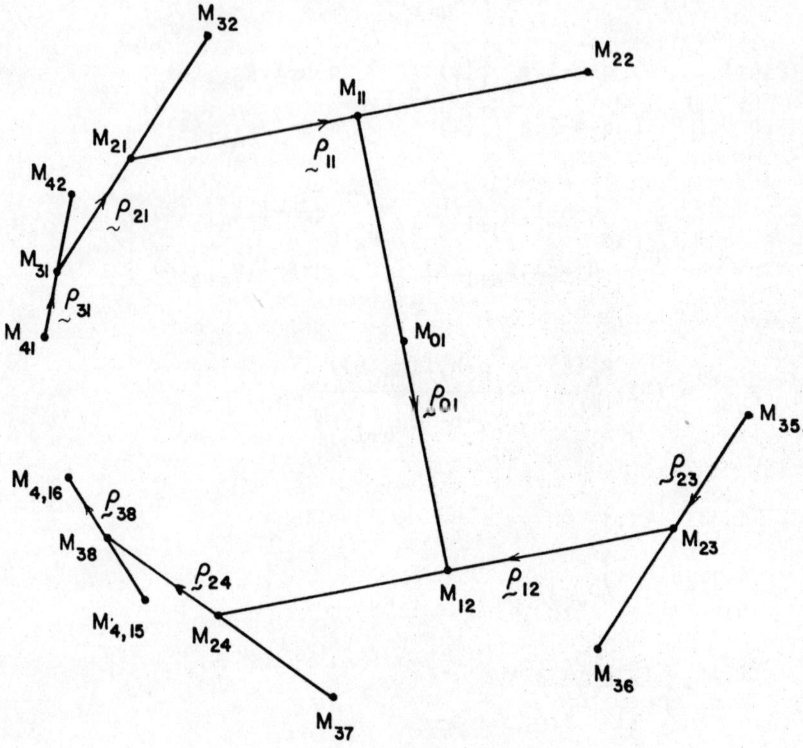

Figure 3. An Unfilled General Hierarchical Dynamical System; n = 9 .

The latter is already a simple HDS but the former can be further broken down into a 3-body ($M_{4,16}$, $M_{4,15}$, M_{37}) and a 2-body (M_{36}, M_{35}) system.

The filled 16-body general HDS of Figure 2 is a 16(8(4(2,2),4(2,2)), 8(4,(2,2),4(2,2))) system while the multiple star Castor is a 6(4(2,2),2) system.

5. THE SIMPLE AND GENERAL 4-BODY HDS.

Although the analysis of Sections 2, 3, and 4 has been carried out for n in general, it is of value to consider the special case where $n = 4$. This case is the lowest possible value of n where the HDS can be simple or general, that is, 4 or 4(2,2) in the nomenclature of Section 4. It is illustrated in Figure 4.

If it is a simple HDS, the equations of motion, to the second order in the expansions, are obtained from (5) by limiting i to the values 2 , 3 , 4 . For the general HDS, equations (10) and (12) are used.

We consider the simple case first. Then the equations are:

$$\begin{aligned}
\ddot{\underset{\sim}{\rho}}_2 &= G(m_1+m_2)\underset{\sim}{\nabla}_2\left[\frac{1}{\rho_2}\{1+\varepsilon_{32}P_2(C_{23})+\varepsilon_{42}P_2(C_{24})\}\right] , \\
\ddot{\underset{\sim}{\rho}}_3 &= G(m_1+m_2+m_3)\underset{\sim}{\nabla}_3\left[\frac{1}{\rho_3}\{1+\varepsilon^{23}P_2(C_{23})+\varepsilon_{43}P_2(C_{34})\}\right] , \qquad (13) \\
\ddot{\underset{\sim}{\rho}}_4 &= G(m_1+m_2+m_3+m_4)\underset{\sim}{\nabla}_4\left[\frac{1}{\rho_4}\{1+\varepsilon^{24}P_2(C_{24})+\varepsilon^{34}P_2(C_{34})\}\right]
\end{aligned}$$

where

$$\begin{aligned}
\varepsilon^{23} &= \frac{m_1m_2}{(m_1+m_2)^2}\alpha_{23}^2 \quad ; \quad \varepsilon_{32} = \frac{m_3}{m_1+m_2}\alpha_{23}^3 \\
\varepsilon^{24} &= \frac{m_1m_2}{(m_1+m_2)(m_1+m_2+m_3)}\alpha_{24}^2 \quad ; \quad \varepsilon_{42} = \frac{m_4}{m_1+m_2}\alpha_{24}^3 \qquad (14) \\
\varepsilon^{34} &= \frac{m_3(m_1+m_2)}{(m_1+m_2+m_3)^2}\alpha_{34}^2 \quad ; \quad \varepsilon_{43} = \frac{m_4}{m_1+m_2+m_3}\alpha_{34}^3
\end{aligned}$$

and $\alpha_{ij} = \rho_i/\rho_j < 1$.

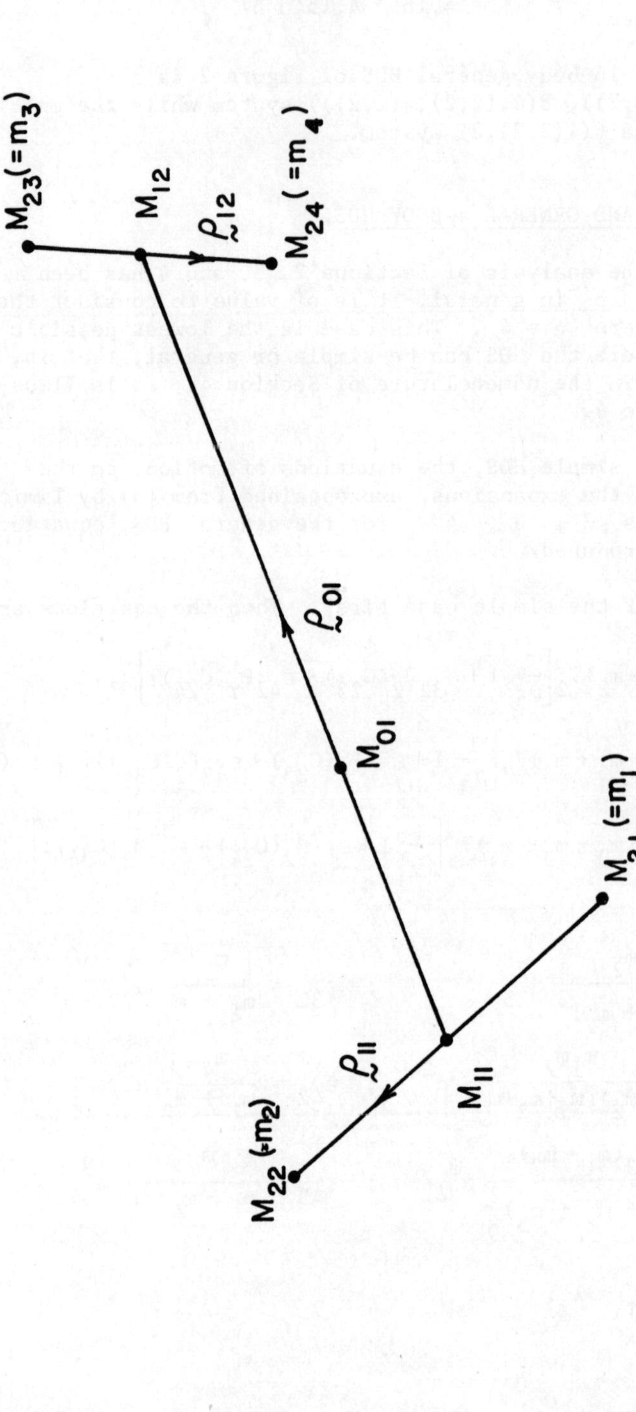

Figure 4. The General Hierarchical Dynamical System; n = 4 .

The corresponding equations for the general 4-body HDS are, correct to the second order in the $\alpha_{k\ell}^{ij}(<1)$:

$$\ddot{\underset{\sim}{\rho}}_{01} = G\, M_{01}\, \underset{\sim}{\nabla}_{01}\left[\frac{1}{\rho_{01}}\{1+\varepsilon_{01}^{11}P_2(C_{01}^{11})+\varepsilon_{01}^{12}P_2(C_{01}^{12})\}\right] ,$$

$$\ddot{\underset{\sim}{\rho}}_{11} = G\, M_{11}\, \underset{\sim}{\nabla}_{11}\left[\frac{1}{\rho_{11}}\{1+\varepsilon_{11}^{01}P_2(C_{01}^{11})\}\right] , \qquad (15)$$

$$\ddot{\underset{\sim}{\rho}}_{12} = G\, M_{12}\, \underset{\sim}{\nabla}_{12}\left[\frac{1}{\rho_{12}}\{1+\varepsilon_{12}^{01}P_2(C_{01}^{12})\}\right] ,$$

where

$$\varepsilon_{01}^{11} = \frac{m_1 m_2}{(m_1+m_2)^2}(\alpha_{01}^{11})^2 \quad ; \quad \varepsilon_{11}^{01} = \frac{m_3+m_4}{m_1+m_2}(\alpha_{01}^{11})^3 ,$$

$$\varepsilon_{01}^{12} = \frac{m_3 m_4}{(m_1+m_2)^2}(\alpha_{01}^{12})^2 \quad ; \quad \varepsilon_{12}^{01} = \frac{m_1+m_2}{m_3+m_4}(\alpha_{01}^{12})^3 \qquad (16)$$

and $\alpha_{01}^{11} = \rho_{11}/\rho_{01}$; $\alpha_{01}^{12} = \rho_{12}/\rho_{01}$.

If we define μ , μ_3 and μ_4 by

$$\mu = \frac{m_2}{m_1+m_2} \quad ; \quad \mu_3 = \frac{m_3}{m_1+m_2} \quad ; \quad \mu_4 = \frac{m_4}{m_1+m_2} ,$$

relations (14) and (16) become, respectively:

$$\varepsilon^{23} = \mu(1-\mu)\,\alpha_{23}^2 \quad ; \quad \varepsilon_{32} = \mu_3\,\alpha_{23}^3 ,$$

$$\varepsilon^{24} = \frac{(1-\mu)}{1+\mu_3}\,\alpha_{24}^2 \quad ; \quad \varepsilon_{42} = \mu_4\,\alpha_{24}^3 , \qquad (14')$$

$$\varepsilon^{34} = \frac{\mu_3}{(1+\mu_3)^2}\,\alpha_{34}^2 \quad ; \quad \varepsilon_{43} = \frac{\mu_4}{1+\mu_3}\,\alpha_{34}^3 ,$$

and

$$\varepsilon_{01}^{11} = \mu(1-\mu)(\alpha_{01}^{11})^2 \quad ; \quad \varepsilon_{11}^{01} = (\mu_3 + \mu_4)(\alpha_{01}^{11})^3 ,$$
$$\varepsilon_{01}^{12} = \mu_3\mu_4(\alpha_{01}^{12})^2 \quad ; \quad \varepsilon_{12}^{01} = \frac{1}{\mu_3 + \mu_4}(\alpha_{01}^{12})^3 . \tag{16'}$$

In both the simple and general HDS, without loss of generality, we can put $\mu \leq 0.5$; while in the general HDS we can also put $\frac{m_4}{m_3 + m_4} = \frac{\mu_4}{\mu_3 + \mu_4} \leq 0.5$. It may also be noted that although we have 3 alphas in (14'), they are not independent since $\alpha_{24} = \alpha_{23} \cdot \alpha_{34}$. Omitting from further consideration the P_2 Legendre polynomials which are of order unity or less, we note that the epsilons provide a measure of the size of the perturbations on the Keplerian orbits in both systems. These perturbations are continually changing in magnitude and direction, aiding and hindering each other in their net effects on the Keplerian orbits. Nevertheless, the magnitudes of the epsilons offer a guide to the size of the perturbations. They also, together with the α sizes, provide guidance to the answer to the question:

"What is the totality of possible 4-body HDS that can occur in nature and have any chance of maintaining their status quo for any length of time?"

For any chance of stability, all alphas must satisfy the condition $0 < \alpha < 1$ so that no cross-over of orbits and possible close encounters leading to high perturbations result. In addition, all epsilons must satisfy the same condition, i.e. $0 < \varepsilon < 1$ since an epsilon as large as unity will produce a perturbation as large as the central force producing the Keplerian orbit.

Consider firstly the simple 4-body HDS expressed by equations (13) and (14').

For a simple 4-body HDS to occur in nature, all μ's must be real and positive, with $\mu \leq 0.5$.

In fact, since two epsilons exist in each equation, the last condition, namely that $0 < \varepsilon < 1$ should be refined.

Let

$$\Sigma_2 = \varepsilon_{32} + \varepsilon_{42} < 1 ,$$

$$\Sigma_3 = \varepsilon^{23} + \varepsilon_{43} < 1 \quad ,$$

$$\Sigma_4 = \varepsilon^{24} + \varepsilon^{34} < 1 \quad .$$

Using (14') we can then write:

$$\mu_3 \, \alpha_{23}^3 + \mu_4 \, \alpha_{34}^3 = \Sigma_2 \quad ,$$

$$\mu(1-\mu) \, \alpha_{23}^2 + \frac{\mu_4}{1+\mu_3} \, \alpha_{34}^3 = \Sigma_3 \quad , \tag{17}$$

$$\frac{\mu(1-\mu)}{1+\mu_3} \, \alpha_{24}^2 + \frac{\mu_3}{(1+\mu_3)^2} \, \alpha_{34}^2 = \Sigma_4 \quad ,$$

where $\alpha_{24} = \alpha_{23} \cdot \alpha_{34}$.

Now within the solar system, and in multiple star systems, the Σ values fall within a wide range. A suitable procedure is to take Σ_2 , Σ_3 , and Σ_4 as given data and then, for each set of Σ values, solve equations (17) to obtain μ , μ_3 and μ_4 as functions of the two independent α's , α_{23} and α_{34} . The functions will, of course, involve the given set of Σ parameter values.

It is found that quadratics in μ , μ_3 and μ_4 are obtained. Curves of the three dimensionless masses may then be plotted in the α_{23} , α_{34} plane within the area $0 < \alpha < 1$ for various mass-values that are real and positive, with $\mu \leq 0.5$. Such curves are limited by the fact that the discriminants in the quadratics must be positive. The curves then outline areas in the α_{23} , α_{34} plane in which all necessary conditions are satisfied. Outside such areas no real 4-body simple HDS can exist for that set of Σ values.

An examination of these plots then enables in principle the totality of all possible 4-body simple HDS to be listed since any point within the 'real' area gives values of μ , μ_3 , μ_4 , α_{23} , α_{34} , Σ_2 , Σ_3 , Σ_4 . The first three values provide the three Jacobian radius vectors, while the last three, together with equations (14) and (17), give the size of perturbations the system experiences.

A recent study (Walker and Roy 1982b) takes all combinations of values,

$$\Sigma_2 = 10^{-\omega_2} , \Sigma_3 = 10^{-\omega_3}, \Sigma_4 = 10^{-\omega_4} ,$$

where the ω's take values 2, 3, 4, 5, 6, so providing a total of 125 such plots. Even if a specific restriction is made to four-body systems where any system is taken to be 'built' from stars, planets and moons (planet mass = 10^{-3} star mass; moon mass = 10^{-3} planet mass) a remarkably diverse variety of four-body systems is found to satisfy the required conditions.

Turning now to the general four-body HDS expressed by equations (15) and (16'), it is seen that a similar though simpler procedure may be used.

Because

$$\alpha_{01}^{12} = (\varepsilon_{11}^{01} \cdot \varepsilon_{12}^{01})^{1/3} (\alpha_{01}^{11})^{-1} , \tag{18}$$

all possible systems, for given ε_{11}^{01} , ε_{12}^{01} , lie on the curve in the ε_{01}^{12} , α_{01}^{11} plane defined by (18). Either one of the expressions

$$\varepsilon_{11}^{01} = (\mu_3 + \mu_4)(\alpha_{01}^{11})^3 \quad \text{or} \quad \varepsilon_{12}^{01} = \frac{1}{\mu_3 + \mu_4} (\alpha_{01}^{12})^3$$

enables the relationship between μ_3 and μ_4 to be found.

Define Σ by

$$\Sigma = \varepsilon_{01}^{11} + \varepsilon_{01}^{12} = \mu(1-\mu)(\alpha_{01}^{11})^2 + \mu_3\mu_4(\alpha_{01}^{12})^2 \quad . \tag{19}$$

Given a set of values of ε_{11}^{01} , ε_{12}^{01} and Σ , relation (19) enables, for every point α_{01}^{12} , α_{01}^{11} in the α-plane lying on the curve defined by (18), the relationship between μ , μ_3 and μ_4 to be found.

Again, as in the study by Walker and Roy 1982b, a range of combinations of values of ε_{11}^{01} , ε_{12}^{01} , Σ enables all possible four-body general HDS to be listed that (a) can exist in nature and (b) have any chance at all of being stable.

In principle, this morphological approach of listing all types of four-body HDS that can occur in nature and have any chance of being stable could be extended to a study of n-body systems where $n > 4$.

We now consider how rather more precise data about the stability of HDS may be obtained. To do so we firstly review some recent work in the general three-body problem.

6. THE ZARE CRITERION.

Putting $n = 3$ in equations (4) we have

$$\frac{m_1 m_2}{m_1 + m_2} \ddot{\underset{\sim}{\rho}}_2 = \underset{\sim}{\nabla}_2 U ,$$
$$\frac{m_3(m_1 + m_2)}{m_1 + m_2 + m_3} \ddot{\underset{\sim}{\rho}}_3 = \underset{\sim}{\nabla}_3 U , \tag{20}$$

where

$$U = \frac{1}{2} G \sum_{i=1}^{3} \sum_{j=1}^{3} \frac{m_i m_j}{r_{ij}} , \quad j \neq i .$$

This three-body HDS consists of a binary $P_1 - P_2$ with P_3 in orbit about the center of mass of P_1 and P_2.

In recent years a number of authors (for example, Marchal and Saari 1975; Szebehely and Zare 1977; Zare 1976, 1977) have shown that it is possible to establish a condition enabling a decision to be made about the permanency or otherwise of the binary. This is analogous to the use of surfaces of zero velocity in the restricted three-body problem to investigate whether or not the massless particle must remain in orbit about one of the massive particles.

Let the energy and angular momentum integrals be formed from equations (20). Let the total energy be E and the total angular momentum vector be $\underset{\sim}{C}$. Then, following Zare 1976, 1977, it may be shown that the stability or otherwise of the binary is controlled by the value of the parameter $S = |\underset{\sim}{C}^2| E$. The value of S is, of course, known from the initial values of the masses and the position and velocity components appearing in the energy and angular momentum relations. If S is smaller than or equal to a critical quantity, S_{cr} , which can be computed, then the binary cannot be broken up by the third mass. If, however, $S > S_{cr}$, then break-up may occur. Zare's criterion $S \leq S_{cr}$

may therefore be usefully applied to any general three-body problem found in nature of the hierarchical type (binary plus third body). Examples of these are triple stellar systems, Planet-Moon-Sun, Sun-Jupiter-Saturn, though in each case, although the general three-body problem model is a close approximation to the system found in nature, the presence of other perturbing bodies cannot be totally disregarded.

The quantity S_{cr} is computed form the values of the three masses, applying the Lagrange collinear solution of the three-body problem (Szebehely and Zare 1977; Zare 1976, 1977).

7. APPLICATION OF EMPIRICAL STABILITY CRITERIA TO THE THREE-BODY HDS

In their statistical empirical approach to the question of the stability of hierarchical dynamical systems ($n \geq 3$), Emslie, Roy and Walker make use of Zare's criterion.

Putting $n = 3$ in equations (5), we have, to the second order in the expansion of the force function U

$$\ddot{\underset{\sim}{\rho}}_2 = G(m_1 + m_2) \underset{\sim}{\nabla}_2 \left(\frac{1}{\rho_2} (1 + \varepsilon_{32} P_2(C_{23}))\right) ,$$

$$\ddot{\underset{\sim}{\rho}}_3 = G(m_1 + m_2 + m_3) \underset{\sim}{\nabla}_3 \left(\frac{1}{\rho_3} (1 + \varepsilon^{23} P_2(C_{23}))\right) , \tag{21}$$

with

$$\varepsilon^{23} = \frac{m_1 m_2}{(m_1 + m_2)^2} \alpha_{23}^2$$

$$\varepsilon_{32} = \frac{m_3}{m_1 + m_2} \alpha_{23}^3 . \tag{22}$$

Thus ε_{32} is a measure of the ratio of the disturbance by P_3 on P_2's orbit about P_1, to the central two-body force between P_2 and P_1. Likewise ε^{23} is a measure of the ratio of the disturbance by P_1 and P_2 on the orbit of P_3 about the centre of mass of P_1 and P_2, to the central two-body force between P_3 on the one hand and P_1 and P_2, assumed to lie at their mass-centre.

Again introduce μ and μ_3 by the relations

$$\mu = m_2/(m_1 + m_2) \quad ; \quad \mu_3 = m_3/(m_1 + m_2) \quad .$$

Then (23)

$$\varepsilon^{23} = \mu(1-\mu)\,\alpha_{23}^2 \quad ; \quad \varepsilon_{32} = \mu_3\,\alpha_{23}^2 \quad .$$

We now examine this picture in the light of Zare's stability criterion (Section 6) based on the quantity $S = |\underset{\sim}{C}^2|E$, where $\underset{\sim}{C}$ and E are the constants appearing respectively in the angular momentum and energy integrals of the general three-body problem. If the three-body system was a hierarchical one (a binary plus a third body in a large orbit about the binary's mass-centre), and $S \leq S_{cr}$, the binary could never be broken up. The critical stability value S_{cr} was derived from the collinear solution of the general three-body problem. To obtain S_{cr} , a ratio X must be found, where X is the solution of Lagrange's quintic equation (Szebehely and Zare 1977; Zare 1976, 1977). In its turn $\alpha_{cr} = (\rho_2/\rho_3)_{cr}$ is related to S_{cr} as follows.

The quantity $\alpha = \rho_2/\rho_3$ is independent of μ and μ_3 ; its value is fixed by the initial configuration of the hierarchical three-body system. $\underset{\sim}{S}$ is a function of the mass parameters μ , μ_3 and also $\underset{\sim}{\rho}_2$, $\dot{\underset{\sim}{\rho}}_2$, $\underset{\sim}{\rho}_3$ and $\dot{\underset{\sim}{\rho}}_3$ (which may be in turn expressed as functions of osculating elements of the disturbed Keplerian orbits of the system), thus

$$S = f(\mu,\mu_3;\ a_2,a_3,e_2,e_3,i_2,i_3,\Omega_2,\Omega_3,\omega_2,\omega_3,f_2,f_3).$$

Let us now consider initially circular, coplanar orbits with the bodies in a straight line $P_1\ P_2\ P_3$. Then

$$S = f(\mu,\mu_3;\ \alpha)$$

where α is now simply the ratio a_2/a_3 .

The critical value of α , α_{cr} , may then be found by equating S and S_{cr} viz.

$$S_{cr} = f(\mu,\mu_3;\ \alpha_{cr}) \quad .$$

Thus to the critical stability criterion of Zare, namely $S \leq S_{cr}$, there corresponds the stability criterion $\alpha \leq \alpha_{cr}$, for a given $\alpha = \rho_2/\rho_3 = a_2/a_3$ (the radii of the initially circular orbits) and a given μ , μ_3 .

Thus for all of possible values of μ and μ_3, plotted on the $\mu - \mu_3$ plane, a surface of values of α_{cr} exists above it in the third dimension α. For a hierarchical three-body problem with initially circular, coplanar orbits, therefore, α is known, as is μ and μ_3. The point μ, μ_3, α can therefore be plotted. If it lies below or on the point μ, μ_3, α_{cr}, the system is stable in the sense that the binary $P_1 - P_2$ cannot be broken up.

From relations (23), it is obvious that a system may be expressed not only as a set of values μ, μ_3, α but also as a set of values ε^{23}, ε_{32}, α. We may therefore readily translate Zare's critical stability surface form the $0\mu\,\mu_3\,\alpha$ parameter space into the $0\varepsilon^{23}\varepsilon_{32}\alpha$ parameter space (see Walker et al., 1980). It is thus possible to use Zare's criterion in relation to the ε-parameters as well as to the μ-parameters.

Walker and Roy have used numerical integration to examine the orbital behaviour of a large number of three-body HDS where the value of S is initially greater than the Zare stability value S_{cr}. All that can be said analytically in this case is that the system may be unstable, that is, the binary may be broken up with P_2 being ejected from the system or P_3 ejected by robbing the binary of some of its energy.

Walker and Roy's numerical experiments with initially circular and coplanar three-body HDS show that Zare's criterion is unnecessarily restrictive.

They took all combinations of pairs of values $\varepsilon^{23} = 10^{-\omega_1}$ and $\varepsilon_{32} = 10^{-\omega_2}$, where ω_1 and ω_2 took the integral values 2, 3, 4, 5, 6, giving 25 sets of pairs.

For a chosen pair of ε^{23} and ε_{32} values, the behavior of a system of a given α was studied.

For values of α far bigger than α_{cr}, but less than unity, where α_{cr} is that value of α corresponding to $S = S_{cr}$ for the system of the same mass values μ and μ_3, the system within a small number of synodic periods showed instability. Such instability may be demonstrated by either a crossover or orbits indicated in Figures 5 and 6 by X, or an approach of P_2 and

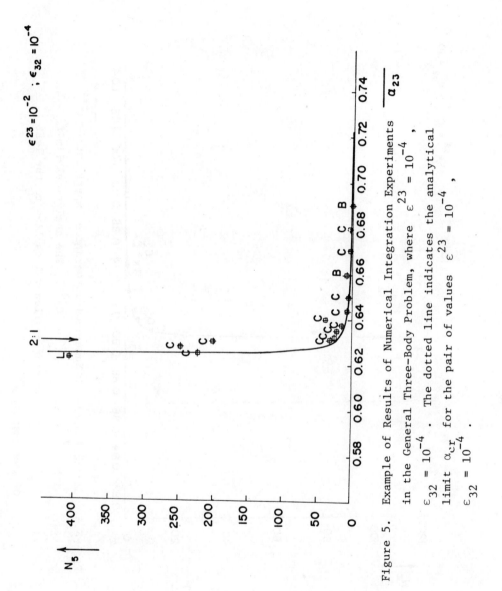

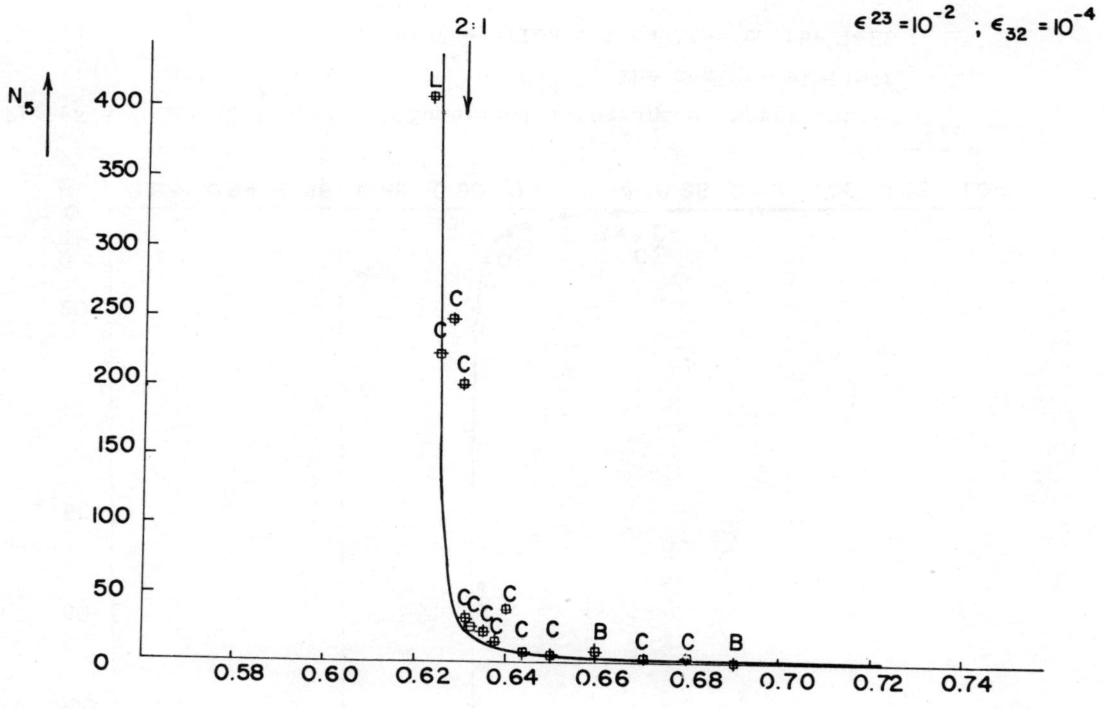

Figure 5. Example of Results of Numerical Integration Experiments in the General Three-Body Problem, where $\varepsilon^{23} = 10^{-4}$, $\varepsilon_{32} = 10^{-4}$. The dotted line indicates the analytical limit α_{cr} for the pair of values $\varepsilon^{23} = 10^{-4}$, $\varepsilon_{32} = 10^{-4}$.

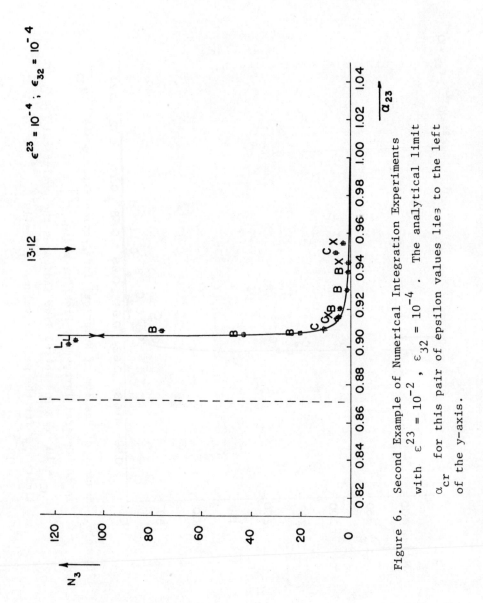

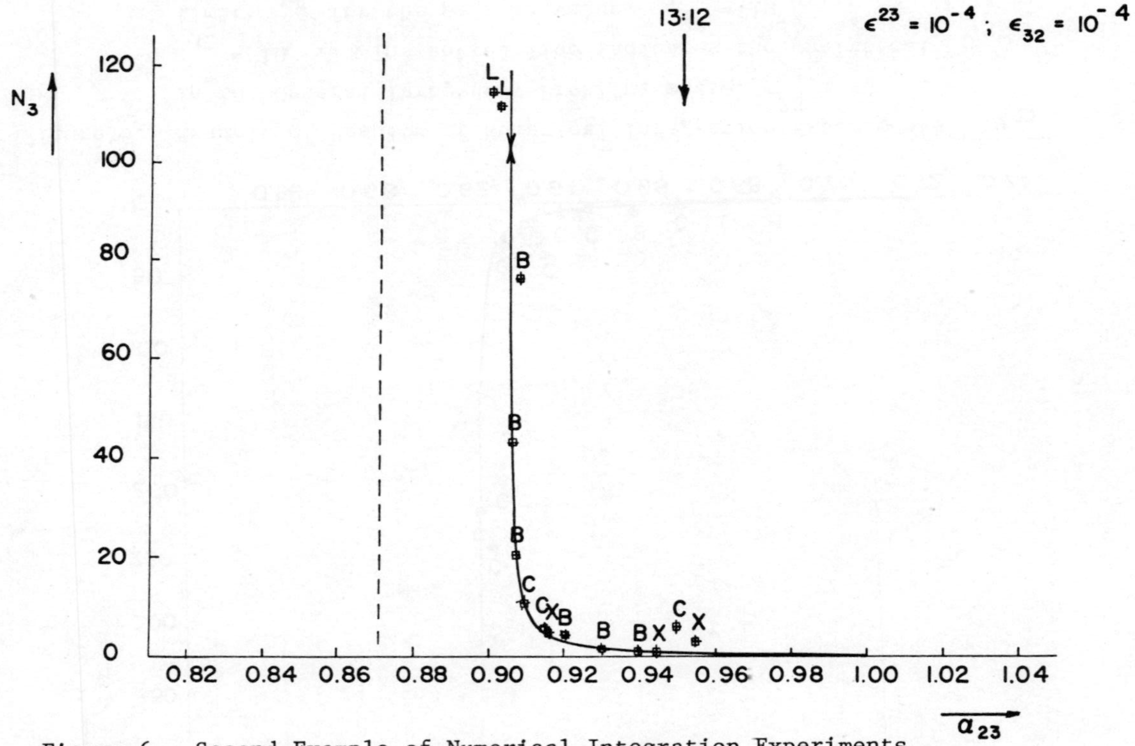

Figure 6. Second Example of Numerical Integration Experiments with $\varepsilon^{23} = 10^{-2}$, $\varepsilon_{32} = 10^{-4}$. The analytical limit α_{cr} for this pair of epsilon values lies to the left of the y-axis.

P_3 so close that an irreversible change takes place in the size and shape of the osculating Keplerian orbits, indicated by C, or the escape of either P_2 or P_3 from the system (termed breakup B).

For smaller values of α, but with values still larger than α_{cr}, the number N_S of synodic periods the system lasted for, before instability showed itself, grew, the growth in N_S following an exponential upward trend with decreasing α and obviously becoming infinite at some value $\alpha_o > \alpha_{cr}$. In cases where the numerical integration was terminated at some value of N_S, the letter L is used to indicate that no instability had showed itself by that time.

It may be noted that in Figures 5 and 6, deviations from the smooth curves exist at commensurabilities in mean motions at 13:12 and 2:1, respectively.

For every pair of ε-values investigated, a value $\alpha_o > \alpha_{cr}$ existed such that, if $\alpha_{cr} < \alpha < \alpha_o$, there is no indication that the mutual perturbations produce instability.

It would appear then that not unexpectedly a zone of stability exists outside the Zare surface capable of being found by numerical integration experiments.

8. APPLICATION OF EMPIRICAL STABILITY CRITERIA TO THE FOUR-BODY HDS.

There does not seem to be a Zare-type analytical stability criterion for the four-body HDS. Nevertheless, it appears possible to approach the stability problem of a four-body HDS in the following way.

A four-body HDS can be thought of as made up of 'disturbed' triple subsets. Thus in Figure 1 the simple four-body HDS has subsets (i) m_1, m_2, m_3 disturbed by m_4, (ii) m_1, m_2, m_4 disturbed by m_3, (iii) $(m_1 + m_2)$, m_3, m_4 disturbed by m_1 and m_2 not being at C_2. Similarly for the general four-body HDS of Figure 4, the subsets are (i) m_1, m_2, $(m_3 + m_4)$ disturbed by m_3 and m_4 not being at M_{12}, (ii) m_3, m_4, $(m_1 + m_2)$ disturbed by m_1 and m_2 not being at M_{11}.

Suppose initially each subset satisfies Zare's criterion, its α value lying well below the α_{cr} calculated for the subset's masses. If the subset was not being disturbed, then its stability would be assured forever by its α being less than α_{cr}. But it is being disturbed and although α_{cr} will not vary, the α value will. If it wanders in a sort of random walk so that it ultimately reaches a situation where $\alpha > \alpha_{cr}$, then the subset may become unstable and certainly will be if it becomes greater than α_o, Walker and Roy's empirical stability value.

The same argument applies to other triple subsets. This line of argument suggests that the epsilons may well be the crucial parameters in a consideration of the long-term stability of the four-body HDS. They are a measure of the disturbances that each body produces on the others' orbits. It is not expected that the computation of the epsilons for a HDS would lead to a determination of the precise time when any two orbits will interchange or a body escape from the system. It should be possible, however, to make a statistical prediction from such data. This statistical, empirical approach to stability is being investigated in a long term programme by Roy and Walker, part of the work involving the numerical integration of general n-body systems (n = 3,4). In the case of $n = 4$, studies have been made to see how different initial sets of starting conditions (the ε, α and μ values) govern the time it takes for such four-body systems (which are, of course, composed of triple subsets) to reach a state where one or more of the Zare stability criteria in the system is violated. From such experiments it is becoming clear that it should be possible from an examination of the 'starting conditions' in any n-body system to provide a statement or statistical estimate of its stability - the dynamical equivalent of the life-time of a planetary atmosphere.

The kinetic theory of gases enables a half-life T (the time it will take half of the molecules in the atmosphere to escape into space) to be calculated from x, the ratio of the mean molecular velocity to the velocity of escape from the planet.

For $x = 1$, the value of T is very small indeed. As x decreases, T grows slowly at first and is measured in minutes, hours, weeks. But quite soon a region of x is reached where T shoots up to durations of astronomical length.

It is possible that the stability of n-body HDS where $n > 3$ can usefully be treated like this. If we begin with a large number of hierarchical dynamical systems where they all have epsilon and alpha values within certain ranges, we may be able to state that the statistical status quo lifetime of these systems is of such and such a duration in the sense that such a lifetime will have to elapse before half the systems will have suffered any change in the status quo of their ordered orbits.

9. APPLICATION TO THE STABILITY OF THE SOLAR SYSTEM.

If the arguments sketched above are sound, then insight into the problem of the stability of the Solar System is achieved by applying them to the Solar System's hierarchical dynamical systems.

If the Solar System is firstly split into triple HDS subsets, for example, Sun-Jupiter-Saturn, or Earth-Sun-Moon, or Sun-Saturn-Uranus, and so on, it is found that with very few exceptions, Zare's stability criterion is well-satisfied (Roy 1980; Walker et al. 1980). Among the exceptions are the retrograde satellites of Jupiter, which is not unexpected since they are probably captured asteroids and may resume an asteroidal role in the future.

These triple subsets are, of course, disturbed by the other members of the HDS and it is possible that the cumulative pseudo-random walk of perturbations will drive some of the subsets outside the Zare limit and even beyond the Walker-Roy empirical limit. If this happens, the status quo of this occurring will depend upon the sizes of the epsilons. If these are small enough, this time interval may be of astronomical length.

Leaving aside the 'hard' commensurability cases in the Solar System such as Neptune-Pluto, or Titan-Hyperion where there is a dynamical locking mechanism, ensuring stability, there would appear on such arguments to be nothing remarkably esoteric about the distribution of solar system orbits or the values of the elements that describe these orbits. In their distribution, near-circularity and near-coplanarity, they merely reflect the sizes of the epsilons and alphas that have reduced the orbits' pseudo-random walks to such small strolls, enabling the Solar System's status quo to be maintained over a long time, perhaps an astronomically long time.

REFERENCES.

Marchal, C. and Saari, D. G.: 1975, Celest. Mech., 12, 155.

Roy, A. E.: 1979, Instabilities in Dynamical Systems, ed. V. Szebehely, Reidel, Dordrecht.

Szebehely, V. and Zare, K.: 1977, Astron. Astrophys., 58, 145.

Walker, I. W.: 1982, Celest. Mech. (in press).

Walker, I. W., Emslie, A. G. and Roy, A. E.: 1980, Celest. Mech., 22, 371.

Walker, I. W. and Roy, A. E.: 1981, Celest. Mech., 24, 195.

Walker, I. W. and Roy, A. E.: 1982a, Celest. Mech. (in press).

Walker, I. W. and Roy, A. E.: 1982b, Celest. Mech. (submitted).

Zare, K.: 1976, Celest. Mech., 14, 73.

Zare, K.: 1977, Celest. Mech., 16, 35.

REFERENCE SYSTEMS FOR EARTH DYNAMICS

R. O. Vicente
Faculty of Sciences
University of Lisbon, Portugal

ABSTRACT. A comparison is presented of reference systems employed by classical and modern techniques of observing the Earth's rotation. Advantages and disadvantages of the systems so far employed are discussed. The observing campaign of project "MERIT" emphasizes the need for consistent systems of reference. Any future system of reference for the Earth's dynamics should be well defined in order to avoid past ambiguities. The future of the Conventional International Origin and proposals about the Conventional Terrestrial System to be adopted by international agreement are presented.

1. INTRODUCTION.

What is the importance of the study of reference systems for Earth dynamics? And for celestial mechanics in general?

Some of the important practical problems are the following: 1) comparison of the observed positions of celestial bodies (in planetary systems in particular) with theory; 2) positions of spacecraft in the solar system; 3) positions of artificial satellites of the Earth.

The importance of a consistent reference system for the Earth dynamics has increased lately. There are several reasons for the increased interest in terrestrial reference systems in fundamental astronomy, terrestrial and space geodesy, and geophysics, namely, the modern techniques of determining distances and station coordinates on the surface of the Earth, and the possibility of measuring displacements of the plates for the researches on global plate tectonics.

V. Szebehely (ed.), Applications of Modern Dynamics to Celestial Mechanics and Astrodynamics, 131–144.

The modern techniques employing Doppler and laser with artificial satellites, lunar laser ranging and radio interferometry (connected and very long base line) claim the precision of decimeters and even a few centimeters.

One of the interesting problems we have to face in the near future is that we might reach levels of determining distances and station coordinates that correspond to the noise level of the system.

The definition of the noise level of the system corresponding to the planet Earth depends on the accuracy of the techniques employed for detecting the geophysical motions of the observatories. Considering the horizontal and vertical motions of the observatories that we know nowadays from geodesy and geophysics, we can imagine there is a permanent instability of the station that will not allow us to reach accuracies greater than this level of instability. One of the future problems will be to find out this noise level of the system will it be of the order of 1 cm or 1 mm?

The study of reference systems for Earth dynamics is closely associated with the dynamics of the rotation of the planet Earth, and we have to consider 3 main aspects: a) time scales; b) structure of the planet; c) reference systems employed. We cannot really separate them since they are interdependent.

2. TIME SCALES.

It is convenient to classify the span of time beginning with the formation of the solar system. We suppose that the age of the solar system is of the order of 4.5×10^9 years and we can consider the following intervals:

$0 - 10^3$ years - very short ,
$10^3 - 10^5$ years - short ,
$10^5 - 10^7$ years - long ,
$> 10^7$ years - very long .

One of the main difficulties in dealing with studies of the dynamics of planets, or any other studies in celestial mechanics, is the fact that our astronomical observations correspond only to intervals of time that we must consider to be very short. One of the longest records of systematic astronomical observations corresponds to the luni-solar precession first observed about 2,000 years ago. We have slightly longer records of observations of total solar eclipses (about 4,000 years) that are useful for studies of the diurnal rotation of the Earth.

On the subject of researches dealing with the Earth's diurnal motion, we shall mention the studies about the growth of coral rings that can be traced to about 450 million years, but these studies imply a number of biological and physical processes which are not yet very well understood.

3. STRUCTURE OF THE PLANET.

We may say nowadays that the Earth is one of the most complex planets of the solar system because the structure of the Earth is composed of a lithosphere, an atmosphere and a magnetosphere. All these components interact and, therefore, make it more difficult to define, without ambiguity, a suitable reference system.

On the other hand, we should not overlook the fact that the simplest mathematical model for studying the Earth's motion still gives a very good approximation. This is when we consider the Earth as a solid rigid body. The adoption of such a model leads to the solution of the problem employing Euler's equations referring to the motion of a rigid body with a fixed point.

Let us now consider the extreme case of a very complicated model which would apply to an Earth changing its shape due, for instance, to elasticity or the transference of heat. One could consider the Earth as formed by a collection of particles having all sorts of motion, but the difficulty would be to represent that model by an adequate set of differential equations which could be integrated. The equations of motion, corresponding to one of such models, are represented by Liouville's equations.

4. EULER'S EQUATIONS.

The vector equation corresponding to Euler's equations is

$$\frac{d\vec{H}}{dt} = \vec{G}$$

which shows that the time derivative of the angular momentum $\vec{H}$ around the centre of mass is equal to the vector moment $\vec{G}$ due to the external forces.

The motion is studied considering the fre motion ($\vec{G} = 0$) and the forced motion ($\vec{G} \neq 0$), following the traditional procedure of celestial mechanics. This procedure has advantages for the integration of the equations of motion. The free motion is called free nutation and the forced motion is designated as forced nutation. The word nutation derives from the Latin "nutatio" which means an oscillation around an axis.

The study and integration of Euler's equations reveals the existence of 4 fundamental axes which complicate the definition of suitable reference systems because they do not coincide. They are the following: 1) the axis OZ fixed in space and directed towards the north pole of the ecliptic; 2) the axis Oz fixed in the body and called the axis of figure; 3) the instantaneous axis of rotation, called the axis of rotation, corresponding to the instantaneous rotation vector $\vec{\omega}$ representing the motion of the Earth at every instant; 4) the axis along the angular momentum vector $\vec{H}$.

There is an important feature in the behavior of the free motion. This type of motion only affects the coordinates of points on the Earth's surface (latitude and longitude), and does not affect, for instance, the equatorial coordinates of a star. This important fact is overlooked quite often. Another interesting property of the free motion is that the axes of rotation of figure and of the angular momentum are in the same plane.

Euler's theory permits the computation of the angles between these axes. The angle between the axes of rotation and of the figure is about 0".3 (or 10 m on the surface of the Earth). The angle between the axes of rotation and of the angular momentum is about 0".001 (or 3 cm on the Earth's surface).

The practical consequence of the fact that the axis of rotation does not coincide with the axis of figure is the existence of the so-called variation of latitude, initially detected about a hundred years ago. Besides this variation of latitude, there is a variation of longitude. The free motion is nowadays designated as polar motion.

There is, unfortunately, a very loose terminology about these phenomena which has been employed since about 1950 when geophysicists became more interested in the explanations of polar motion. This terminology does not conform to the basic dynamics of the Earth's motion and it has been criticised by Melchior (1980). A working group on nutation of the International Astronomical Union has just published a report where a set of definitions is given, and it is hoped that it will be universally adopted (report to be published in Celestial Mechanics).

The main fact to keep in mind is that the axis of rotation practically coincides with the axis of angular momentum from the point of view of the observations because the classical techniques (visual and photographic zenith telescopes, astrolabes) cannot attain a precision of the order of 0".001, and the modern techniques have not yet been able to reach an accuracy of 0".001.

We know that in the free motion ($\vec{G} = 0$) the axis of angular momentum is fixed in space

$$\frac{d\vec{H}}{dt} = 0 \quad \vec{H} = \text{const.}$$

This important property has not yet been employed for writing down the equations of motion in terms of the angular momentum vector. The advantages, derived from the fact that the axis of rotation practically coincides with the axis of angular momentum, were pointed out by Oppolzer (1886) who was the first to see the benefits of employing the axis of rotation to obtain the equations of motion of the Earth.

Another fact that has been overlooked recently concerns the observations which are referred to the axis of rotation when we employ any of the classical techniques. This axis of rotation intersects the surface of the Earth at the points called geographical poles, and this definition is given in any classical treatise on positional astronomy (Woolard and Clemence, 1966).

Considering such model for the Earth's structure, all the circular paths described in the diurnal motion of the stars are centered in the celestial rotation pole, that is, the intersection of the axis of rotation with the celestial sphere.

It is obvious that the pole of figure cannot be determined from the observations, considering the previous definitions. In the theoretical case, it would be the centre of the circle defined by the motion of the axis of rotation (called the polhode), but the observations show that the polhode is an irregular curve. All the past directors of the Central Bureau of the International Latitude Service called it the barycenter and have determined it, since 1900, for different intervals of the observations. We can say that these barycenters are near to the pole of figure.

The above considerations show the advantages of employing the axis of rotation as a convenient axis for the body-fixed system of reference not only from the theoretical point of view but also from the practical aspects of astronomical observations.

5. LIOUVILLE'S EQUATIONS.

We can generalize Euler's equations by the employment of a tensor notation and the application of the summation convention as follows:

$$\frac{dH_i}{dt} = \varepsilon_{ijk}\,\omega_j\,H_k = L_i$$

where H_i are the components of angular momentum, ε_{ijk} are the elements of the unit tensor and L_i are the components of the moments of the external forces acting on the system.

We can separate the angular momentum H_i into two parts,

$$H_i = C_{ij}\,\omega_j + h_i$$

where the inertia tensor is

$$C_{ij} = \int_V \rho(x_k x_k \delta_{ij} - x_i x_j)dv$$

The relative angular momentum due to the motion of particles u_i , relative to the coordinate system x_i , is

$$h_i = \int_V \rho \, \varepsilon_{ijk} \, x_j \, u_k \, dv.$$

The quantities C_{ij} and h_i depend on the fields of density $\rho(x_k,t)$ and relative velocity $u_i(x_k,t)$.

Introducing these expressions in the generalization of Euler's equations we obtain

$$\frac{d}{dt}(C_{ij} \, \omega_j + h_i) + \varepsilon_{ijk} \, \omega_j (C_{k\ell} \, \omega_\ell + h_k) = L_i$$

which are Liouville's equations.

The integration of such a system of differential equations is a difficult problem and can only be attempted by successive approximations. In such a model, where every particle of the body moves, it is extremely difficult to define any set of axes linked to the body which does not move and can be employed as a reference system.

Fortunately, the behavior of the Earth can be considered in a way which does not correspond to such an extreme model. Actually, the observations show that the deviation from a rigid model is of the order of 0".01 in the main term of the forced nutation in obliquity. This is called the constant of nutation (Jeffreys, 1959). Therefore, we should keep in mind the rigid Earth model and consider the actual departures from this model as small perturbations, a technique which is widely employed in celestial mechanics.

Unfortunately, this subject sometimes is confused (the departure of the Earth from a rigid model) and systems of reference are introduced which are difficult to visualize and to construct from the point of view of the observations.

6. REFERENCE SYSTEMS.

The ideal solution for the problem of reference systems for the Earth needs only two systems:
1) a space-fixed system X_i with unit vectors $\vec{E}_i$
2) a body-fixed system x_i with unit vectors $\vec{e}_i$ (i = 1,2,3).

These two systems define respectively the position of the Earth as a planet in space and the positions of points on the Earth's surface. They correspond to a Conventional Inertial System (CIS) and to a Conventional Terrestrial System (CTS).

The mathematical transformation between any of the possible reference systems X_i and x_i is represented by a matrix.

Let us consider some of these matrices which are necessary to take account of the main motions of the Earth:

Diurnal Motion: $D_{ij}(t)$, $\vec{e}_i(t) = D_{ij}(t)\vec{E}_j$

Precession: $P_{ij}(t)$, $\vec{e}_i(t) = P_{ij}(t)D_{jk}(t)\vec{E}_k$

Nutation: $N_{ij}(t)$, $\vec{e}_i(t) = N_{ij}(t)P_{jk}(t)D_{kl}(t)\vec{E}_l$

The nutation matrix should be separated into the Free $N^{FR}_{ij}(t)$ and Forced $N^{FO}_{ij}(t)$ nutations in agreement with the ideas previously explained and in order to avoid the mistakes mentioned.

All the computer programs used with the modern techniques incorporate these rotation matrices but it is not yet certain if they employ them in a consistent reference system.

7. DETERMINISM IN CELESTIAL MECHANICS.

We have to keep in mind that our theoretical models, in spite of the advances made, have to be considered as asymptotic approximations to the physical reality of the Earth or the universe in general.

It is very important to point out that the value of the constant of nutation was obtained from the observations and not from theoretical expressions, therefore, an inconsistency exists. A theory based on a simplified model of the Earth where we insert a value obtained from observations which correspond to the real behavior of our planet.

This is the problem which is mentioned by Szebehely (1982), and we cannot, unfortunately, avoid such a difficulty. This is related to the question if celestial mechanics is deterministic or not. It is known that 19th century mechanics considered as one of their great triumphs the possibility of forecasting the position of celestial bodies. But they did not define clearly what they meant by forecasting in space and time and they did not specify the errors of the observations.

Considering the time scales previously mentioned, even for such a short interval as 10^5 years, we cannot discriminate in favor or against determinism in dynamics because our interval of observations is hopelessly very short. At most, we have only 100 years of fairly accurate observations of the bodies of the planetary system.

Considering now the space position of the different bodies of the planetary system, we have to take into account the erros of the observations. If one finds a planetary body within, say, 5° of the position forecasted by the theory--is that a confirmation or a rejection of the applicable theory? Can we consider that as an error of observation or a genuine disagreement with the theory?

If we relax our requirements, we may consider a precision of 5° a good confirmation of the theory. But if we wish to obtain higher accuracy, say 1' or 1", then one may say that the theory is hopelessly wrong.

These two aspects of forecasting the position in space and time of celestial bodies was mentioned in order to point out that it is difficult to argue in favor (or against) determinism in mechanics. The application of the principles of mechanics to celestial bodies (celestial mechanics) does not seem to support or reject the idea that dynamics is a deterministic science.

8. OBSERVATIONAL RESULTS.

We must remember that the only way to study the motions of the Earth is through a set of well-planned observations.

The classical techniques employing meridian transit circles, visual and photographic zenith telescopes, and astrolabes, have been able to determine the values of the main components of polar motion and the irregularities of universal time. The fact that there have been regular, long and well-planned programs of observations enabled us to discover some interesting irregularities in the Earth's dynamics in spite of the lower precision (about 1.5 m) of these techniques. These discoveries were aided by the existence of international computing centers, that is, the International Polar Motion Service (IPMS) and the Bureau International de l'Heure (BIH).

The observations corresponding to the International Latitude Service (ILS) have been determining variations of latitude permanently since 1900, and have started a number of projects, aimed at the geophysical explanations for the path of the instantaneous pole of rotation (the polhode). One important fact that has been often overlooked is that the barycenter of the polhode is near to the pole of figure of the Earth, corresponding to the axis of the greatest moment of inertia. There seems to be no other known way of determining a point on the Earth's surface near to the pole of figure.

This important property led to the definition of the Conventional International Origin (CIO), based on the well-known latitudes of the ILS observatories, all situated on the 39°08' N parallel, and it was officially adopted by the International Astronomical Union [Transactions IAU, Vol. 13B, p. 111 (1967)]. Unfortunately, there are still numerous statements ignoring this definition and leading to wrong conclusions. The adoption of the CIO as the origin of the pole coordinates led to a better and more consistent way of expressing the pole coordinates determined by the IMPS. Another great advantage of this definition is the fact that it is obtained from well-known station coordinates and, therefore, it can be compared and recovered any time so long as these observatories remain active at the same sites.

The long term studies about the behavior of the polhode and fluctuations in the length of the day can only be undertaken and have some meaning if we keep the same origin as a reference. The long and consistent ILS series, in spite of its lower precision in comparison with present day techniques, originated a data set (Yumi and Yokoyama, 1980) that permitted not only the identification of a 30 year period (Vicente and Currie, 1976) in the polar motion, but also offered the best determination of the period of Chandler's component (Wilson and Vicente, 1980).

It is well known that long and regular data sets, for instance, solar eclipses and lunar occultations (Morrison, 1979) have furnished very valuable elements about the Earth's dynamics in spite of their low precision. The low precision observations obtained by the ILS chain of observatories, and the consistent employment of the CIO, will prove useful in the future to measure possible displacements in the plate tectonics investigations. We can arrive at the same conclusion about Chandler's period. It will be necessary to obtain many decades of regular and continuous observations with the modern techniques before we can improve the values obtained for this period.

One of the main difficulties encountered in the search for a convenient reference system is associated with the statistical analysis of the results obtained by the IPMS and BIH, and the adoption of a proper statistical model. Jeffreys (1940, 1968) mentions that no standard statistical method can be satisfactory on account of correlation of errors and variation of weights. Maximum likelihood methods have shown advantages, giving estimates and valid uncertainties of Chandler's period (Wilson and Vicente, 1981).

Another difficult problem is the definition of the meridian of zero longitude. The BIH has adopted an algorithm for the definition of the fundamental meridian based on the longitudes of certain observatories but, unfortunately, the observatories considered vary from time to time; some observatories are eliminated and others are added (Feissel, 1980). This implies changes in the computational methods to define the fundamental meridian, and it is difficult to know if we are keeping always the same meridian within the precision of nanoseconds required by some of the present day techniques. We can even ask the question if it is possible to define a fundamental meridian, within the precision obtained by atomic clocks, employing a set of observatories that are not always the same in the course of the years.

The philosophy underlying the definition of the fundamental meridian is different from the conception that led to the definition of the CIO. The fundamental meridian is defined in terms of dozens of observatories distributed world-wide in the hope that the geophysical motions of the individual observatories will be attenuated, but the total number of observatories is not kept constant. The CIO definition is based on a fixed number of observatories of well-known coordinates, determined since the beginning of the century, but the world coverage is not very good and, therefore, the geophysical motions of the observatories are not well compensated.

9. REFERENCE SYSTEMS EMPLOYED.

The modern techniques of lunar laser ranging and radio interferometry used to observe celestial bodies (the Moon and radio sources), also employ an astronomical reference system which is defined in relation to the axis of rotation. This is an important advantage of these modern techniques because we can relate their observations to a reference system which has been employed by the classical techniques for nearly a century. This is important for long term studies about the behavior of the Earth's motions around the center of mass.

The two other modern techniques of satellite laser ranging and Doppler do not use a reference system connected with the celestial sphere because they employ artificial satellites. But the connection with astronomical reference systems appears through the need of taking account, for instance, of precession and nutation.

Many of the computer programs employed for the reduction of observations, obtained by the modern techniques, are not yet consistent in the utilization of reference systems, that is, in one part of the program they employ a reference system which might not be consistent with the reference system employed in another part of the program.

The geopotential series adopted by some of the modern techniques (Doppler and laser with artificial satellites) have been referred to the CIO, and any change in the origin of the pole coordinates might bring inconsistencies in the results, if no proper precautions are taken. The reference systems employed in several investigations, including Doppler techniques, have not been consistent in the past, and various satellite solutions yield values for the position of the pole of the z-axis which differ by several meters at the Earth's surface (Anderle and Tanenbaum, 1974).

This lack of consistency is not important for determining the period of Chandler's component of the polar motion. This has been demonstrated by Vicente (1981). The existence of systematic errors in two long series of classical observations did not affect the value of the period. But, on the other hand, the study of possible variations in Chandler's component for shorter intervals, for instance, several months or years, is strongly dependent on the reference systems adopted and the number of observing sites.

10. POSSIBLE REFERENCE SYSTEMS.

After defining a suitable Conventional Inertial System (CIS) which does not concern us at the moment, we have to consider adequate reference systems linked to the Earth. We have previously mentioned the main axes connected with the Earth's dynamics and, therefore, we have several options. We can suppose, for instance, the following systems:

1) linked with the dynamical motion of the Earth, where the axis of rotation assumes an important role;
2) based on certain points defined on the surface of the Earth, and keeping in consideration the rotation of the Earth;
3) defined by a set of reference points situated, for instance, in different continents whose distances have been determined by methods employed in space geodesy (the minimum number of points would be three).

We can point out some of the uncertainties which affect the systems described:

1) The dynamical theory of the motion of the Earth around its center of mass does not consider all the forces acting on the Earth, and the known physical properties of the structure of our planet. Some of the problems are: secular deceleration of the Earth's rotation, seasonal variations of the Earth's rotation, irregular variations (probably random) of the rotation of the Earth, periodic terms (for instance, annual and Chandlerian) of the motion of the pole, and other irregularities of the motion of the pole.

   This reference system has the advantage of providing a link between purely terrestrial systems and heliocentric reference systems because our fundamental plane is the equator of the Earth.
2) The adoption of certain points on the Earth's surface introduces more unknown parameters, for instance, we need a theory of the motions of the Earth's crust in space related to the rotation of the Earth. That is, besides the difficulties mentioned for the previous system, we have to consider the motions of the terrestrial crust. Some parts of the crust move in relation to other parts, and this is to be related to the axis of rotation. Does the crust, as a whole, move in relation to the mantle?
3) This system is purely terrestrial, and, in analogy with the Conventional Inertial System already mentioned, we could consider it the ideal one if the points on the Earth's crust were perfectly at rest. But, as we know, some of the motions we have to consider are: a) the points adopted are situated in different tectonic plates, and we must have a mathematical

theory of the motions of these plates; b) portions of the crust might sink or rise in relation to other portions; c) good values of earth tides and ocean load factors are needed at the chosen fundamental points. This system has also the inconvenience that it cannot be immediately related to the Conventional Inertial System.

There does not seem to exist at the present time a unique reference system that would satisfy all the requirements and all of the above described reference systems can be criticized.

We can say it is not possible to have a good Conventional Terrestrial System within the precision of the order of a few centimeters, because we have not yet a mathematical theory describing all the phenomena mentioned. But we must keep in mind that the departures of the structure of the Earth from a rigid body model are really small, and, therefore, we probably should adopt a Conventional Terrestrial System which does relate well to the rotation of the Earth.

11. THE MERIT CAMPAIGNS.

The International Astronomical Union set up a working group in order to promote a comparative evaluation of the techniques utilized for the determination of the rotation of the Earth. This is called the MERIT (Monitoring Earth Rotation by Intercomparison of Techniques) working group.

The MERIT "short campaign" was accomplished between August 1 and October 31, 1980. It was designed to test the possibility of acquiring data from classical and modern techniques with special concern about the possibility of exchanging and disseminating the results obtained and testing the techniques of transmitting data quickly.

The results obtained in three months' observations cannot be employed to check which technique is more convenient because they refer to a short time interval and, also, the constants and standards employed by the different techniques were not consistent.

We have mentioned before the difficulties and pitfalls of studying polar motion and, therefore, it is premature to arrive at any conclusions obtained by the several techniques during the MERIT short campaign.

It is hoped that during the MERIT "main campaign" (scheduled for 1983-84) the group in charge of constants and standards will be able to define clearly a set of constants and reference systems which will be uniformly employed by all the techniques used. This implies a great amount of work to make all the computer programs consistent, but without such effort, it will be meaningless to try to compare the results.

Even the smoothing process employed by the several techniques should be standardized because it introduces uncertainties in the values obtained (Wilson and Vicente, 1981). We are referring to two types of smoothing: 1) from the raw observational data into the data to be used in the computer programs; 2) the smoothing employed for obtaining polar motion coordinates (x,y) at 2, 3, or 5-day intervals. This is especially important as we are hoping to get precisions of at least one order of magnitude better than the classical techniques, that is, of the order of decimeters.

12. THE FUTURE CONVENTIONAL TERRESTRIAL SYSTEM.

Considering that we are near the noise level of the system, it is difficult to envision a set of observatories, within the present day capabilities of scientific technology, which satisfies all the best theoretical requirements.

The main point we must keep in mind is the reason of comparing observations. It does not matter if the reference pole is not the best available, because the important thing is to keep the same standard for comparison purposes. We can demonstrate the advantages of such a procedure from the well-known example of the fundamental gravity value at Potsdam, which was known to be in error, but was used as a standard of comparison by all scientists determining gravity values in their geodetic work.

The important goal is to keep the continuity of records for long series of data that were started at the beginning of the century. It is essential to avoid discontinuities and interruptions that cannot be recovered in the future once observations stop at the same sites. Any Conventional Terrestrial System (CTS) to be adopted, has, therefore, to take into account the advantages of the CIO and define a system well-related to the reference systems adopted in the past.

Having a small number of observatories observing permanently is feasible. This has the advantage of making the comparisons easier, but also there are no difficulties in the definition of the fundamental meridian because they will not drop out of the observing program. This world network will not avoid completely the variations due to the geophysical motions of the stations which have been mentioned. This possible network combines the advantages of the present ILS chain with some of the good features of the BIH set of observatories.

A possible approach to the definition of the CTS would be the adoption of the well determined ILS observatories together with a small set of world observatories. These should be well distributed, and should observe regularly for at least several decades.

REFERENCES.

Anderle, R. J. and Tanenbaum, M. G. (1974), Nav. Weapons Lab. Techn. Rep. TR-3161.

Feissel, M. (1980), Bull. Géodésique, 54, p. 81.

Jeffreys, H. (1940), Mon. Not. Roy. Astron. Soc., 100, p. 139.

Jeffreys, H. (1959), Mon. Not. Roy. Astron. Soc., 119, p. 75.

Jeffreys, H. (1968), Mon. Not. Roy. Astron. Soc., 141, p. 255.

Melchior, P. (1980), in "Nutation and the Earth's Rotation" (ed. E. P. Fedorov, M. L. Smith and P. L. Bender), Reidel Publishing Company, Holland, p. 17.

Morrison, L. V. (1979), Geophys. Journ. Roy. Astron. Soc., 58, p. 349.

Oppolzer, T. (1886), "Traité Determination des Orbites", 1, Gauthier Villars, Paris.

Szebehely, V. (1982), in "Applications of Modern Dynamics to Celestial Mechanics and Astrodynamics" (ed. V. Szebehely), Reidel Publishing Company, Holland, in this volume, p. 321.

Vicente, R. O. (1981), Bull. Géodésique, in press.

Vicente, R. O. and Currie, R. G. (1976), Geophys. Journ. Roy. Astron. Soc., 46, p. 67.

Wilson, C. R. and Vicente, R. O. (1980), Geophys. Journ. Roy. Astron. Soc., 62, p. 605.

Wilson, C. R. and Vicente, R. O. (1981), Astron. Nachrichten, 302, p. 227.

Woolard, E. W. and Clemence, G. M. (1966), "Spherical Astronomy", Academic Press, New York.

Yumi, S. and Yokoyama, K. (1980), "Results of the International Latitude Service in a Homogeneous System (1899.9-1979.0)", Central Bureau Int. Polar Motion Service, Mizusawa, Japan.

RECENT PROGRESS IN THE THEORY OF THE TROJAN ASTEROIDS

Boris Garfinkel

Yale University, New Haven, Connecticut

ABSTRACT. This paper summarizes the author's previous publications on the subject in the light of the numerical integrations by Deprit and Henrard. The period of libration is expressed as a function of the mass parameter and the normalized Jacobian constant. Brown's conjecture regarding the termination of the tadpole branch of the family at L_3 is refined, and a heuristic proof of its validity is offered.

In the previous publications the author has constructed a formal long-periodic solution of the restricted problem of three bodies in a state of 1:1 resonance, to $O(m^{3/2})$, where m is the mass parameter of the system.

The solution carries a small divisor,

$$D_k = \omega_1 - k\omega_2 ,$$

where ω_1 and ω_2 are the characteristic frequencies of the motion.

Thus, the solution is local, rather than global, as it must exclude the neighborhood of the exact k:1 commensurability between ω_1 and ω_2, where $D_k = 0$. From the results of numerical integrations carried out by Deprit and Henrard (1970), we know that at $D_k = 0$ our long-periodic family $\mathcal{L}$ is interrupted by bifurcations and short-periodic "bridges". Accordingly, our family, parametrized by the normalized Jacobian constant α^2, must be defined in the Strömgren sense as the union of the admissible intervals, of the form $\mathcal{L} = U_j \{|\alpha-\alpha_j| > \varepsilon_j\}$, $j = k, k+1, \ldots\infty$ for a given value

V. Szebehely (ed.), Applications of Modern Dynamics to Celestial Mechanics and Astrodynamics, 145–152.

of m. Here $\{\alpha_j(m)\}$ is the sequence of the critical α_j corresponding to the exact j:1 commensurability.

Inasmuch as the intervals $|\alpha-\alpha_j| < \varepsilon_j$ of "avoidance" can be shown to be disjointed, it follows that, despite the clustering of the sequence $\{\alpha_j\}$ at $j = \infty$, it is always possible to find a member of our family that lies as close to the separatrix $\alpha=1$ of the "tadpole" branch as we wish.

A notable feature of our method of solution is the expansion of the disturbing function R about the circle r=1, rather than about the Lagrangian point L_4 with the coordinates $r = 1$, $\theta = \pi/3$. This mode of expansion is equivalent to analytic continuation, for it replaces the circle of convergence centered at L_4 by an annulus, $|r-1| < \varepsilon$, of convergence with $0 \leq \theta \leq 2\pi$. Inasmuch as the Trojan orbits generally lie within this annulus, our solution admits large amplitudes of libration, including the entire "tadpole" branch, with $0 \leq \alpha \leq 1$ and most of the horseshoe branch, with $1 < \alpha < (2m)^{-1/6}$. Here an upper bound was placed on α in order to avoid a close approach to Jupiter, located at r=1, θ=0, where the solution is singular. It may be noted, however, that an expansion about $r=1 + \beta(\theta)$, where β is conveniently chosen, would remove the latter singularity. This would open the possibility of investigating the termination of the horseshoe branch in the vicinity of L_2, as well as the circulating orbits of the outer and the inner "ovals".

The parameter α is related to the Jacobian constant C as defined in Szebehely (1967) by the formula

$$\alpha^2 = \frac{C-3}{2m} + \psi(\lambda_2).$$

Here $\psi(\lambda)$ is the so-called regularizing function, and λ is the mean synodic longitude, which plays the role of the critical argument librating in the interval $\lambda_1 \leq \lambda \leq \lambda_2$.

The function $\psi(\lambda)$ originating the second-order Hamiltonian F_2, has been transferred to the unperturbed Hamiltonian F_0 in order to forestall a singularity arising from the 1:1 resonance. This procedure I call Poincaré's transfer, for it has been recommended by Poincaré in similar situations in resonance problems.

The canonical variables employed are the set $(G,\lambda\ ;\ \xi,\eta)$. Here ξ and η are Poincaré's eccentric variables, defined by $\xi + i\eta \equiv z = \sqrt{2\Gamma}\, e^{i\ell}$, $\Gamma \equiv L - G$, with ℓ the mean anomaly and L and G the usual variables of Delaunay. In terms of these variables the Hamiltonian F becomes $F = F_0 + F_1 + \dots$, with

$$F_0 = \frac{1}{2}G^{-2} + G + mf_0(\lambda) - \frac{1}{2}\omega_2(\xi^2 + \eta^2) - \frac{1}{2}m ,$$

$$F_1 = m[-2f_1\rho + (f_1 + 1)\xi + f_2\eta] + \dots,$$

where f_0, f_1, and f_2 are known functions of λ. In particular, Poincaré's transfer function, $f_0(\lambda)$ is defined by

$$f_0 = \frac{1}{2s} + 2s^2 - \frac{3}{2} + m[\psi(\lambda) - \psi(\lambda_2)] , \quad s = \sin(\lambda/2).$$

It combines the dominant term of the 1:1 resonance transferred from the original F_1 and the regularizing function $\psi(\lambda)$ transferred from the original F_2. Clearly, F_0 decomposes into the sum

$$F_0 = F_0^{(1)} + F_0^{(2)} ,$$

with

$$F_0^{(1)} = \frac{1}{2}G^{-2} + G + mf_0(\lambda)$$

$$F_0^{(2)} = \frac{1}{2}\omega^2(\xi^2 + \eta^2) - \frac{1}{2}m .$$

The corresponding integrals of "energy" are

$$\frac{1}{2}G^{-2} + G + mf_0(\lambda) = h = \text{const.}$$

$$\frac{1}{2}(\xi^2 + \eta^2) = \Gamma = \text{const.}$$

Furthermore, $F_0^{(1)}$ is in the form of the Ideal Resonance Problem, while $F_0^{(2)}$ represents the Simple Harmonic Oscillator.

The Ideal Resonance Problem was defined by the author (1976) as

$$F = B(y) + 2\mu^2 A(y)f(x), \quad f(x + 2\pi) = f(x), \quad \mu << 1,$$

and solved for the case of libration. The identifications

$$x = \lambda \quad , \quad y = G-1 \equiv \rho \quad , \quad f = f_0$$

$$A = 1 \quad , \quad B = \frac{1}{2}G^{-2} + G \quad , \quad \mu = \sqrt{\frac{1}{2}m}$$

permit us to write down immediately the solution of the undisturbed problem to O(m) as

$$\begin{cases} \rho = -\frac{1}{3}\sqrt{6m(\alpha^2-f_0)} + \frac{4}{9}m(\alpha^2-f_0) + \dots \\ t-t_1 = \int \frac{d\lambda}{\sqrt{6m(\alpha^2-f_0)}} - \frac{4}{9}(\lambda-\lambda_1) + \dots \\ z = \sqrt{2\Gamma}\, e^{i(\omega_2 t+\phi)} \end{cases}$$

with α^2 defined by $h = \frac{3}{2} + m\alpha^2$.

The perturbations arising from F_1 have been calculated by the method of Lie series (Hori 1966). [1] In order to remove the short periodic terms, we choose the initial conditions of the motion so that $\Gamma = 0$. Then the long-periodic family of orbits of period $T = 2\pi/\omega_1$, appears in the parametric form,

$$\rho = \rho(\lambda,\alpha), \qquad z = \bar{z}(\lambda,\alpha) + g(t,\alpha), \qquad t = t(\lambda,\alpha). \tag{1}$$

Here $g(t,\alpha)$ are the epicyclic terms exhibiting waves, cusps, and loops of high frequency $k\omega_1$, confirming the results of the numerical integration by Deprit and Henrard (1970). In our solution, the dominant epicyclic term is of the form

$$g = kc_k e^{ik\omega_1}/D_k + \dots, \tag{2}$$

where c_k is the corresponding Fourier coefficient of the function $f \equiv f_1 + if_2$, and D_k is the critical divisor. The term $\bar{z}$ refers to the "mean"orbit obtained by averaging out the high frequency terms, and thus freed of the singularity.

The algorithm consists of the following formulas:

$$\begin{cases} \rho = \rho_1 + \frac{2}{3}\rho_1^2 + \frac{5}{18}\rho_1^3 + \dots \quad \frac{2m}{3}(1+m)\, f_1 + \dots, \\ z = m(m+1)[f(\lambda) + g(t)], \\ t - t_1 = \int_{\lambda_1}^{\lambda} (1+mh)[6m(\alpha^2-f_0)]^{-\frac{1}{2}} d\lambda - \frac{4}{9}(\lambda-\lambda_1) + \dots, \end{cases} \tag{3}$$

with

$$\left\{\begin{array}{l}
\rho_1(\lambda) = -\frac{1}{3}\sqrt{6m(\alpha^2-f_0)}\ \operatorname{sgn}\dot{\lambda} \\
f_0(\lambda) = \frac{1}{2s} + 2s^2 - \frac{3}{2} + m[\psi(\lambda) - \psi(\lambda_2)] \\
\psi(\lambda) = \frac{1}{96}\, s^{-4}\ (85^3 - 1)^2(12 - 13s^2) \qquad (4) \\
f(\lambda) = (\frac{1}{4s^2} - 2s)(1-2ic/s) \equiv f_1 + if_2 \\
h(\lambda) = \frac{1}{36}\ (3s^{-3} - 4s^{-1} + 20\alpha^2 + 78 - 160s^2) + 0(m) \\
s = \sin\lambda/2 \quad , \qquad c = \cos\lambda/2 \ .
\end{array}\right.$$

The bounds λ_1 and λ_2 of the libration are the roots of $f_0(\lambda) = \alpha^2$.

The full expression for $g(t)$ is

$$g(t) = \sum_1^\infty j\omega_1 c_j e^{ij\omega_1 t}/D_j$$

$$D_j \equiv \omega_2 - j\omega_1 \qquad (5)$$

$$c_j \equiv \frac{1}{T}\int_o^T e^{-ij\omega_1 t} f(\lambda(t))dt \ .$$

It would be tempting to remove the singularity at $D_k = 0$ by means of the quadrature

$$g = -e^{i\omega_2 t}\int_o^t e^{-i\omega_2 t} df,$$

$$df = f'(\lambda)\ \dot{\lambda}\ dt, \qquad (6)$$

where the function $\lambda(t)$ is known from the inversion of the function $t(\lambda)$, carried out by the author in 1980. Unfortunately, this procedure seems to introduce short-periodic terms into $g(t)$, thereby destroying the periodicity of the solution. As Brouwer remarked, "a global solution of the problem of double resonance is an outstanding unsolved problem of celestial mechanics".

Recent progress in the theory includes the calculation of the period $T(\alpha,m)$ of libration. From the first of equation 3 we deduce the hyperelliptic integral,

$$T = \oint (1 + mh)[6m(\alpha^2-f_0)]^{-\frac{1}{2}} \, d\lambda, \tag{7}$$

where $h(\lambda)$ and $f_0(\lambda)$ are known functions defined in Equation 4. A table of $T(\alpha)$ for $m = 10^{-3}$ was published in Garfinkel 1978, p. 270 on the basis of the quadrature of Equation 7 carried out by J. Mengel.

A formal expansion of the function $T(\alpha,m)$ in a power series involves a representation of T as a triple product

$$T(\alpha,m) = T(o,m)\,\tau_1(\alpha,o)\,\tau_2(\alpha,m), \tag{8}$$

where $\tau_1(\alpha,m)$ and $\tau_2(\alpha,m)$ are the first and the second normalizations, respectively, defined by

$$\begin{aligned} \tau_1(\alpha,m) &= T(\alpha,m)/T(o,m), \\ \tau_2(\alpha,m) &= \tau_1(\alpha,m)/\tau_1(\alpha,o). \end{aligned} \tag{9}$$

The first factor in Equation 8 is Lagrange's period of small oscillations given by the well-known expression

$$\left\{ \begin{aligned} T(o,m) &= 2\pi/\omega_0 \\ \omega_0^2 &= \frac{1}{2}\left[1 - \sqrt{1-27m(1-m)}\right]. \end{aligned} \right. \tag{10}$$

This is the principal part of the period-mass dependence, carrying a singularity at m = o, with

$$T(o,m) = O(1/\sqrt{m}), \quad (m \sim o). \tag{11}$$

The second factor in Equation (8) can be written as

$$\begin{aligned} \tau_1(\alpha,o) &= \frac{3\sqrt{2}}{8\pi} \oint (\alpha^2-f_0)^{-\frac{1}{2}} \, d\lambda \\ &= \frac{3}{4\pi} \oint \left[\frac{z}{(1-z^2)(z-z_1)(z_2-z)(z-z_3)}\right]^{\frac{1}{2}} dz, \end{aligned} \tag{12}$$

where $z = \sin \lambda/2$, and $z_3 < z_1 \leq z_2$ are the roots of the cubic equation, $4z^3 - (2\alpha^2 + 3)z+1 = 0$. The expression (12) is a hyper-elliptic integral of class two, named after Hagihara. It represents the principal part of the period-amplitude dependence, carrying a logarithmic singularity at $\alpha=1$, with $\tau(\alpha,o) = - O[\log(1-\alpha)]$, $(\alpha \sim 1)$. The formula can be either evaluated by quadrature or represented as a power series involving standard elliptic integrals (Garfinkel, 1980). The following asymtotic expansions are useful:

$$\tau(\alpha,o) = 1 + \frac{1}{6}\alpha^2 + \frac{169}{1296}\alpha^4 + \ldots, \quad (\alpha \sim o),$$

$$\tau(\alpha,o) = \frac{6}{\pi\sqrt{14}}[(1+0.18878\beta^2 +..)\log 1/\beta + 1.74732 + 0.13314\beta^2+..],$$

$$(\alpha \sim 1), \qquad \beta \equiv \sqrt{1-\alpha^2}. \tag{13}$$

The third factor in Equation 12 is a small correction to the period, and it can be conveniently given by the asymptotic expansion

$$\tau_2(\alpha,m) = 1 + 3.23m - 1.64m\beta^2 + \ldots$$
$$-(4.248m - 3.61m\beta^2 + ..)/\tau_1(\alpha,o), \quad (\beta \sim o), \tag{14}$$

valid for $\alpha \sim 1$. The corresponding expansion for $\alpha \sim o$ has not yet been derived. The error of formula (14) does not exceed 5×10^{-5} at $\lambda_2 = 150°$, as shown by quadrature of Equation 12.

As an example of the use of Equation 8, we calculate for $m = 10^{-3}$, $\alpha^2 = 0.98672$:

$$\tau(o,m) = 76.252, \quad \tau_1(\alpha,o) = 1.796, \quad \tau_2(\alpha,m) = 1.001,$$

$$T(\alpha,m) = 137.10, \quad \lambda_2 = 170° .$$

For Jupiter-Sun, $\dot{T}(\alpha,m) = 258.83$ years .

This paper is concluded with a discussion of E. W. Brown's (1911) conjecture that the tadpole branch of the long-periodic family terminates at L_3 with $\alpha=1$, and evolves into the horse-shoe branch for $\alpha > 1$.

The mathematical existence of the horseshoe-shaped orbits has been proved by the results of the numerical integrations carried out by Rabe (1961) and D. Taylor (1980) as well as by the author's formal solution of the problem. As to the first part of the conjecture, it had been widely accepted and "confirmed" by Rabe (1961), only to be finally refuted by the more precise numerical integration carried out by Deprit and Henrard (1970). The occurence of epicyclic terms in our solution supports this refutation. In fact, the separatrix of the tadpole branch, corresponding to $\alpha=1$, encloses L_3 in an asymptotic loop, as can be shown by a detailed analysis.

An attempt to rehabilitate the first part of the Brown conjecture involves the construction of the "mean" orbit obtained by the averaging out the epicyclic terms of frequency $j\omega_1$, thus dropping g(t) from our solution (1). The conjecture can then be refined in two alternative forms:

a) "The separatrix $\alpha=1$ of the mean tadpole branch approaches L_3 asymptotically as $t \to \infty$ ".

b) "The separatrix $\alpha=1$ of the tadpole branch spirals asymptotically toward a limit cycle, centered on L_3".

The center of the limit cycle has the coordinates

$$\bar{r} = 1 - \frac{7}{12} m + \dots , \quad \bar{\theta} = \pi \qquad (15)$$

which differs from L_3 by a quantity of $O(m^2)$. Although our $O(m^{3/2})$ theory is not sufficient to establish the coefficient of $O(m^2)$ in Equation 15, a heuristic proof of b) is being offered here. Indeed, if L_3 and the center of the limit cycle do not coincide, then there exist two remarkable points in close proximity. This conclusion would contradict the Einstein Principle that "God does not play dice".

REFERENCES.

Brown, E. W. (1911), Mon. Not. RAS, 71, p. 438.

Deprit, A. and Henrard, J. (1970), in "Periodic Orbits, Stability, and Resonances", p. 1, ed. G.E.O. Giacaglia, (Reidel).

Garfinkel, B. (1976), Celestial Mechanics, 13, p. 229.
(1977), Astron. J., 82, p. 368.
(1978), Celestial Mechanics, 18, p. 259.
(1980), Celestial Mechanics, 22, p. 267.

Hori, G. (1966), Publ. Astron. Soc. Jpn., 18, p. 287.

Rabe, E. (1961), Astron. J., 66, p. 500.

THE ADIABATIC INVARIANT : ITS USE IN CELESTIAL MECHANICS

Jacques Henrard

Dept. of Mathematics, Facultés Universitaires de Namur, B-5000 Namur, Namur, Belgium

ABSTRACT

Many problems in the dynamical evolution of the Solar System can be modelized by some pendulum like Hamiltonian system with one degree of freedom and slowly varying parameters. The adiabatic invariant introduced in the context of quantum mechanics and of physics of nuclear particles is a very effective tool for the study of such problems.

In this paper, we describe the basic ideas of this theory and apply it to the problem of capture into resonance of Titan and Hyperion.

1. INTRODUCTION

It seems that many of the particular features in the dynamics of the Solar System (like resonances) can be explained by the effect over long period of time of small non conservative forces. If this is so, we can deduce from the present state of the Solar System some information about these forces (and thus the physics of the planets and satellites) and some information about the primeval Solar System.

Very often, the study of these dynamical models leads to some pendulum like Hamiltonian system with one degree of freedom and slowly varying parameters.

In such situations, the theory of the adiabatic invariant is a very useful tool. With very general and realistic assumptions on the variation of the parameters, it can provide very precise in-

V. Szebehely (ed.), Applications of Modern Dynamics to Celestial Mechanics and Astrodynamics, 153–171.

formation about the evolution of the dynamical system.

It has not been widely used in Celestial Mechanics, perhaps because it is not well-known in this context having been studied mainly in the context of nuclear physics.

We would like in this paper to describe its basic concepts and results and to apply them to an example to show how they can very easily and precisely unravel the implications of a given physical model.

We begin with the description of the widely used (and sometimes abused) action-angle variables. If they are well-known in the context of Celestial Mechanics, they are not always defined precisely and we thought it worthwhile to do it even in the simple context of one degree of freedom systems.

In Section 3, we introduce the basic concept and result of the adiabatic invariant theory. This basic first order result has been generalized to n-th order by several authors. These generalizations do not agree with each other as various authors use various definitions for the adiabatic invariant and make various assumptions on the dynamical systems they consider. We plan to review these generalizations in a forthcoming publication.

In Section 4, we give a brief summary of an earlier paper analyzing an extension of the theory to the jump over a critical curve. This jump is instrumental in some capture into resonance.

The following sections are devoted to the Titan-Hyperion resonance. We describe the conservative approximation we use (an averaged restricted planar and circular three body problem) and the tidal dissipation effects which introduce a slowly varying parameter (the semi-major axis of Titan).

We can then apply the adiabatic invariant theory and it extension to this dynamical model. Of course, the dynamical model can be critisized. At the meeting, Dr. Message has pointed out to us that the eccentricity of Titan should not be neglected right away and Dr. Milani that the averaged problem differs considerably from the non averaged one for orbits which come close to Titan (orbits with moderate to high eccentricity). We point out also that it is not certain at all that the tidal dissipation is strong enough in this case to allow the capture to take place within the age of the Solar System.

Nevertheless, we believe that the dynamical model we use is more realistic than most of the ones which have been analyzed previously (we take into account terms up to the fifth order in the eccentricity) and that the results we get within that model are more general (we cover escape as well as capture into resonance for quite large range of initial eccentricity) and more precise (we can tabulate for instance the present libration amplitude versus the initial eccentricity).

2. THE ACTION-ANGLE VARIABLES

Let us consider an autonomous Hamiltonian system with one degree of freedom

$$H(\phi, I) = h \tag{1}$$

defined on a domain D of the phase space. Let us assume then H is twice continuously differentiable and that all trajectories of the system are periodic.

We shall take as an example of such a system the pendulum the Hamiltonian of which is

$$H = \frac{1}{2} I^2 - b \cos \phi . \tag{2}$$

The domain D under consideration will be the domain inside the critical curve (the libration domain), so that all trajectories are periodic (see Figure 1). With some minor modifications, we could also choose one of the outside domains (positive or negative circulation domain).

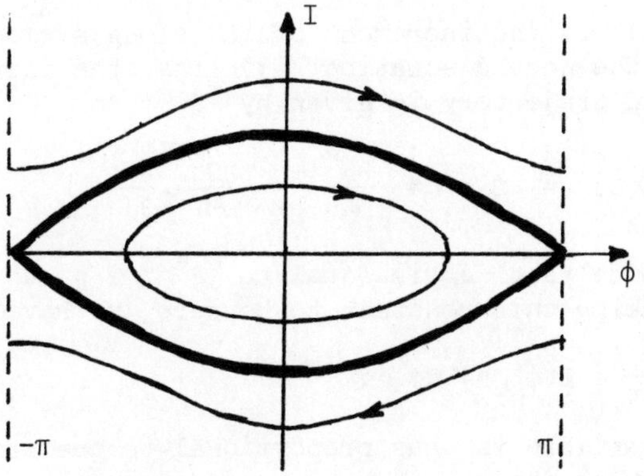

Figure 1. The phase space of the pendulum (for $b > 0$). The phase space is actually a cylinder (the lines $\phi = -\pi$ and $\phi = \pi$ should be identified). The critical curves are shown in heavy lines.

We introduce the action-angle variables (ψ, J). These variables are particularly well-suited to describe the problem. This is why in many problems of this kind, the first step is to define them.

We introduce them by means of a canonical transformation

$$I = \frac{\partial}{\partial\phi} S(\phi, J) \qquad , \qquad \psi = \frac{\partial}{\partial J} S(\phi, J) \tag{3}$$

generated by a function $S(\phi, J)$ to be defined in such a way that the Hamiltonian function (1) is transformed by (3) into a function $K(J)$ independent of the variable ψ.

The action J will thus be a constant along a trajectory. Furthermore, we ask that the variable ψ is actually an angle and increases by 2π along each trajectory. Hence ψ will measure the position along a trajectory and each trajectory will be labelled by a value of J.

From the Hamiltonian-Jacobi equation

$$H(\phi, \frac{\partial S}{\partial \phi}) = K(J) \tag{4}$$

we find

$$S = \int_{\phi^{\star}}^{\phi} I(\phi, K(J))\, d\phi \tag{5}$$

where the function $I(\phi, h)$ is implicitely defined by (1).

If $S^{\star}(h)$ is the increment of S along a closed trajectory, we find from the second equation (3) that the increment of ψ along a closed trajectory is given by

$$\psi(T) - \psi(0) = 2\pi = \frac{\partial S^{\star}}{\partial J} = \frac{\partial S^{\star}}{\partial h}\frac{\partial K}{\partial J} . \tag{6}$$

We conclude that J is equal to $S^{\star} / 2\pi$ up to an additive constant. Taking this constant to be zero, we have

$$J = \frac{1}{2\pi} \oint I(\phi, h)\, d\phi \quad . \tag{7}$$

This action variable is thus proportional to the area enclosed by the periodic trajectory.

In the example of the pendulum mentioned above, we have that

$$I(\phi, h) = [2h + 2b \cos\phi]^{1/2} \quad . \tag{8}$$

In order to perform the circuit integration in (5) and (7), we introduce the auxiliary angle l defined by

$$\sin \frac{\phi}{2} = \sqrt{\frac{h+b}{2\,b}}\ \sin l \quad . \tag{9}$$

The angular variable is such that it goes from zero to 2π along a closed trajectory. We then have with $\alpha^2 = (h+b)/2\,b$

$$I(l\,,h) = 2\,b^{1/2}\ \alpha \cos l \quad , \tag{10}$$

$$S = 2\,(2\,b)^{1/2}\,\{\mathbb{E}(l\,;\alpha^2) + (\alpha^2-1)\,\mathbb{F}(l\,;\alpha^2)\} \quad , \tag{11}$$

$$J = \frac{4\,(2\,b)^{1/2}}{\pi}\{\mathbb{E}(\alpha^2) + (\alpha^2-1)\,\mathbb{K}(\alpha^2)\} \ , \tag{12}$$

$$\psi = \frac{\pi}{2\,\mathbb{K}(\alpha^2)}\,\mathbb{F}(l\,;\alpha^2) \tag{13}$$

where $\mathbb{E}(l\,;\alpha^2)$, $\mathbb{F}(l\,;\alpha^2)$, $\mathbb{E}(\alpha^2)$ and $\mathbb{K}(\alpha^2)$ are the usual elliptic integrals (see, for instance, Abramovith and Segan, 1964).

The new Hamiltonian function $K(J)$ is given implicitely by equation (12) where we take

$$\alpha^2 = (K+b)\,/\,2\,b \quad . \tag{14}$$

As it can be seen from this example, the introduction of action-angle variables leads to complex analytical manipulations. Therefore, it is often useful to think in terms of action-angle variables (because they are theoretically the best suited) but to actually perform the computation in some other set of variables (see, for instance Deprit, 1982, for a good illustration of this fact).

3. THE ADIABATIC INVARIANT

We assume now that the Hamiltonian function (1) depends actually on a parameter λ which changes slowly with time. On a short time scale, the system will behave almost like the system with a constant λ . Any trajectory will be well-approximated by one of the periodic trajectories described by the equation $H(\phi\,,I\,;\lambda) = h$.

But this is not true on a longer time scale. This means that at any time the trajectory will be close to one of the curves mentioned above but that this curve itself changes slowly (actually λ and h change slowly).

The problem is to describe how this curve (or the level of energy h) changes on a very long time scale.

The key to the answer is the adiabatic invariant. By an adiabatic invariant of order one, we mean a function of the phase space $A(\phi, I; \lambda)$ the variation of which, along a trajectory of length $T < 1/\varepsilon$, goes to zero with ε, where ε is an upper bound of the derivatives of λ with respect to time. Once we know an adiabatic invariant, it is easy to follow the variation of h and thus of the curve which approximates the trajectory on a short time scale.

The beauty of it is that we do not even have to know in detail the variations of h or λ with time. Knowing that, at time t_1, $\lambda = \lambda_1$ and $h = h_1$, we can deduce the value h_2 reached at the time when $\lambda = \lambda_2$ without knowing the path between λ_1 and λ_2 and of course the path between h_1 and h_2.

This concept of the adiabatic invariant has been investigated mainly in the context of quantum mechanics and of the physics of nuclear particles (see, for instance, Lenard, 1959, and Kruskal, 1962). It can be found also in some (Russian) books on classical mechanics (Landau et Lipschitz, 1960; Arnold, 1978). It has been introduced in the context of Celestial Mechanics by Goldreich and Toomre (1969).

The terminology and the results on the adiabatic invariant are not quite settled yet. Even the definition is not agreed upon by all the authors. We follow here the definition and the result given by Arnold (1978). A refinement of this result and some comments on related results will appear in a further publication.

Theorem 1

Let us assume that the Hamiltonian function $H(\phi, I; \lambda)$ is defined and twice continuously differentiable on a domain D of the phase space (ϕ, I) and an interval E of the parameter λ. We assume also that for each fixed value λ in E the curves

$$H(\phi, I; \lambda) = h \tag{15}$$

are close and that the corresponding periodic orbits have their period uniformly bounded.

With respect to the variation of λ, we assume that the two first derivatives of λ are continuous and that there exists a small quantity ε such that

$$|\dot{\lambda}| \leqslant \varepsilon \quad , \quad |\ddot{\lambda}| \leqslant \varepsilon^2 . \tag{16}$$

We shall show that the action J is an adiabatic invariant of order one. More precisely, if t_1 and t_2 are two instants such that

$$|t_1 - t_2| \leqslant 1/\varepsilon \tag{17}$$

and if, at time t_i, the value of J (computed from formula (7) for the value of λ and the position in phase space at time t_i) is J_i, then

$$|J_1 - J_2| < M\,\varepsilon \tag{18}$$

where M is some constant independent of ε.

To show this, we go back to the definition of the action-angle variable and remark that the generating function $S(\phi, J; \lambda)$ defined in (5) now depends on the time through the parameter λ. The new Hamiltonian function describing the motion is no longer $K(J)$ but

$$K_1(\phi, J; \lambda) = K(J; \lambda) + \dot{\lambda}\frac{\partial}{\partial\lambda} S(\phi, J; \lambda) \tag{19}$$

where ϕ has to be expressed in function of $(\psi, J; \lambda)$ after the partial differentiation.

This Hamiltonian function is no longer independent of the angular variable ψ and thus the action J is no longer constant although his variation is obviously small because the second term of (19) is small with $\dot{\lambda}$.

To get a better idea of the variation of J and to show that *on the mean* it does not vary, we perform a first order averaging of (19) by means of the Lie transform method (Deprit, 1969).

If we define

$$\langle \partial S / \partial\lambda \rangle = \frac{1}{2\pi}\int_0^{2\pi} \frac{\partial}{\partial\lambda} S(\phi, J; \lambda)\, d\psi \tag{20}$$

the averaged value of $\partial S / \partial\lambda$ and

$$(\partial S / \partial\lambda)_P = \partial S / \partial\lambda - \langle \partial S / \partial\lambda \rangle \tag{21}$$

the purely periodic part of the same function, we can take as generating function of the Lie transform method the function

$$W(\psi, J) = \dot{\lambda}\left(\frac{\partial K}{\partial J}\right)^{-1} \int_0^{\psi} \left(\frac{\partial S}{\partial\lambda}\right)_P d\psi \quad . \tag{22}$$

Note that W is periodic in ψ and truly of the order of $\dot{\lambda}$ because $(\partial K / \partial J)^{-1}$ is bounded by our assumption on the period of the periodic orbits of the system with λ = cte.

The function W defines a canonical transformation from the phase space (ψ, J) to the averaged phase space $(\overline{\psi}, \overline{J})$ with

$$\psi = \exp(D_W)\,\overline{\psi} = \overline{\psi} + \frac{\partial W}{\partial J} + \theta(\varepsilon^2) \quad , \tag{23}$$

$$J = \exp(D_W)\,\overline{J} = \overline{J} - \frac{\partial W}{\partial \psi} + \theta(\varepsilon^2) \quad , \tag{24}$$

where D_W is the Lie derivative associated with the function W, i.e. the Poisson Bracket with the function W.

The new Hamiltonian function in $(\overline{\psi}, \overline{J})$ is given by

$$K_2(\overline{\psi}, \overline{J}; \lambda) = \exp(D_W)\,K_1 - \int_0^1 \exp(\varepsilon\, D_W)\,\frac{\partial W}{\partial t}\,d\varepsilon \tag{25}$$

or

$$K_2(\overline{\psi}, \overline{J}; \lambda) = K(\overline{J}, \lambda) + \dot{\lambda} <\partial S / \partial \lambda> + \theta(\dot{\lambda}^2, \ddot{\lambda}) \quad . \tag{26}$$

This new Hamiltonian is now independent of $\overline{\psi}$ up to term of order ε^2. This means that the derivative of $\overline{J}$ is at most of order ε^2 and thus that the change of $\overline{J}$ over a time of the order of $1/\varepsilon$ is at most of the order of ε.

Hence $\overline{J}$ is an adiabatic invariant of the first order. As J differs from $\overline{J}$ by terms of at most the order of ε, J itself is an adiabatic invariant of the first order.

As a simple example of the use of the adiabatic invariant, let us consider the pendulum described by the Hamiltonian function (2). Let us assume that the length of the pendulum decreases slowly so that the parameter b in (2) decreases slowly.

As b decreases, the curves $H(\phi, I; b) = h$ will become more flat in the phase space (ϕ, I) (see Figure 2).

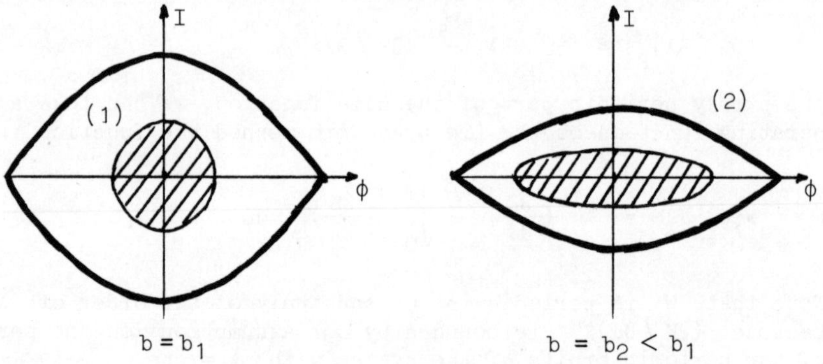

Figure 2. The phase space for different values of b.

Let us assume that, at time t_1 , the value of b is b_1 and the initial conditions are along the curve marked (1) in Figure 2. What will be the behaviour of the trajectory a long time after that when the value of b reaches the value b_2 ?

We know that on a short time scale around the time t_2 , the trajectory will stay close to a curve $H(\phi, I, b_2) = h_2$ labelled (2) in Figure 2. The value of h_2 is such that the value of J_1 equals the value of J_2 , i.e. the area enclosed by the curve (1) is equal to the area enclosed by the curve (2) .

We see at once that the amplitude of the oscillation of the pendulum has to increase.

With the help of formulae (12) and (13) , it is not difficult to compute by exactly how much this amplitude has increased.

For instance, we have computed that if we start from an amplitude of 30° and if we wait until b has decreased by a factor of two, the final amplitude will be $35^\circ 46'$. Also if we want the final amplitude to be close to 180° , we have to wait until b has decreased by a factor of 355 .

4. TRANSITION OR CAPTURE INTO RESONANCE

Let us consider again the example of the pendulum. What will happen if we let b decrease beyond a factor of 355 ?

The theory we have just developped will be of no help. Indeed, as we approach oscillations of 180° , the periods of oscillation are unbounded and the scheme proposed fails (see the remark after equation (22)).

Other methods used to show the adiabatic invariance of J fail as well. The classical one leads to the following estimation of the variation of J over a period T of oscillations

$$|J(t_1 + T) - J(t_1)| \leqslant M \lambda^2 T^3 \tag{27}$$

(see Henrard, 1981). The right hand member is unbounded with T and the estimation is of no use. We need a much finer analysis especially near the unstable equilibrium position where most of the time is spent.

The problem is important because it is precisely in this situation that interesting things happen. This is the situation where a transition can be made from one type of motion to another (for instance, from positive circulation of a pendulum to libration ... or will it be to negative circulation ?). This covers also the problem of capture into resonance of dynamical systems as we know that resonance leads to systems related to the pendulum.

Following the suggestion of Goldreich (1965) that tides are responsible for the capture into resonance of pair of satellites or of the rotational motion of a satellite, many authors have investigated this mechanism (references can be found in the introduction of Henrard, 1982). Most of them use either numerical integration or pecularities of the model they are using.

For instance, to cite the most mathematically minded papers on this topic, Yoder (1979) and Burns (1979) consider a pendulum described by

$$H = \frac{1}{2}(I + c)^2 - b \cos \phi \tag{28}$$

with the assumption that $\varepsilon_1 = b^{-3/2}\dot{b}$ and $\varepsilon_2 = b^{-1}\dot{c}$ are constant. It can be transformed into an autonomous system in two dimensional phase space. Neishtadt (1974) considers the Hamiltonian function

$$H = 4I^2 - 2\lambda I + \mu\sqrt{2I}\cos\phi \tag{29}$$

with $\dot{\lambda}$ constant (this is a simplified model of the problem of capture into resonance of pair of satellites).

An analysis of the general case is proposed in (Henrard, 1982). We shall give here only a general idea of the result. The reader interested in more details can consult the paper just quoted.

The neighborhood of the critical curves is divided into three regions : one (region III) which is close to the unstable equilibrium and the other two (regions I and II) close to each branch of the critical curve (see Figure 3).

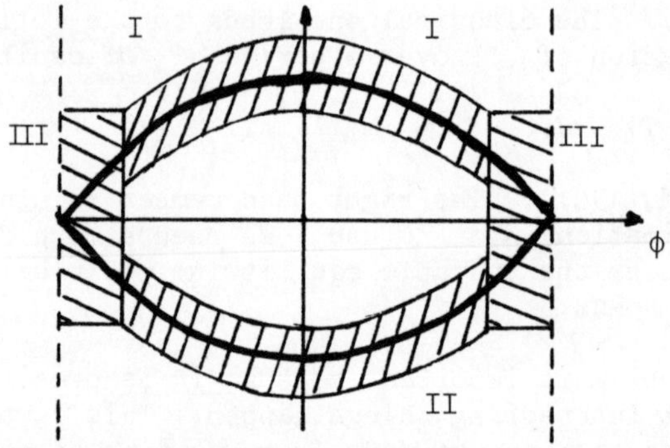

Figure 3. The three regions forming a neighborhood of the critical curve.

When the sizes of the three regions are carefully chosen, the following results can be shown to hold.

(1) The gain in energy (with respect to the energy of the unstable equilibrium) along the trajectory which traverses completely region I or II (i.e. which goes from region III to region III) is a first order constant in ε (let us say $B_1\varepsilon$, and $B_2\varepsilon$) plus quantities of the second order in ε . This first order quantity can be computed by a quadrature.

(2) The gain in energy (again with respect to the energy of the unstable equilibrium) for a traverse of region III is zero at the first order in ε when the trajectory is not too close to the critical curve.

(3) When a trajectory coming from region I (resp. II) enters region III , it comes out in region I (resp. II) or II (resp. I) according to the sign of the energy when it enters. We assume again that the trajectory is not too close to the critical curve.

The evolution of the system through the critical region can be deduced from these results. We shall illustrate this in a particular case.

Let us assume that H (the energy relative to the energy at the unstable equilibrium) is negative in the libration region and positive in the two circulation regions. Let us assume furthermore that the motion is initially in the positive circulation region and approaching from above the critical curve (thus $B_1 < 0$). The interesting case appears when $B_2 > 0$ and $B_1 + B_2 < 0$.

If the motion crosses the level $H = 0$ early enough in one of his traverses of region I , it will reach at the end of it, a negative level large enough so that the following traverse of region II cannot bring it back to $H = 0$. After that, it will loose energy at every cycle, eventually ending up in libration.

On the other hand, if the motion crosses the level $H = 0$ late enough in one of his traverses of region I , the following traverse of region II will bring it back to positive value of H but this time in the region of negative circulation. Each following traverse of region II will increase its energy and it will eventually escape from the critical region in the negative circulation domain (see Fig. 4).

The motions between those two behaviours are those coming too close to the critical curve in region III and which we have excluded. Very likely they are captured by an invariant set inside region III . We cannot say what that invariant set might be without describing more precisely the system (especially the function $\lambda(t)$) which we do not want to do.

Hence everything depends on the phase at which the system reaches the level $H = 0$. This is directly related to the phase at which the system enters the region I or the energy level reached at the last passage through region III before transition. In most of the applications, we have no information about this phase because it depends critically on initial conditions or unmodeled small perturbations. It thus seems reasonable to introduce "probability" consideration and to ascribe (why not ?) equal probability to all possible energy level (between 0 and $-B_1$) at the last passage through region III .

With these definitions, it is easy to show that the probability of capture in libration is $(B_1 + B_2) / B_1$ and the probability of escape in negative rotation if $-B_2 / B_1$.

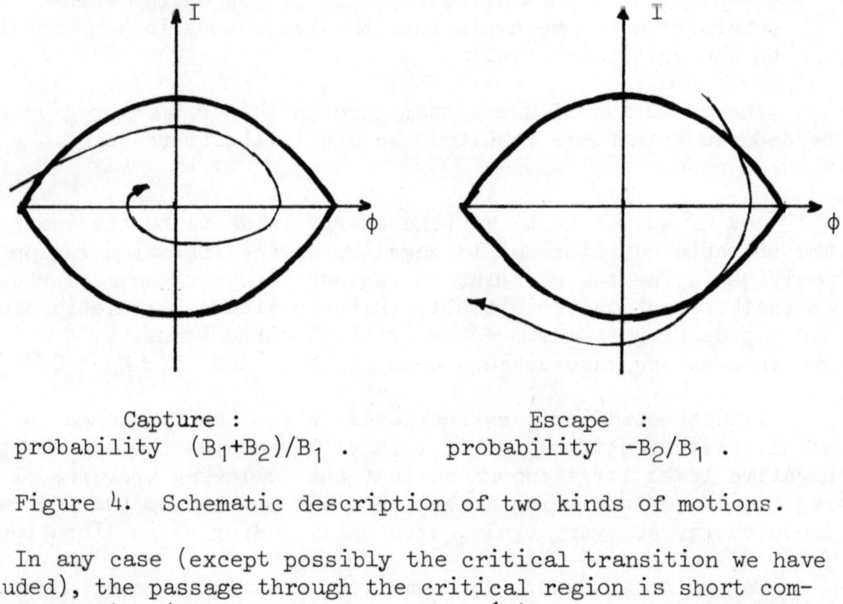

Capture : probability $(B_1+B_2)/B_1$. Escape : probability $-B_2/B_1$.

Figure 4. Schematic description of two kinds of motions.

In any case (except possibly the critical transition we have excluded), the passage through the critical region is short compared to $1 / \varepsilon$ (of the order of $\log \varepsilon^{-1}$). Hence the adiabatic invariant does not change its zero order value except possibly for the discontinuity due to the fact that the topology of the solutions of the unperturbed problem changes across the critical curve (see Fig. 5).

From this we conclude that the adiabatic invariant is still invariant across a critical curve (provision being made for the possible discontinuity).

But two paths are usually possible for the evolution of the

system depending on its phase at the time of transition.

The fact that this phase cannot be predicted because we do not know the system and its initial conditions well enough is expressed in terms of probabilities.

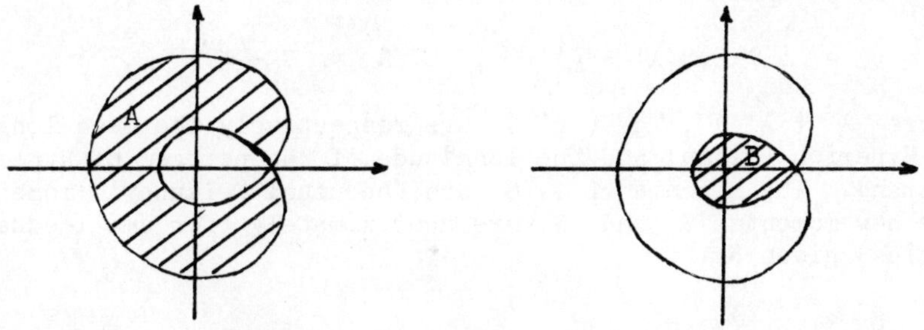

Figure 5. Discontinuity of the adiabatic invariant during a transition from region A to region B .

5. THE TITAN-HYPERION RESONANCE

As an illustration of the previous theory, we shall examine the (possible) capture into resonance of the two satellites of Saturn, Titan and Hyperion through tidal dissipation effects.

We choose this example not because of its planetological significance. It is believed that in the case of Titan, the tidal dissipation force is too weak for the capture to have taken place during the age of the solar system.

We choose it because the mechanism of capture has been studied by several authors (let us mention Greenberg, 1973, and Colombo et al, 1974) through other techniques and we believe that we can show that the use of the adiabatic invariant leads to an analysis which is at the same time more general and more sharp.

As model of the gravitational interaction of Saturn, Titan and Hyperion considered as point masses, we use the planar circular restricted three body problem. Titan is assumed to be on a circular orbit around Saturn and we study the motion of Hyperion.

The Hamiltonian function of the problem is given by

$$H = -\frac{1}{2a} - \mu \left[\frac{1}{\Delta} - \frac{\vec{s}/\vec{r}}{s^3}\right] \tag{30}$$

where a is the semi-major axis of Hyperion, μ the mass of Titan, Δ the distance Titan-Hyperion and $\vec{s}$ and $\vec{r}$ the position vectors of Titan and Hyperion with respect to Saturn.

In order to bring forward the 3 : 4 resonance, we use the following canonical variables

$$\begin{aligned} \sigma &= 4\lambda - 3\lambda' - g \quad , \quad & S &= L - G \quad , \\ \nu &= 4\lambda - 3\lambda' - g' \quad , \quad & N &= G - \tfrac{3}{4} L \quad , \end{aligned} \qquad (31)$$

where λ (λ') , g (g') are respectively the mean longitude of Hyperion (Titan) and the longitude of the perigee of Hyperion (Titan). The momenta L , G are the usual Delaunay's momenta. The new momenta S and N are approximately (for small eccentricities) given by

$$\begin{aligned} S &\simeq \frac{1}{2} \sqrt{a}\; e^2 \quad , \\ N &\simeq \frac{1}{2} \sqrt{a} \left(\frac{1}{2} - e^2\right) \quad . \end{aligned} \qquad (32)$$

At the conjunction of Titan and Hyperion, the angle σ (resp. ν) measures the mean anomaly of Hyperion (resp. Titan).

After performing an averaging over the fast frequency λ' , the Hamiltonian function becomes

$$H = -3\,(S+N)\,n' - \frac{1}{32\,(S+N)^2} + \frac{\mu}{a'} \sum_{(k)} P^{(k)}\left(\frac{a'}{a}\right) e^{k_1} \cos k_2 \sigma \qquad (33)$$

which can be considered as an Hamiltonian of a problem with one degree of freedom (σ , S) depending upon the parameters a' and N .

For the computations which follow, we have truncated the perturbation function at the sixth power in e and we have expanded the coefficients P^k around a nominal value of a'/a up to the second power in the increment of (a'/a) .

In the plane, the coordinates of which are

$$x = \sqrt{2S}\, \cos\sigma \quad , \qquad y = \sqrt{2S}\, \sin\sigma \quad , \qquad (34)$$

the trajectories are the curve H = constant given in Fig. 6 for the four typical values of N .

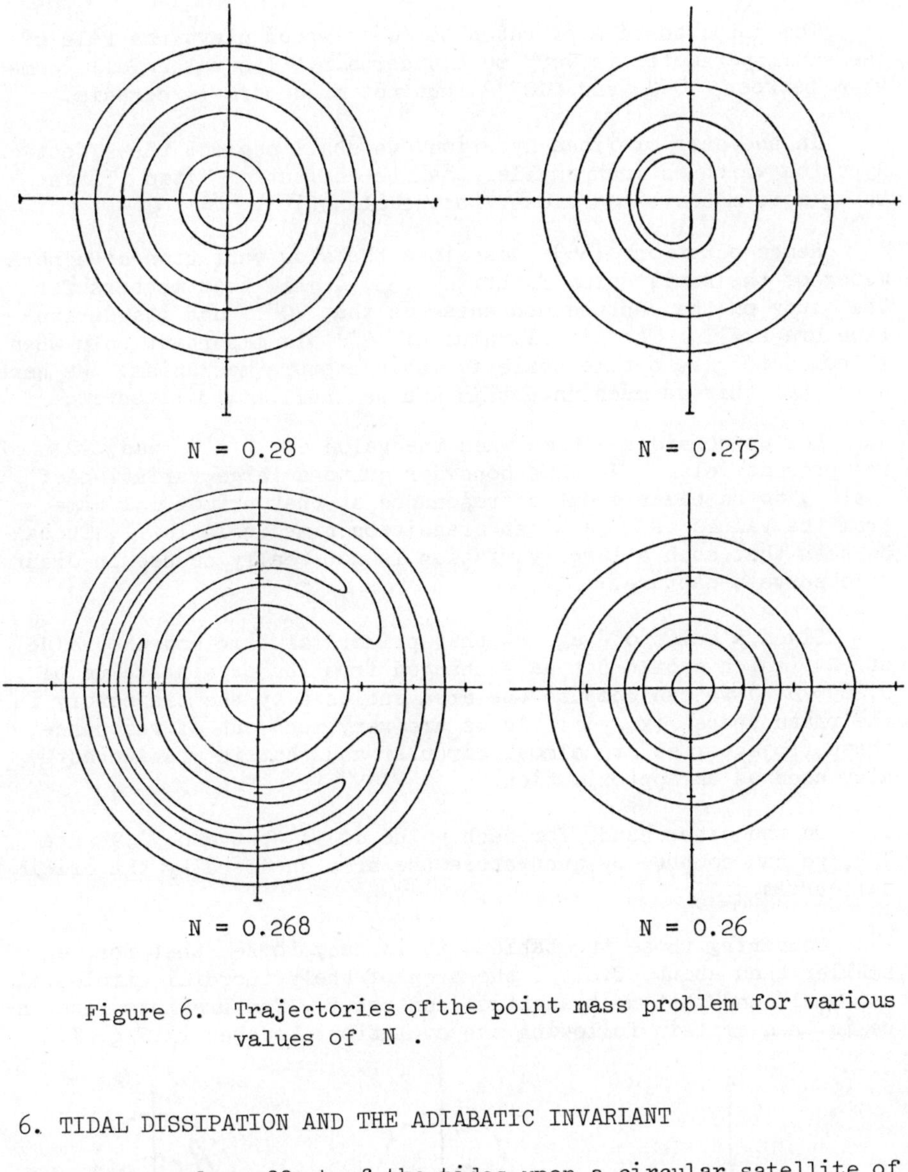

Figure 6. Trajectories of the point mass problem for various values of N .

6. TIDAL DISSIPATION AND THE ADIABATIC INVARIANT

The secular effect of the tides upon a circular satellite of semi-major axis " a' " and mean motion " n' " can be modelled by

$$\dot{a}' = \frac{9}{2} k_2 \mu \left(\frac{R_e}{a}\right)^5 Q^{-1} (a' n') \tag{35}$$

where μ is the mass ratio (mass of satellite / mass of planet) and k_2 , Q , R_e the love number, dissipation parameter and radius of the planet.

The value to be attributed to Q^{-1} which plays the role of the small parameter is very much uncertain. Its value falls somewhere between 10^{-4} and 10^{-13} but not much more is certain.

In the case of Titan-Hyperion, we shall neglect the effect on Hyperion which is much smaller than the effect on Titan (Titan being more massive and closer to the planet).

Hence equation (35) describes the slow variation of a parameter of the Hamiltonian function (33). All that matters for the study of the capture mechanism is that Q^{-1} and its derivative are small. The actual value of Q^{-1} is important only when it comes to give a time scale for this capture mechanism. We have seen that this is much uncertain and we shall not discuss it.

Let us consider a time when the value of " a'" was 90 % of its present value. We thus consider quite a large variation of " a'", to be clearly out of resonance at that primordial time. From the values of " a'" at transition time in Table II, it can be seen that such a large variation is not really needed in order to observe a capture.

Given a value of e_0 at that primordial time and the value of N (which should not have changed from its present value of $0.269\sqrt{a'}$) we can compute the area enclosed by the trajectory in the phase space (x, y). As we are very much out of resonance, these trajectories are almost circular and that is what we actually used as an approximation.

On the other hand, for each value of a' between 0.9 and 1., we can compute by quadrature the area enclosed by the critical curves.

Comparing those two tables, it is easy to see that for e_0 smaller than about 0.075, the area of the primordial circle is too small and that no transition can occur. Capture into resonance is then certain following the evolution sketched in Fig. 7.

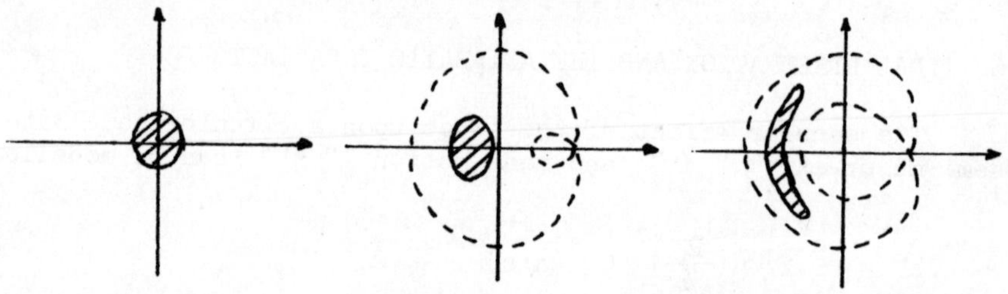

Primordial time Present time

Figure 7. Capture into resonance without transition.

The area of the bean shaped trajectories at the right of Fig. 7 can be related numerically to the amplitude of libration. As it is the same as the primordial area, we can relate the primordial eccentricity and the amplitude of libration as in Table I.

e_0	0	0.0174	0.05	0.075
Amplitude	0	35°	98°	143°

Table I. Primordial eccentricity and present amplitude of libration.

As the present amplitude is about 35° , we can infer that the primordial eccentricity of Hyperion was 0.0174 .

In case the primordial eccentricity is larger, two paths of evolution are possible. They are sketched in Figs. 8 and 9.

a) Capture into resonance.

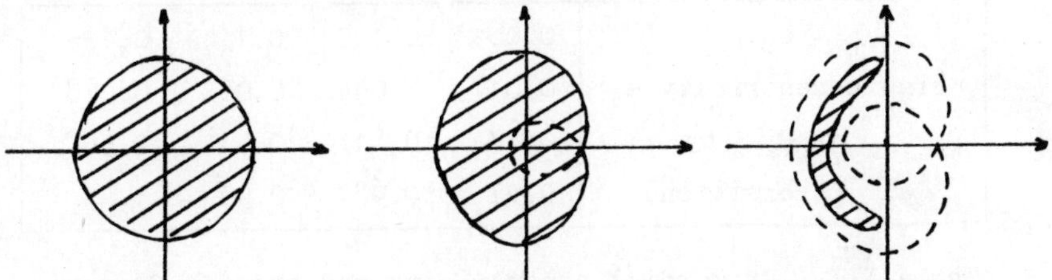

Primordial time. Transition time. Present time.

b) Escape from resonance.

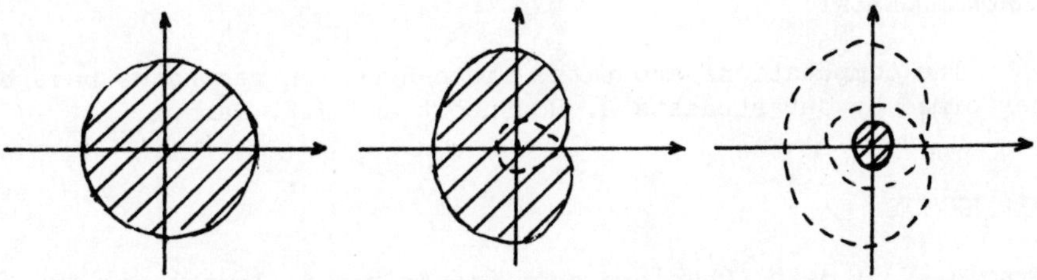

Primordial time. Transition time. Present time.

Figure 8. Two evolution paths after transition.

In case of capture, we compute in Table II the same elements as in Table I plus the probability of such an event which is no longer certain and the semi-major axis of Titan at transition.

e_0	0.081	0.100	0.114	0.125
Amplitude	148°	160°	169°	180°
Probability	0.83	0.59	0.49	0.43
a' (transition)	0.97	0.98	0.99	1.

Table II. Primordial eccentricity and present amplitude of libration in case of capture.

In case of escape, the present state of the satellite is best represented by its "free eccentricity". The present trajectory in the plane (x , y) is close to a circle the center of which corresponds to a small "forced eccentricity" (0.004) and the radius of which corresponds to the free eccentricity. The results are summarized in Table III. We see that the passage through resonance is the cause of a diminution of the eccentricity from e_0 to e_1 .

e_0	0.081	0.100	0.114	0.125
Free eccentricity e_1	0.017	0.044	0.063	0.078
Probability	0.17	0.41	0.51	0.56
a' (transition)	0.97	0.98	0.99	1.

Table III. Primordial eccentricity and present free eccentricity in case of escape.

AKNOWLEDGEMENT

The computations about the Titan-Hyperion resonance have been performed by our students G. Roothooft and J.F. Pardon.

REFERENCES

Arnold, V. : 1978, Chapitre supplémentaires de la théorie des équations différentielles ordinaires, Edition MIR, Moscou.

Burns, T.J. : 1979, On the rotation of Mercury, Celest. Mech. 19, pp. 297-313.

Colombo, G., Franklin, F., and Shapiro, I.I. : 1974, On the formation of the orbit-orbit resonance of Titan and Hyperion, Astr. J. 79, pp. 61-72.

Deprit, A. : 1969, Canonical transformation depending on a small parameter, Celest. Mech. 1, pp. 12-30.

Deprit, A., and Richardson, D. : 1982, Disemcumbering transformations for ideal resonance problems, submitted for publication in Celest. Mech.

Henrard, J. : 1982, Capture into resonance : an extension of the use of the adiabatic invariant, Celest. Mech., in print.

Goldreich, P. : 1965, An explanation of the frequent occurence of commensurable mean motions in the Solar System, M.N.R.A.S. 130, pp. 159-181.

Goldreich, P., and Toomre, A. : 1969, Some remarks on polar wandering, J. of Geoph. Res. 74, pp. 2555-2567.

Greenberg, R. : 1973, Evolution of satellite resonances by tidal dissipation, Astro. J. 78, pp. 338-346.

Kruskal, M. : Asymptotic theory of Hamiltonian and other systems with all solutions nearly periodic, J. of Math. Phys. 3, pp. 806-828.

Lenard, A. : 1959, Adiabatic invariance to all orders, Annals of Physics 6, pp. 261-276.

Landau, L., and Lipschitz, E. : 1966, Mécanique, Edition MIR, Moscou.

Neishtadt, A.I. : 1975, Passage through a separatrix in a resonance problem with a slowly-varying parameter, PMM 39, pp. 621-632.

Yoder, C.F. : 1979, Diagrammatic theory of transition of pendulum like systems, Celest. Mech. 19, pp. 3-30.

MODERN AND OLD VIEWS IN THE DYNAMICAL FOUNDATIONS OF CLASSICAL STATISTICAL MECHANICS

Luigi Galgani

Istituto di Fisica dell'Universita, Milan, Italy

ABSTRACT. A short review is offered giving dynamical foundations to some ideas in statistical mechanics which go back to L. Boltzmann and W. Nernst, concerning the distribution of energy in classical systems of weakly coupled oscillators.

1. The problem of understanding the dynamical foundations of statistical mechanics for systems of weakly coupled oscillators, which was considered to have been settled around the turn of the century, was unexpectedly brought to new life in the year 1954 with two quite different contributions, namely the work of Kólmogorov and the work of Fermi, Pasta, and Ulam.

Considering for example, a system of N harmonic oscillators coupled by some small non-linear interaction, the main problem under discussion is how energy is shared, in time-average, among the oscillators. The "classical" answer is that one should have equipartition.

On the other hand, the mathematical theorem of Kólmogorov, proving the existence of a set of invariant tori of considerable Lebesque measure in phase space (at least for not too high energies), showed that the systems under consideration were in general non-ergodic so that equipartition does not necessarily follow. On the other hand, the distribution of energy among oscillators was studied numerically by Fermi, Pasta and Ulam in a particular model (64 particles on a line with some non-linear nearest-neighbor coupling) and the startling result was found that at least for the special initial conditions considered and for the times of evolution taken into account, energy was shared only by few of the oscillators of

V. Szebehely (ed.), Applications of Modern Dynamics to Celestial Mechanics and Astrodynamics, 173–184.

the system. In the light of the present understanding of the Kólmogorov theorem, the result of Fermi can be interpreted by thinking that he had taken initial conditions leading to quasi-periodic motions on Kólmogorov tori.

In the meantime, the numerical result of Fermi was confirmed by Contopoulos in the field of celestial mechanics in the years around 1958, when systems of two degrees of freedom were intensively studied by finding essentially invariant tori which were just slightly deformed with respect to the tori of the corresponding unperturbed systems. One had then to wait until the year 1964 for the next fundamental discovery, namely the rather abrupt transition to stochasticity as energy is increased above some threshold, which was made by Hénon and Heiles(1964).

The concept of a stochasticity threshold was then extended by Izrailev and Chinikov (1966) to systems of more than two degrees of freedom, such as typically the Fermi-Pasta-Ulam model, by conceiving that one could define for each oscillator $i = 1, \dots, N$ a critical energy ε_i having in some sense a role analogous to the critical energy of the Hénon-Heiles model. The main idea is still that one should find chaotic motions when the initial conditions in phase space are changed by going to higher energies along a given direction, for example, by assigning energy to just one oscillator in the form of purely kinetic energy. In such ways, if the transition to stochasticity is rather abrupt, one thus defines N critical energies $\varepsilon_1, \dots, \varepsilon_N$; but the problem remains whether such critical energies still have an analogous meaning when several oscillators are initially excited.

The problem can be illustrated by making reference to Figure 1, where a system of two oscillators is considered just for graphical convenience,

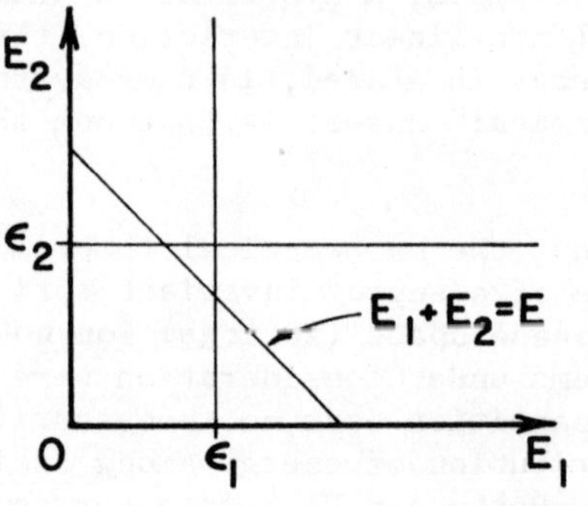

Figure 1.

while one should think of a system of N oscillators with $N > 2$. Here, E_1, E_2 are the energies of the oscillators and the "energy surface" is $E_1 + E_2 = E$ in the harmonic approximation, while ε_1 and ε_2 are the critical energies in question. The rectangle $\{E_1, E_2; E_1 < \varepsilon_1, E_2 < \varepsilon_2\}$ should correspond to initial conditions giving motions on invariant tori; the region with $E_1 > \varepsilon_1$ and $E_2 > \varepsilon_2$ should correspond to completely chaotic motions, and the problem is that of understanding the dynamical role of the two "lateral strips" in which only one oscillator is above threshold, namely the strip $\{E_1, E_2 ; E_1 < \infty, E_2 < \varepsilon_2\}$ and the strip $\{E_1, E_2 ; E_1 < \varepsilon_1, E_2 < \infty\}$. Clearly the analogs of such strips for $N > 2$ are very important in the thermodynamic limit $N \to \infty$, $E \to \infty$, $\frac{E}{N} \to u < \infty$, because they should have non-vanishing measure, while the analog of the rectangle corresponding to tori should have vanishing measure, as was particularly stressed by C. Cercignani.

We thus come to a central point, related to what I like to call the Froeschlé effect. In the year 1971 this author, just in order to understand the problem under discussion, studied a related problem for mappings. Indeed, as it is well known, a system of two oscillators is related to a mapping of a plane into itself, having typically a fixed point with invariant curves around it and a chaotic region beyond a certain "last invariant curve" (which is the analog of a critical energy). In order to mimic a system of three oscillators he considered a mapping of a plane into itself as described above and a replica of it, then defining correspondingly by direct product a mapping of R^4 into itself; non-trivial results were then expected to occur by introducing a small interaction between the two original mappings.

Making reference to a figure analogous to Figure 1, ε_1 and ε_2 corresponding to the radii of the last invariant curves, the results found for not too large interactions are the following. For initial conditions in the finite rectangle there were no apparent differences with respect to the uncoupled case (completely ordered motions; on tori) apart from very few cases; for initial conditions with $E_1 > \varepsilon_1$, $E_2 > \varepsilon_2$ completely chaotic motions were found; for initial conditions in the lateral strips, the oscillator below threshold would move apparently on invariant curves for short times, but in fact its radius would slightly fluctuate performing some kind of slow random motion until after a long time it would possibly go beyond its critical energy, thus performing motions of the same chaotic type as if it were isolated. In the course of time that oscillator, or the other one, could then come below threshold, and the one being below threshold would behave as described above. This is a typical effect of intermittency, where one oscillator performs an ordered motion for some time, a chaotic motion for another thime interval, and so on.

Obviously, the important problem remains that for a system of harmonic oscillators, at variance with a system of coupled mappings, there is no natural set of critical energies built in; in other words, in such case the interaction should produce both the critical energies $\varepsilon_1, \ldots, \varepsilon_N$ and the Froeschlé intermittency effect, but this was never clearly shown up to now.

2. Having thus given a short review of the modern advances in the dynamics of systems of weakly coupled oscillators, I come now to the problem of a possible interpretation of them in connection with the foundations of classical statistical mechanics. But first of all, let me mention that I am well aware of the completely speculative character of the following considerations; nevertheless, I dare to present them. This is because I am sure that they follow a line of thought that can clearly be traced back to Boltzmann and Nernst, in whose works many premonitions of the dynamical facts described above can indeed be found. And, I can even quote the words of Boltzmann himself, written in the main work I will refer to, where having presented his speculations, he adds: "It may be objected that the above is nothing more than a series of imperfectly proved hypotheses, but granting its improbability, it suffices that this explanation is not impossible. For then I have shown that the problem is not insoluble, and nature will have found a better solution than mine."

So I have to illustrate the main idea of Boltzmann, which was presented in a very short and vivid paper published in beautiful english in Nature in the year 1895. The problem was that of solving the paradox that, while relying on the Maxwell-Boltzmann distribution law he found the equipartition law, such law however, did not always work well, in particular for gas molecules; neither does it work well, as we know today, for solids at low temperature and for the black-body radiation. In a sense, things appeared to be such that some degrees of freedom were, so to say, frozen, and did not take part in the energy sharing; in fact, the frozen oscillators appeared to be those of higher frequencies (typically, internal vibrations in molecules). So, in a first, very crude approximation the thought of a system as composed of two subsystems, namely the one having low frequencies and the one having high frequencies, the oscillators of the first subsystem being characterized by quite unstable motions, while the other ones should have quite stable motions; this is the first essential insight in connection with the dynamical foundations of statistical mechanics. Indeed, while according to the equipartition theorem the internal energy is NkT (N = number of degrees of freedom, k = Boltzmann's constant, T = absolute temperature), such mechanical or "ideal" internal energy does not coincide with the thermodynamic internal energy, because in thermal exchanges only the fraction of energy owned by the unstable (low frequency) oscillators is relevant, the remaining one being frozen because of its stability character. In Boltzmann's words "...the vis viva of the internal motion is transformed into progressive and rotatory motion so slowly, that when a gas is brought to a lower temperature the molecules may retain for days, or even for years, the higher vis viva of their internal vibrations corresponding to the original temperature. This transference of energy, in fact, takes place so slowly that it cannot be perceived amid the fluctuations of temperature of the surrounding bodies."

This is then the main idea of Boltzmann, namely of discriminating between the ideal internal energy NkT (Equipartition principle) and the thermodynamic internal energy, which is the fraction of it, having quite unstable character and thus being relevant for thermal exchanges. Here however, several problems remain. First of all, there is the problem that one should then in principle be in presence of a non-equilibrium situation; to this problem we will come back below. Furthermore, the distinction of the two groups of oscillators is in a sense too sharp.

This latter problem can be overcome by making reference to an idea of W. Nernst, which by the way, fits very well with the Froeschle effect. The main idea of Nernst can be found in a long, very difficult paper that he wrote in the year 1916 and which was highly influenced by Planck's second theory published in the year 1912, in which Planck conveived that an oscillator of frequency ν should have an energy $(\frac{1}{2})h\nu$ at zero temperature,

h being Planck's constant. By his insights in the principles of thermodynamics, in particular in connection with his own third principle, which is just related to zero-point energy, Nernst conceived that this energy $(\frac{1}{2})h\nu$ should be of ordered type. More

properly, he conceived that any oscillator should have a critical energy $\varepsilon(\nu)$ discriminating between ordered motions below it and disordered motions above it. The reading of pages 91-92 of his(1912) paper is a really shocking experience in the light of the Froeschlé's effect mentioned above, and indeed it occurred to me to understand the results of Froeschlé and those pages of Nernst at the same time after several years of attempts at understanding them separately.

3. Now it is easy to implement the main idea of Boltzmann through the idea of Nernst in the light of the Froeschle effect. Consider a system of N oscillators, all of the same frequency ν (which is now a parameter), at equilibrium at a given absolute temperature T. The energy E of a single oscillator is then a (positive) random variable with probability density given by the Maxwell-Boltzmann law

$$p(E) = \frac{1}{kT} e^{-E/kT} \tag{1}$$

and one has

$$\int_0^\infty p(E)\, dE = 1 \tag{2}$$

$$\overline{E} = \int_0^\infty E\, p(E)\, dE = kT, \tag{3}$$

so that the average energy $\overline{E}$ is independent of the frequency ν according to the equipartition principle, and the total internal energy is the "ideal" energy NkT.

Assume now that there exists a critical energy $\varepsilon = \varepsilon(\nu)$ in the sense of Froeschlé. Thus we have in the so-called μ space (the phase space on one oscillator) a circle of energy $\varepsilon(\nu)$ such that the oscillators that are inside it perform ordered motions (on circles, with slowly varying radii), while the oscillators outside the circle of energy $\varepsilon(\nu)$ perform quite chaotic motions. At any time the fractions of oscillators below and above threshold are

$$n_0 = \int_0^\varepsilon p(E)\, dE \quad \text{and} \quad n_1 = 1 - n_0 \text{ respectively, with}$$

$$n_0 = 1 - e^{-\varepsilon(\nu)/kT}, \quad n_1 = e^{-\varepsilon(\nu) kT} \tag{4}$$

and the ideal internal energy NkT is divided into two parts, namely NU_0 and NU_1 respectively, where

$$U_1 = \int_{\varepsilon(\nu)}^\infty E\, p(E) dE = \frac{\varepsilon(\nu) + kT}{e^{\varepsilon(\nu)/kT}} \tag{5}$$

and $U_0 = kT - U_1$. In the spirit of the Boltzmann idea, only NU_1 is the thermodynamic internal energy.

As one sees, if ν is considered as a parameter, all frequencies contribute to the internal energy, each with a fraction $n_1(\nu,T)[\varepsilon(\nu) + kT]$. However, if one assumes that $\varepsilon(\nu)$ is an increasing function of ν (for example, a linear function of ν), then at a fixed temperature T the high frequencies carry only a negligible contribution to the thermodynamic internal energy, as conceived by Boltzmann.

In such a way we can conclude that, if the Froeschle effect is a real one for systems of coupled oscillators, then, implementing the Boltzmann idea through the Nernst's idea in classical statistical mechanics the high frequencies do not contribute to the thermodynamic internal energy and the so-called ultraviolet catastrophe is avoided.

4. Having obtained this qualitative result, how much should one then insist on the quantitative estimates given by the classical theory sketched above ? Such a "theory" is indeed a rather crude one, based on a highly idealized phenomenon (the existence of sharp energy thresholds) and on a very rough approximation, namely that of excluding the contribution of all frozen oscillators. One could imagine instead that by depleting the chaotic oscillators by thermal interaction with a reservoir at a lower temperature, a new equilibrium will be reached, at longer times, in which the frozen oscillators will also take part.

However, it is a very curious fact that even in such a crude approximation the quantitative estimates are not so bad, and actually much better than one would imagine at first sight. Some considerations along these lines follow.

Let us restrict our attention to the problem of the black-body radiation. In such case, if one assumes that there is thermal equilibrium, the general Wien's law requires that the internal energy per unit volume and per unit frequency be (Planck, 1923)

$$u(\nu,T) = \nu^2 kT\, f(\nu/T), \tag{6}$$

where f is an undetermined function. Now, recalling that the number of oscillators per unit volume with frequencies between ν and $\nu+d\nu$ is $N(\nu)d\nu$, where

$$N(\nu) = \frac{8\pi}{c^3}\,\nu^2 \tag{7}$$

c being the velocity of light, the theory proposed above gives

$$u(\nu,T) = N(\nu)\ \frac{\varepsilon(\nu) + kT}{e^{\varepsilon(\nu)/kT}} = \frac{8\pi\nu^2}{c^3}\ \frac{\varepsilon(\nu) + kT}{e^{\varepsilon(\nu)/kT}}\ . \quad (8)$$

By comparison with (6) one thus deduces that ε is proportional to frequency,namely, $\varepsilon(\nu) = h\nu$, where h is a constant, to be possibly identified with Planck's constant (which should in principle be determined by the non-linear interaction among the oscillators of the electromagnetic field produced by electric charges). As a consequence we may say that in the very crude approximation described above, we have for the energy density of the electromagnetic field the law

$$g(\nu,T) = \frac{8\pi}{c^3}\ \nu^2\ \frac{h\nu + kT}{e^x} \qquad (9)$$

where

$$x = \frac{h\nu}{kT}\ , \qquad (10)$$

as a classical analog to Planck's law

$$f(\nu,T) = \frac{8\pi}{c^3}\ \nu^2\ \frac{h\nu}{e^x - 1}\ . \qquad (11)$$

5. Now, everyone has seen the famous isotherms of Lummes and Pringsheim of the year 1900. Their fundamental figure was of the type of Figure 2, where

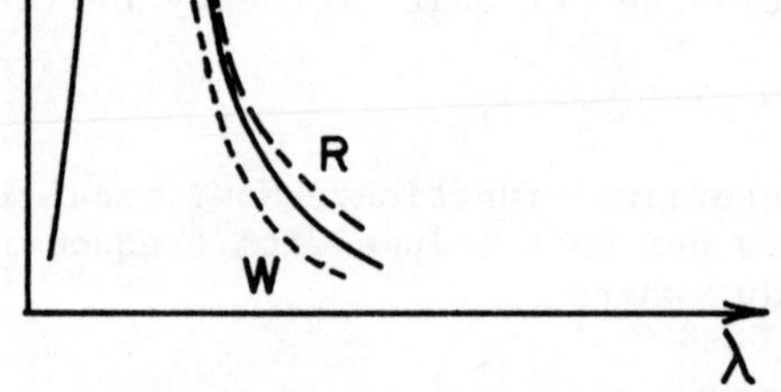

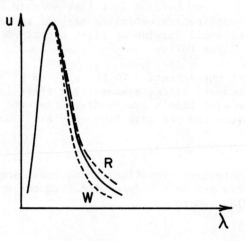

Figure 2

the continuous line gives the experimental values, while the other two lines represent two analytical formulae by Wien and Rayleigh. In terms of frequencies these can be written as

$$u_W(\nu,T) = \frac{8\pi}{c^3}\nu^2\frac{h\nu}{e^x}\ , \ u_R(\nu,T) = \frac{8\pi}{c^3}\ \nu^2\ \frac{kT}{e^x}\ , \qquad (12)$$

where h and k are to be considered as two parameters to be fitted. As is seen from the figure, these two curves deviate from opposite sides with respect to the experimental curve. Now the theoretical curve proposed here, namely the law (11), is just an arithmetic mean of such two laws, so that clearly it will not deviate much from the experimental curve. As a matter of fact, in the same year, 1900, another interpolation formula between those of Wien and of Rayleigh was proposed by Thiesen, namely a geometric mean

$$u_T(\nu,T) = \alpha\ \frac{8\pi}{c^3}\ \nu^2\ \frac{\sqrt{\nu T}}{e^x}\ , \qquad (13)$$

α being a parameter (actually $\alpha = \sqrt{hk}$). This law is also reported in the original figure of Lummes and Pringsheim, and it fits the data quite well.

Usually one assumes that the decision in favor of Planck's law, which was proposed shortly afterwards, was made through the results of Rubens and Kurlbaum (1901) who produced three curves of isochromatics (curves of energy versus temperature at fixed ν). In two of them Planck's law fits much better than Thiesen's, Theisen's, but Thei sen's was a little better in the third one. It is a fact that the law proposed here seems to fit the data of these three curves a little better or as well as Planck's, as shown by the following three tables (notice that the intensities are experimentally determined up to a multiplicative constant and an additive constant).

TABLE I. $\lambda = 24\mu$ and $\lambda = 31.6\mu$.

T	u observed	f	g
85	-15.5	-12.8	-14.4
193	-9.4	-8.8	-10.0
293	0	0	0
523	30.3	27.6	30.2
773	64.3	61.5	64.7
1023	98.3	96.4	98.6
1273	132	132	132
1523	167	167.8	164.9
1773	201.5	203.7	197.5

TABLE 2. $\lambda = 51.2\mu$.

T	u observed	f	g
85	-20.6	-22.5	-26.5
193	-11.8	-12.2	-14.1
293	0	0	0
523	31.0	30.2	32.1
773	64.5	63.9	65.6
1023	98.1	97.9	99.1
1273	132	132	132
1523	164.5	166.1	164.6
1773	196.8	200.3	197.2

TABLE 3. $\lambda = 8.85\mu$.

T	u observed	f	g
100	-1.6	-1.3	-1.2
200	-1.5	-1.2	-1.1
373	3.4	3.1	3.0
500	13.5	12.5	12.3
800	53.5	50.7	51.2
1100	102	100.1	101.2
1273	132	132	132
1400	154	155.8	154.9
1700	212.5	213.9	209.4

In fact, the law proposed here differs as much as 30% with respect to that of Planck; typically, for example, the Stefan-Boltzmann constant turns out to be 23% higher. As a consequence, as Planck's law is usually assumed to be well verified, one has to presume that the law proposed here fails at a quantitative level. It is however, a curious fact that such law, which has some theoretical foundation, turns out to be the closest one to Planck's among the other ones which were proposed with no theoretical foundation.

As a piece of information, I like to add here that in fact the situation concerning the experimental verification of Planck's law is not quite clear. A systematic review of all experiments up to 1919 was made by Nernst and Wulf who found systematic discrepancies of the order of 7% (or even of 25%, according to the value attributed to the so-called second radiation constant $C_2 = hc/k$). By the way, they could also determine graphically a variant of Planck's law fitting the data better, and it is a very strange, fortuitous fact that the form they proposed agrees quite well with the one proposed here. An answer to Nernst and Wulf was given two years later by Rubens and Michel who determined a number of isochromatics and maintained to have found agreements within 1% of Planck's law. Things are, however, not so clear as shown for example, by a rather disconcerting result of Michel in the following year on the same experimental data (namely, a fit for C_2 gave a value going towards the wrong direction). After a review given in the year 1929, no systematic research on the experimental verification of Planck's law was made, (Muller, 1929).

Without entering into further details, I would like to mention that it is not impossible that systematic disagreements with respect to Planck's law may be found; moreover, it is also quite conceivable that one realize that nonequilibrium phenomena, as envisaged above, may turn out to be relevant. Furthermore, the possible discrepancies with respect to Planck's law which were recently reported in connection with the cosmic background radiation could also be of interest, (Woody and Richards, 1979).

6. I thank Professor Victor Szebehely for his kind invitation to give the present lecture at this conference. I hope that the rather speculative character of the considerations reported above was sufficiently balanced by the other lecture I gave, concerning a new proof of the Kolmogorov theorem on invariant tori by use of the Lie transform technique, the details of which will be reported elsewhere.

REFERENCES.

Boltzmann, L., (1895), Nature, **51**, p. 413.
Contopoulos, G., (1963), Astron. J., **69**, p. 73.
Fermi, E., J. Pasta and S. Ulam, (1954), Los Alamos Report, reprinted in E. Fermi, Collected Works, (1965), Roma and Chicago, p. 978.
Froeschlé, C., (1971), Astrophys. Space Sci., **14**, p. 110; see also L. Tennison, M. A. Lieberman and A. J. Lichtenberg (1979), Nonlinear Dynamics and the Beam-Beam Interaction, M. Month and J. C. Herrera (eds.) Am. Dust. Phys. Conf. Proc., **57**, p. 272.
Galgani, L., (1981), Lett. Nuovo Cim., **31**, p. 65; Nuovo Cim. (1981), 62, p. 306.
Hénon, M. and C. Heiles, (1964), Astron. J., **69**, p. 73.
Izrailev, F. M. and B. V. Chinikov, (1966), Dokl. Akad. Nauk USSR, **166**, p. 57.
Kangro, H., (1976), Early History of Planck's Radiation Law, London.
Kolmogorov, A. N., (1954), Dokl. Akad. Nauk USSR, **98**, No. 4, p. 527.
Lummer, O. and E. Pringsheim, (1900), Ver. Dtsch. Phys. Ges., **2**, p. 163.
Michel, G., (1922), Z. Phys., **9**, p. 285.
Muller, C., (1929), Die experimentelle Prüfung der Strahlungsgesetze und die Bestimmung der Konstanten, In Handbuch der Experimentalphysik, **9**, Leipzig.
Nernst, W., (1916) Ver. Dtsch. Phys. Ges., **18**, p. 83.

Nernst, W. and T. Wulf, (1919), Ver. Dtsch. Phys. Ges., 21, p. 294 and 332.
Planck, M., (1912), Ann. Phys., 37, p. 642.
Planck, M., (1923), Vorlesugen über die Theorie der Wärmestrahlung, Leipzig, (English translation, The Theory of Heat Radiation, (New York, NY, 1959).
Rubens, H. and F. Kurlbaum, (1901), Ann. Phys., 309, p. 649.
Rubens, H. and G. Michel, (1921), Phys. Zeitschr., 22, p. 569.
Woody, D. P. and P. L. Richards, (1979), Phys. Rev. Lett., 42, p.925.

THE KAM INVARIANT AND POINCARÉ'S THEOREM

T. Petrosky and I. Prigogine*

Center for Studies in Statistical Mechanics
University of Texas, Austin, Texas, U.S.A.
*Also Service de Chimie-Physique II
Université Libre de Bruxelles, Belgique

ABSTRACT

The relation between the KAM invariants and Poincare's classical theorem on nonexistence of uniform invariants is discussed, and their compatibility is pointed out. It is also shown that the KAM invariants are regular invariants in the sense defined by Prigogine and his coworkers. Then, the relation between the ergodic problem and the KAM invariant is discussed.

(1) INTRODUCTION

The study of invariants of motion is an interesting subject which is closely connected with a fundamental problem of stability in classical dynamics, that is, the problem of existence of quasi-periodic motion and of ergodicity in non-integrable systems.

One of the cornerstones of this field has been set by Poincare in his celebrated theorem on the nonexistence of uniform integrals of the motion in non-integrable systems (see Whittaker, E., 1964; Prigogine, I., 1962). While Poincare's theorem has been widely quoted, its interpretation and importance have been the subject of some controversy. Indeed, some authors have taken this theorem as a statement about metrical indecomposability which ensures the ergodicity of the system (Fermi, E., 1923; Ter Harr, D., 1955; Brilluoin, L., 1966). It is, however, clear that Poincare's negative result is only a necessary but not a sufficient condition for ergodicity. After Poincare's theorem, thus, the challenge to find new types of in-

V. Szebehely (ed.), Applications of Modern Dynamics to Celestial Mechanics and Astrodynamics, 185–199.

variants for nonintegrable systems still remained (see, for example, the work of Whittaker (1964) on the adelphic integral).

A crucial development in this field is the Kolmogorov-Arnold-Moser theory, i.e., the so-called KAM theory. In a fundamental paper, Kolmogorov (1954) pointed out that for a large class of systems there exist invariant tori on sets of positive measure. This statement was proved later by Arnold (1963) and, independently, by Moser (1962) (see also Barrar, R., 1970). This result is quite important, because it gives a first rigorous analytic method for predicting the long time ($t \to \infty$) behavior of non-integrable systems, which is impossible to achieve through numerical methods. In spite of their significance, the properties of the KAM invariants, including their compatibility with Poincare's theorem and their relation to ergodicity, have not been completely clarified.

Independently of the above developments, recent development in nonequilibrium statistical mechanics has led to an interesting classification of the invariants of the motion. Through the investigation of the asymptotic solution of the Liouville equation, Resibois and Prigogine discovered that for systems with an infinite number of degrees of freedom to which kinetic theory applies, one may construct additional classes of invariants, the so-called "singular invariants" (Prigogine, I., 1962; Resibois, P. & Prigogine, I., 1960). The remaining class of invariants, which includes the Hamiltonian as a typical example, was called the class of "regular invariants," and is closely related to the uniform integrals in Poincare's theorem (see Prigogine, I., 1962). The significance of this classification is that for suitable initial conditions we may show that singular invariants do not affect the ergodic limit, i.e., the long time averages of physical quantities.

The purpose of this paper is to discuss the relation of the KAM invariants to Poincare's theorem and to show that they correspond to regular invariants. This result gives more precise information about the physical role of the KAM invariants in the ergodic problem. As will be shown in Section 3, the KAM invariants are expressed by singular "functions" -- distributions -- with an argument which is a highly complicated function in phase space. This fact enables them to establish the compatibility of the KAM invariants with Poincare's theorem. However, because of their singular structure, the question of whether the existence of the KAM invariants are compatible with the mean ergodic theorem (see Michael, R. & Simon, B., 1972), assuming the system starts with a smooth initial condition, still remains an open problem.

In the next section, we will briefly summarize Poincare's theorem and introduce the classification of invariants. In section 3 we will construct explicitly the KAM invariants and show their compatibility with Poincare's theorem. Section 4 will be devoted to the proof that the KAM invariants are "regular" invariants. In the last section, we will present a few concluding remarks.

(2) INVARIANTS OF MOTION AND POINCARE'S THEOREM

The invariant of the motion $\Phi(p,q)$ is defined as the stationary solution of the Liouville equation

$$i\frac{\partial\rho}{\partial t} = L\rho \ , \tag{2.1}$$

i.e., the solution of

$$L\Phi = 0 \ . \tag{2.2}$$

Here, L denotes the Liouville operator, i.e., the Poission bracket $i\{H,\cdot\}$ with the Hamiltonian $H(p,q)$,

$$L = i\sum_{\alpha=1}^{N}\left(\frac{\partial H}{\partial q_\alpha}\frac{\partial}{\partial p_\alpha} - \frac{\partial H}{\partial p_\alpha}\frac{\partial}{\partial q_\alpha}\right) . \tag{2.3}$$

The function $\rho(p,q:t)$ is a distribution function in phase space. N is the number of degrees of freedom of the system. The variables (p,q) are the canonical variables, say, the action-angle variables, and p and q are abbreviations for the set $\{p_1,p_2,\ldots p_N\}$ of action variables and the set $\{q_1,q_2,\ldots q_N\}$ of angle variables. Properties of invariants of the motion may, therefore, be studied through the solution of the Liouville equation (2.1), which is formally expressed by the contour integral

$$\rho(t) = e^{-iLt}\rho(0) = \frac{1}{2\pi i}\int_c dz e^{-izt}\frac{1}{z-L}\rho(0) \ , \tag{2.4}$$

where the path c of the integration is parallel to the real axis on the upper half-plane for $t>0$ and goes to $-\infty$ from $+\infty$.

It may be noted that the distribution function that we consider may not be directly observed itself in the experiment but is used to compute average values of observable quantities $A(p,q)$ which are in general smooth functions of the phase point p and q,

$$\langle A(t)\rangle = \langle A,\rho(t)\rangle \equiv \int dpdq\ A(p,q)\ \rho(p,q:t) . \tag{2.5}$$

This fact allows us to consider wide classes of functions ρ for solutions of the equations (2.1) and (2.2), including the distribution. For such singular cases, we must, of course, interpret Eq. (2.2) as

$$\langle G, L\Phi \rangle = 0, \tag{2.6}$$

for any functions of rapid decrease (the so-called test functions) $G(p,q)$ (see Michael, R. & Simon, B., 1972). This extension of the class of functions enables us to construct new types of invariants even for non-integrable systems to which Poincare's theorem applies.

Let us now briefly summarize Poincare's theorem. We consider the system with the following Hamiltonian

$$\begin{aligned} H(p,q) &= H_o(p) + \lambda V(p,q) \\ &= H_o(p) + \lambda \Sigma_n' v_n(p) e^{i(n,q)} \; . \end{aligned} \tag{2.7}$$

Here n is the integer vector, $n = \{n_1 n_2, \ldots, n_N\}$, and we use the notation $(n,q) \equiv \Sigma_{\alpha=1}^{N} n_\alpha q_\alpha$ for simplicity. The parameter λ is introduced to denote the magnitude of the interaction $V(p,q)$. We have written $V(p,q)$ as a Fourier series, taking into account its periodic dependence on the angles. The prime on the summation sign means that $n \neq 0$, because the angle independent component of $V(p,q)$, if any, will be included into the free Hamiltonian $H_o(p)$. We further assume that the Hessian of H_o does not vanish.

$$\det \left| \frac{\partial^2 H_o(p)}{\partial p_\alpha \partial p_\beta} \right| \neq 0 \; . \tag{2.8}$$

Following the decomposition of the Hamiltonian Eq. (2.7), the Liouvillian (2.3) is written by

$$\begin{aligned} L &= L_o + \lambda L_1 \\ &= -i \sum_\alpha^N \omega_\alpha(p) \frac{\partial}{\partial q_\alpha} + i\lambda \Sigma_n' e^{i(n,q)} \sum_\alpha^N \{ i v_n(p) n_\alpha \frac{\partial}{\partial p_\alpha} - v_n^{(\alpha)}(p) \frac{\partial}{\partial q_\alpha} \}, \end{aligned} \tag{2.9}$$

where $\omega_\alpha(p) \equiv H_o^{(\alpha)}(p)$ are frequencies, and we use the following abbreviation to denote the derivative on the actions

$$F^{(\alpha_1, \alpha_2, \ldots \alpha_k)}(p) \equiv \frac{\partial^k F(p)}{\partial p_{\alpha_1} \partial p_{\alpha_2} \cdots \partial p_{\alpha_k}} \; . \tag{2.10}$$

Poincare's theorem asserts that if the interaction of the system satisfies the condition

$$\lambda v_n(p) \neq 0 \; , \tag{2.11}$$

at the "resonance" points which are determined by the equation,

$$(n,\omega(p)) \equiv \sum_{\alpha}^{N} n_\alpha \omega_\alpha (p) = 0, \tag{2.12}$$

there is no invariant $\Phi(p,q,\lambda)$ which is uniform, that is, single-valued and regular in p,q, and λ except for the Hamiltonian (and functions of it). The proof is sketched as follows (see Whittaker, E., 1964; Prigogine, I., 1962). We consider an invariant of the form

$$\Phi(p,q) = \Phi^0(p) + \sum_{r=1}^{\infty} \lambda^r \Phi^{(r)}(p,q)$$

$$= \Phi^0(p) + \sum_{r=1}^{\infty} \lambda^r \sum_n \phi_n^{(r)}(p) e^{i(n,q)} , \tag{2.13}$$

where the coefficients are analytic functions of their arguments and $\Phi(p,q)$ is assumed to be periodic in the angles. The unperturbed pert Φ^0 is a function only of the actions p due to the condition (2.8).

If Φ is not a function of H, Φ^0 cannot be a function of H_o. But this is impossible for the analytic function Φ^0 since from Equation (2.2), we have $\lambda L_1 \Phi^{(0)} + L_0 \lambda \Phi^1 = 0$ to the first order in λ. Equivalently, for the Fourier coefficients, we have

$$\lambda(n\ \omega(p))\phi_n^{(1)}(p) = \lambda v_n(p) \sum_{\alpha}^{N} n_\alpha \frac{\partial \Phi^0(p)}{\partial p_\alpha} . \tag{2.14}$$

This relation implies that $(n,(\partial\Phi^0/\partial p)) = 0$ holds at the resonance point, and it leads to the condition

$$\frac{1}{\omega_\alpha(p)} \frac{\partial \Phi^0}{\partial p_\alpha} = \frac{1}{\omega_\beta(p)} \frac{\partial \Phi^0}{\partial p_\beta} , \tag{2.15}$$

for all points p, because the resonance points are densely distributed in any domain of p. Therefore, Φ^0 must be a function of H_o.

If $\Phi(p,q)$ is defined by a distribution (such as the Dirac δ-function), we may not conclude that $(n,(\partial\Phi^0/\partial p))=0$ holds <u>at</u> the resonance point, because Equation (2.14) must be interpreted in a way similar to that of Equation (2.6). Thus, if we remove the condition of analyticity, we still have a possibility of constructing new types of invariants for the system described by the Hamiltonian (2.7) with the conditions of (2.8) and (2.11) with (2.12).

The problem of invariants of the motion has been repeatedly discussed in the framework of nonequilibrium statistical mechanics (Prigogine, I., 1962; Balescu, R., 1970; Grecos, A.P., 1971; Prigogine, I. & Grecos, A.P., 1973), since this problem is important in connection with the ergodic hypothesis. It leads to an interesting classification of the invariants which is closely related to Poincare's theorem. In order to introduce this classification scheme, we first notice that the free Hamiltonian $H_0(p)$ is just the average value of H in Equation (2.7) over the angles,

$$H_o(p) = \left(\frac{1}{2\pi}\right)^N \int_0^{2\pi} dq_1 \ldots dq_N H(p,q) \equiv PH(p,q). \tag{2.16}$$

This relation defines the orthogonal projection P. We introduce also $Q = 1-P$ and we have

$$P = P^2 \ , \ Q = Q^2, \ PQ = QP = 0 \ . \tag{2.17}$$

The interaction $\lambda V(p,q)$ is, therefore, the Q projection of the Hamiltonian, i.e., $\lambda V = QH$.

According to the decomposition of the Hamiltonian Eq. (2.7) into the P and Q components, we decompose also the distribution function ρ into the Fourier series,

$$\begin{aligned}\rho(p,q:t) &= \rho_0(p:t) + \rho_c(p,q:t) \\ &= \rho_0(p:t) + \sum_n{}' \rho_n(p:t) e^{i(n,q)} \ ,\end{aligned} \tag{2.18}$$

where $\rho_0 \equiv P\rho$ and $\rho_c \equiv Q\rho$.

Note that ρ_0 is an invariant of the motion for the free Hamiltonian system. We will, therefore, study the influence of the interaction on the P-component of ρ separately from the Q-component. This separation may be done by using the following identity for the resolvent operator of L in Eq. (2.4) (see e.g. Balescu, R., 1975).

$$\frac{1}{z-L} = \{P + \mathcal{C}(z)\}\frac{1}{z-\psi(z)}\{P + \mathcal{D}(z)\} + \frac{1}{z-QLQ} \ , \tag{2.19}$$

where

$$\psi(z) = PLQ\frac{1}{z-QLQ}QLP, \tag{2.20}$$

$$\mathcal{D}(z) = PLQ\frac{1}{z-QLQ} \tag{2.21}$$

and

$$\mathcal{C}(z) = \frac{1}{z-QLQ}QLP. \tag{2.22}$$

To obtain the identity of Equation (2.19), we have used the fact that PLP = 0. The above operators, which we call kinetic operators, are the bssic quantities in our formalism. Of special importance is the "collision operator" $\psi(z)$ which determines the evolution of $\rho_0(p;t)$ through intermediate states in the Q-subspace. The operator $\mathcal{D}(z)$ and $\mathcal{C}(z)$ are called the "destruction" and the "creation" operators, respectively. More details of these concepts are found in the textbook by Prigogine (1962).

The collision operator $\psi(z)$ gives us a criterion as to whether or not the system is "dissipative." To see this, let us consider the "ergodic limit" of the function $\rho(p,q;t)$,

$$\bar{\rho}(p,q) = \lim_{T\to\infty} \frac{1}{T}\int_0^T dt\rho(p,q;t) = \lim_{z\to+io} \frac{z}{z-L}\rho(p,q;0). \qquad (2.23)$$

Using the decomposition of the resolvent operator Eq. (2.19), we obtain

$$\bar{\rho}_0(p) = \lim_{z\to+io} \frac{z}{z-\psi(z)}\{\rho_0(p;0)+\mathcal{D}(z)\rho_c(p,q;0)\}, \qquad (2.24)$$

and

$$\rho_c(p,q) = \lim_{z\to+io} \{\mathcal{C}(z)\frac{z}{z-\psi(z)}\left[\rho_0(p:0)+\mathcal{D}(z)\rho_c(p,q;0)\right] + \frac{z}{z-QLQ}\rho_c(p,q;0)\}. \qquad (2.25)$$

A characteristic feature of integrable systems is that there is a one-to-one correspondence between the initial value $\rho_0(p;0)$ and its ergodic limit $\bar{\rho}_0(p)$. However, this is not true if

$$\psi(+io)\neq 0. \qquad (2.26)$$

Eq. (2.25) is called the "dissipativity condition." We may show that the resonance condition (2.11) with (2.12) is necessary to ensure the dissipativity condition (Prigogine, I., 1962).

We now introduce a classification of the invariants of the motion. Combining Equation (2.19) with the definition of the invariants Eq. (2.2), we obtain

$$\psi(z)\Phi_0+z\mathcal{D}(z)\Phi_c=0 , \qquad (2.27)$$

and

$$\Phi_c = C(z)\Phi_0 + \frac{z}{z-QLQ}\Phi_c \tag{2.28}$$

where $\Phi_0 \equiv P\Phi$ and $\Phi_c \equiv Q\Phi$. The limit $z \to +io$ corresponds to the asymptotic limit, $t \to \infty$. In this limit we may classify the invariants as follows. If the invariant Φ satisfies

$$\lim_{z \to +io} z\mathcal{D}(z)\ \Phi_c = 0,\ \text{i.e.,}\ \psi(+io)\Phi_0 = 0\ , \tag{2.29}$$

and

$$\lim_{z \to +io} \frac{z}{z-QLQ}\Phi_c = 0,\ \text{i.e.,}\ \Phi_c = C(+io)\Phi_0\ , \tag{2.30}$$

then we call Φ a "regular invariant". If either one or both conditions Equations (2.29) and (2.30) fail to hold, we call it a "singular invariant." We further call Φ_0 a "collisional invariant", if it satisfies Equation (2.29).

The regular invariant is closely related to the uniform invariants in Poincare's theorem. Indeed, we may prove that the Hamiltonian is a regular invariant and, following a discussion similar to that between Equations (2.11) to (2.15), we may show that there is no regular invariant except for the Hamiltonian (and functions of it), if the invariant $\Phi(p,q,\lambda)$ is analytic in its arguments (Prigogine, I., Grecos, A., George, Cl., 1977). The above classification, however, gives in addition to Poincare's theorem, interesting information about the ergodic limit, Equations (2.24) and (2.25). We consider, for example, a simple initial condition, $\rho(p,q{:}0)=\rho_0(p;0)$, i.e., $\rho_c(p,q{:}0)=0$. For this case we see from Equation (2.24) that only the component $\rho_0^{eq}(p)$ in $\rho_0(p{:}0)$ which satisfies the equation $\psi(+io)\rho_0^{eq}(p)=0$, gives a contribution to Equation (2.24), and we get $\bar{\rho}_0(p)=\rho_0^{eq}(p)$. Similarly we can prove from Equation (2.25) that $\bar{\rho}_c(p,q)=C(+io)\times \rho_0^{eq}(p)$ holds. This fact shows that only regular invariants may affect the ergodicity even when singular invariants exist. Since the KAM invariants are highly singular functions (see next section), it is interesting to identify the class to which the KAM invariants belong, to understand their physical significance.

(3) CONSTRUCTION OF THE KAM INVARIANTS

We will construct the KAM invariants to the first order approximation following Kolmogorov's procedure (1954). To ensure the existence of the KAM invariants, we assume that the Hamiltonian Eq. (2.7) is analytic and bounded for $|p_\alpha| \le \bar{p}$ and $|\mathrm{Im} q_\alpha| \le r$, and λ is sufficiently small. We expand the Hamiltonian in a Taylor series around a point p^0 which satisfies the Diophontine inequality

$$|(n,\omega(p^0))| \geq \frac{K}{\|n\|^\mu} , \tag{3.1}$$

for all integer vector n with $\|n\|\equiv\sum_{\alpha=1}^{N}|n_\alpha|\geq 1$ and for a certain choice of constants $K>0$ and $\mu>N$. Then, moving the origin of the action variables to p^0_α (which can be done by a canonical transformation, $p_\alpha - p^0_\alpha = p'_\alpha$ and $q_\alpha = q'_\alpha$), we obtain a new Hamiltonian

$$H=H_0(p^0)+\sum_{\alpha}^{N}\omega_\alpha(p^0)\,p'_\alpha+\lambda\{A_0(p^0,q')+\sum_{\alpha}^{N}B_0^{(\alpha)}(p^0,q')p'_\alpha\}$$

$$+\tfrac{1}{2}\sum_{\alpha\beta}^{N}C_0^{(\alpha\beta)}(p^0,q,\lambda)p'_\alpha p'_\beta + D_0(p',q',p^0,\lambda) , \tag{3.2}$$

with

$$A_0(p^0,q')=V(p^0,q')=\sum_n{}' v_n(p^0)e^{i(n,q')} , \tag{3.3}$$

$$B_0^{(\alpha)}(p^0,q') = V^{(\alpha)}(p^0,q')=\sum_n{}' v_n^{(\alpha)}(p^0)e^{i(n,q')} , \tag{3.4}$$

$$C_0^{(\alpha\beta)}(p^0,q',\lambda)=H^{(\alpha\beta)}(p^0,q',\lambda)=H_0^{(\alpha\beta)}(p^0)+\lambda\sum_n{}' v_n^{(\alpha\beta)}(p^0)e^{i(n,q')} . \tag{3.5}$$

$D_0(p',q',p^0,\lambda)$ stands for terms of third and higher power in p'. Hereafter we drop primes on the variables, for simplicity. To remove the "perturbation" terms λA_0 and $\lambda B_0^{(\alpha)}$, we use Kolmogorov's canonical transformation from (p,q) to (P,Q),

$$Q_\alpha=q_\alpha+\lambda X_\alpha(q) , \tag{3.6}$$

$$p_\alpha=\sum_{\beta}^{N}\{\delta_{\alpha,\beta}+\lambda\frac{\partial X_\beta(q)}{\partial q_\alpha}\}P_\beta+\lambda\eta_\alpha+\lambda\frac{\partial Y(q)}{\partial q_\alpha} , \tag{3.7}$$

where η_α, $X_\alpha(q)$ and $Y(q)$ are unknown constants and functions which will be determined later. Substituting Equation (3.7) into Equation (3.2) and putting

$$A_0(q) + \sum_{\alpha}^{N}\omega_\alpha(p^0)\{\eta_\alpha + \frac{\partial Y(q)}{\partial q_\alpha}\} - \zeta = 0 \tag{3.8}$$

$$B_0^{(\alpha)}(q)+\sum_{\beta}^{N}C_0^{(\alpha\beta)}(q)\{\eta_\alpha+\frac{\partial Y(q)}{\partial q_\alpha}\}+\sum_{\beta}^{N}\omega_\beta(p^0)\frac{\partial X_\alpha(q)}{\partial q_\beta} = 0 \tag{3.9}$$

We obtain a new Hamiltonian $H^{(1)}$ with new perturbations $\lambda^2 A_1$ and $\lambda^2 B_1$,

$$H^{(1)}=H_0(p^0)+\lambda\zeta+\sum_\alpha^N \omega_\alpha(p^0)P_\alpha+\lambda^2\{A_1(p^0,q,\lambda)+\sum_\beta^N B_1^{(\alpha)}(p^0,q,\lambda)P_\alpha\}$$
$$+\tfrac{1}{2}\sum_{\alpha\beta}^N C_1^{(\alpha\beta)}(p^0,q,\lambda)\; P_\alpha P_\beta + D_1(P,q,p^0,\lambda) \tag{3.10}$$

where ζ is an unknown constant and $D_1(P,q,p^0,\lambda)$ stands for terms of third and higher power in P. If we repeat a similar procedure for Equation (3.10), we obtain new perturbations $\lambda^4 A_2$ and $\lambda^4 B_2$. Iterating the operation infinitely, we arrive at a Hamiltonian $H^{(\infty)}$ which has a form similar to Equation (3.10) but without the perturbations A and B. Then, it is easy to see from the canonical equation for $H^{(\infty)}$ that $P_\alpha^{(\infty)}(t)=0$ is a solution, where $P_\alpha^{(\infty)}$ is the new canonical variable which is obtained after an infinite iteration of the canonical transformation. Therefore, we may write the KAM invariant by

$$\Phi_K(p,q)=\prod_{\alpha=1}^N \delta(P_\alpha^{(\infty)}(p,q)). \tag{3.11}$$

To determine the explicit form for p_α in the first order approximation, we should solve the set of equations (3.8) and (3.9). By Fourier expanding X_α and Y,

$$X_\alpha(q) = \sum_n{}' x_\alpha(n)e^{i(n,q)}, \tag{3.12}$$
$$Y(q) = \sum_n{}' y(n)e^{i(n,q)}, \tag{3.13}$$

under the condition Eq. (2.8), we can obtain the solutions of $x_\alpha(n)$, $y(n)$, η_α and ζ (Kolmogorov, A. N., 1954) which are expressed by the Fourier coefficients of the interaction and its derivatives on p (see Equations (3.3)-(3.5)). Combining these solutions with Equation (3.7) and going back to the old variables in the original Hamiltonian Eq. (2.7), we obtain

$$p_\alpha-p_\alpha^0=\sum_\beta^N \{\delta_{\alpha\beta}-\lambda\sum_n{}' \frac{n_\alpha}{(n,\omega(p^0))}\left[v_n^{(\beta)}(p^0)-\frac{(n,\omega^{(\beta)}(p^0))}{(n,\omega(p^0))}\right]e^{i(n,q)}\}P_\beta$$
$$-\lambda\sum_n{}' \frac{n_\alpha}{(n,\omega(p^0))}v_n(p^0)\;e^{i(n,q)} + O(\lambda^2). \tag{3.14}$$

After inverting Equation (3.14) on P_β, we substitute it into Equation (3.11). Let us now consider a λ expansion adopted by Poincare. (see Equation (2.13)). The KAM invariant in the first order approximation is then

$$\Phi_K=\Phi^0+\lambda\Phi_K^{(1)}$$
$$=\prod_\alpha^N \delta(p_\alpha-p_\alpha^0)+\lambda\sum_n{}' e^{i(n,q)}\frac{1}{(n,\omega(p^0))}\{v_n(p^0)+$$

$$+\sum_{\beta}^{N}\left[v_n^{(\beta)}(p^0) - \frac{(n,\omega^{(\beta)}(p^0))}{(n,\omega(p^0))} v_n(p^0)\right](p_\beta - p_\beta^0)\}$$

$$\times\sum_{\alpha}^{N} n_\alpha \frac{\partial}{\partial p_\alpha} \prod_{\gamma}^{N} \delta(p_\gamma - p_\gamma^0) \ . \tag{3.15}$$

This construction could be extended to arbitrary order in λ. It is easy to prove that Equation (2.14) holds in the distribution sense. Therefore, Φ_K is an invariant in the first order approximation on λ. This result also shows why Poincare's theorem cannot be applied to the KAM invariants. Indeed, it is easy to check from the direct calculations that Equation (2.15) does not hold in the distribution sense, even though Equation (2.14) holds.

(4) KAM INVARIANTS ARE REGULAR INVARIANTS

We now prove that the KAM invariants are regular invariants in the first order approximation. It is convenient to introduce at this point certain "matrix" elements

$$<m|L|n> = (\frac{1}{2\pi})^N \int_0^{2\pi} dq_1 \ldots dq_N e^{-i(m,q)} L e^{+i(n,q)} \ . \tag{4.1}$$

By the definition of L, Equation (2.9) , we have

$$<m|L_0|n> = (n,\omega(p))\delta_{m,n} \ , \tag{4.2}$$

and

$$<m|L_1|n> = -v_{m-n}(p) \sum_{\alpha}^{N} (m_\alpha - n_\alpha) \frac{\partial}{\partial p_\alpha} + \sum_{\alpha}^{N} n_\alpha v_{m-n}^{(\alpha)}(p) \ . \tag{4.3}$$

It is easy to see from the definition of the collision operator, Equation (2.20), that the first non-trivial term in λ is

$$\psi^{(2)}(z) = \lambda^2 <0|L_1 \frac{1}{z-L_0} L_1|0>. \tag{4.4}$$

Using Equations (4.2) and (4.3), we obtain

$$\psi^{(2)}(z) = -\lambda^2 \sum_n{}' \sum_{\alpha}^{N} n_\alpha \frac{\partial}{\partial p_\alpha} \frac{|v_n(p)|^2}{z-(n,\omega(p))} \sum_{\beta}^{N} n_\beta \frac{\partial}{\partial p_\beta} \ . \tag{4.5}$$

Similarly, from Equations (2.21) and (2.22), we obtain for $n \neq 0$

$$\mathcal{D}_{0,n}^{(1)}(z) = +\lambda \sum_{\alpha}^{N} n_\alpha \frac{\partial}{\partial p_\alpha} \frac{v_{-n}(p)}{z-(n,\omega(p))} \ , \tag{4.6}$$

and

$$\mathcal{C}_{n,0}^{(1)}(z) = -\lambda \frac{v_n(p)}{z-(n,\omega(p))} \sum_{\alpha}^{N} n_\alpha \frac{\partial}{\partial p_\alpha} \ . \tag{4.7}$$

We first prove that the P-component of Φ_K^0, i.e., $P\Phi_K^0=\prod_{\alpha=1}^N \delta(p_\alpha-p_\alpha^0)$, is a collisional invariant. Since $P\Phi_K^0$ is distribution, we must consider the produce of $\psi^{(2)}(z)P\Phi_K^0$ with a test function $g(p)$. Using Equation (4.5), we obtain

$$I=\langle g,\psi^{(2)}(+i\varepsilon)P\Phi_K^0\rangle = \int dp\, g(p)\psi^{(2)}(+i\varepsilon)P\Phi_K^0(p)$$

$$=2\lambda^2 i\sum_{n>0}^{\infty}\frac{\varepsilon}{\varepsilon^2+(n,\omega(p^0))^2}\sum_\beta^N n_\beta\left\{\frac{\partial}{\partial p_\beta}\sum_\alpha^N n_\alpha g^{(\alpha)}(p)|v_n(p)|^2\right\}_{p=p^0}$$

$$-2\lambda^2 i\sum_{n>0}^{\infty}\frac{\varepsilon\,(n,\omega(p^0))}{\left[\varepsilon^2+(n,\omega(p^0))^2\right]^2}\sum_\beta^N n_\beta(n,\omega^{(\beta)}(p^0))\sum_\alpha^N n_\alpha g^{(\alpha)}(p^0)|v_n(p^0)|^2, \tag{4.8}$$

where ε is a small positive number. The summation n must be taken over all non-negative integers $n_\gamma \geq 0$ for $\gamma = 1,\ldots,N$ with the condition $\|n\|\geq 1$. Since the Hamiltonian is assumed to be bounded for $|p_\alpha|\leq p$ and $|\mathrm{Im}q_\alpha|\leq r$, its derivatives on p are also bounded because of Cauchy's inequality (see Cartan, H., 1963). To estimate Equation (4.8), we introduce the notation

$$\|f_k\| = \sup_{|p_\alpha|\leq p}\ \sup_{|\mathrm{Im}q_\alpha|\leq r}\left|\frac{\partial^k f(p,q)}{\partial p_{\alpha_1}\cdots\partial p_{\alpha_k}}\right|, \quad (k\geq 0), \tag{4.9}$$

where k=0 in the right-hand side means $f(p,q)$ itself. Because of the boundedness of the interaction for $|\mathrm{Im}q_\alpha|\leq r$, we can prove that (Arnold, V. I., 1963),

$$|\lambda v_n(p)|\leq\|\lambda V_0\|e^{-r\|n\|}. \tag{4.10}$$

Using the Diophantine inequality (3.1) and the above notations we estimate the first term I_1 in Equation (4.8) by

$$|I_1|\leq 2\varepsilon M\sum_{n>0}^{\infty}\frac{\|n\|^2}{\varepsilon^2+\dfrac{K^2}{\|n\|^{2\mu}}}e^{-2r\|n\|}\leq 2\varepsilon\frac{M}{K^2}\sum_{n>0}^{\infty}\|n\|^{2(\mu+1)}e^{-2r\|n\|}, \tag{4.11}$$

where $M\equiv\|\lambda V_0\|(\|g_2\|+2\|g_1\|\cdot\|\lambda V_1\|)$. The right-hand side of Equation (4.11) goes to zero when $\varepsilon\to 0$ for a system with finite degrees of freedom N.

Similarly, the second term I_2 in Equation (4.8) is estimated by

$$|I_2| \leq 2\varepsilon \frac{M'}{K^4} \sum_{n>0}^{\infty} \|n\|^{4(\mu+1)} e^{-2r\|n\|} \xrightarrow[\varepsilon\to 0]{} 0 \ , \qquad (4.12)$$

where $M' \equiv \|\omega\| \cdot \|\omega_1\| \cdot \|g_1\| \cdot \|\lambda V_0\|^2$. Therefore, we have $|I| \leq |I_1| + |I_2| \to 0$ for $\varepsilon \to 0$, which implies that the P-components of the KAM invariants are collisional invariants in the first order approximation in λ.

Another condition for the regular invariant, i.e., Equation (2.30) is proved quite similarly with a test function $G(p,q)$ as

$$|\langle G, \frac{\varepsilon}{i\varepsilon - L_0} Q\Phi_k^{(1)} \rangle| \xrightarrow[\varepsilon\to 0]{} 0 \ . \qquad (4.13)$$

Therefore, the KAM invariants are regular invariants.

(5) CONCLUDING REMARKS

We have constructed the KAM invariants to first order in λ. They are expressed by a singular function, that is, the δ-function having as an argument the parameter p^0 which is distributed in a highly complicated way in phase space as a result of the restriction of the Diophontine inequality (3.1). In addition, we have shown that the KAM invariants satisfy the stationary solution of the Liouville equation in the distribution sense. This is the very reason why Poincare's theorem cannot be applied to the KAM invariants. We have then proved that the KAM invariants are classified as regular invariants in the first order approximation. It should be noted that the convergence of Equation (4.8) to zero is ensured by the fact that the factor $\lambda v_n(p^0)/(n,\omega(p^0))$ and its derivatives converge to zero. The proof may be extended to higher order approximations in λ, because in the higher order terms additional potentials λv_n always appear in combination with (n,ω) as above. We may, therefore, conclude that if we start with a singular distribution function which is a function of the KAM invariants, we then obtain, in the ergodic limit, not only a function of the Hamiltonian, but also a function of the KAM invariants. In other words, the KAM invariants prevent the system from reaching the microcanonical distribution. However, a question still remains with regard to the mean ergodic theorem, which asserts that if we start with an initial distribution function which belongs to the L^2 space, then in the ergodic limit, the function stays also in the L^2 space (Michael, R. & Simon, B., 1972). Concerning the problem of determining suitable classes of initial distribution functions which result in ergodic behavior, it might be interesting to ask about the possibility of whether we may construct a L^2 function as a combination of the KAM invariants by introducing a suitable measure. This question is still open. But we believe that the kinetic description explained here which deals with the time evolution of the distribution functions rather than with each trajectory might give us a new way of predicting the long time behavior of non-integrable systems.

ACKNOWLEDGEMENT

The authors would like to thank Professors Cl. George, A. Grecos, G. Nicolis, and L. Reichl for interesting comments and discussions. One of the authors (T.P.) also wishes to thank Professor V. Szebehely for his kind hospitality at the NATO Advanced Study Institute at Cortina d'Ampezzo, during the summer of 1981. This work has been supported by the U.S. Air Force (Texas A&M Research Foundation Contract F33617-78-D0629) and partly by the Robert A. Welch Foundation.

REFERENCES

Arnold, V. I.: 1963, Usp. Mat. Nauk 18 (Eng. Trans. Russ. Math. Surv. 18: 1963, p.9; 18: 1963, p. 85).

Balescu, R.: 1975, "Equilibrium and Non-Equilibrium Statistical Mechanics" (John Wiley, New York).

Balescu, R., Clavin, P., Mandel, P., and Turner, J.: 1970, Bull. Cl. Sci. Acad., Belg. 55, p. 1055.

Barrar, R.: 1970, Celestial Mech. 2, p. 494.

Brilluoin, L.: 1966, Arch. Rat. Mech. Anal. 5, p. 76.

Cartan, H.: 1963, "Elementary Theory of Analytic Functions of One or Several Complex Variables" (Addison-Wesley).

Fermi, E.: 1923, Zeit. Physik 24, p. 261.

Grecos, A. P.: 1971, Physica 51, p. 50.

Kolmogorov, A. N.: 1954, Dokl. Akad. Nauk 98, p. 527 (Eng. trans. Los Alamos Scientific Laboratory translation LA-TR-71-67).

Michael, R. and Simon, B.: 1972, "Methods of Modern Mathematical Physics" (Academic Press, New York).

Moser, J.: 1962, Nach. Akad. Wiss. Gottinger II Math. Phys. Kl. 1, p. 1.

Prigogine, I.: 1962, "Non-Equilibrium Statistical Mechanics" (Interscience, New York).

Prigogine, I. and Grecos, A.P.: 1973, in "Cooperative Phenomena, by H. Haken and M. Wagner (eds.) (Springer-Verlag, Berlin).

Prigogine, I., Grecos, A. and George, Cl.: 1977, Celestial Mech. 16, p. 489.

Resibois, P. and Prigogine, I.: 1960, Bull. Cl. Sci. Acad. Belg. 46, p. 53.

Stey, G.: 1978, "Proceedings of NATO Advanced Study Institute", Instabilities in Dynamical Systems, V.G. Szebehely (ed), D. Reidel Publ. Co., Dordrecht, Holland/Boston, U.S.A., p. 103.

Ter Harr, D.: 1955, Rev. Mod. Physics 27, p. 289.

Whittaker, E.: 1964, "A Treatise on Analytical Dynamics of Particles and Rigid Bodies" (Cambridge Univ. Press, London).

REGULARIZATION OF THE SINGULARITIES OF THE N-BODY PROBLEM

C. Marchal

D.E.S. - ONERA
92320 Chatillon, France

ABSTRACT. The n-body problem has two types of singularities: the collisions of two or several bodies and "the infinite expansions in a bounded interval of time."

The regularization of singularities answer to the question "How to extend naturally a solution after a singularity" has a theoretical interest and it is also useful for numerical integrations.

There are two types of regularization: the analytical regularization (Siegel's regularization) and the topological regularization or regularization by continuity (Easton's regularization); here we consider especially the latter type.

If we consider either plane motions or three-dimensional motions, the collisions of two infinitesimal masses are not Easton-regularizable and the only Easton-regularizable singularities are the other binary collisions. An adjacent result is the existence of new, non-regularizable, types of triple or multiple collisions (implying infinitesimal masses).

For rectilinear motions the binary collisions are not Easton-regularizable (because all neighboring motions have such a collision) but they are Siegel-regularizable and thus the problem can be handled with the usual picture of rectilinear motions and their invariable succession of bodies.

As for two or three-dimensional motions the "infinite expansion in a bounded interval of time" are not regularizable but some multiple collisions are regularizable (with at most one

V. Szebehely (ed.), Applications of Modern Dynamics to Celestial Mechanics and Astrodynamics, 201–236.

non-infinitesimal body). A complete list of these regularizable collisions is given.

1. INTRODUCTION.

Since Newton's discoveries, the problem of the motion of n masses moving under the influence of their mutual attraction has been known as the n-body problem.

That problem has two types of singularities: the collisions and the infinite expansion in a bounded interval of time; we will deal with the following question, "Are these singularities regularizable?".

There are two types of regularization: the analytical regularization (or Siegel-regularization) and the topological regularization or regularization by continuity (also called Easton regularization [1]); we will essentially use the topological regularization.

The interest of the regularization is not only theoretical. Indeed, the singularities and their vicinity are a source of large errors and strong instabilities in the numerical integrations; these difficulties are removed when the singularities are eliminated.

2. THE REGULARIZATION.

MacGehee [2] outlined the differences between the two types of regularization.

In the analytical regularization a function such as $x = t^{1/3}$ is extended into the past (i.e., for $t < 0$) by $x = -[-t]^{1/3}$ while $y = t^{2/3}$ is extended into $y = [-t]^{2/3}$ because they both correspond to the general equations $x^3 = t$ and $y^3 = t^2$.

By various methods Siegel extended his definition of regularization, but he could not extend $x = t^{p/q}$ to the past when p is odd and q is even.

Easton uses an independent method. If the differential system $d\vec{x}/dt = \vec{f}(\vec{x},t)$ has a singularity at the point $(\vec{x}_s, t_s)$ the solution leading to $\vec{x}_s, t_s$ is called singular. However, generally most of the neighboring solutions are regular and if they remain together after t_s , the extension of the singular solution can be defined by continuity.

If the regular neighboring solutions diverge, the singularity is not Easton-regularizable.

The examples given by McGehee [2] show that the two types of regularization are independent, however, they generally lead to the same extension when a singularity is regularizable according to both meanings.

We will especially consider the Easton-regularization which is more natural and independent of analytical conventions.

3. THE N-BODY PROBLEM.

3.1 Notations

We will use the usual notations: $m_1, m_2, \ldots, m_n$ are the n mass points of interest, G is the constant of gravitation, and $\vec{r}_j$, $\vec{V}_j$, $j = \{1, \ldots, n\}$ are the radius-vectors and the velocity vectors (referred to in the axes of the center of mass).

$$\sum_{j=1}^{n} m_j \vec{r}_j \times \vec{V}_j = \vec{c} = \text{integral of angular momentum}, \tag{1}$$

$$\frac{1}{2} \sum_{j=1}^{n} m_j V_j^2 - G \sum_{1 \le j < k \le n} \frac{m_j m_k}{r_{jk}} = h = \text{integral of energy}, \tag{2}$$

$$\vec{r}_{jk} = \vec{r}_k - \vec{r}_j \ , \quad r_{jk} = ||\vec{r}_{jk}|| \ , \quad |\vec{V}_j| = V_j \ . \tag{3}$$

The equations of motion are:

$$\vec{r}_j'' = d^2\vec{r}_j/dt^2 = \sum_{k=1;k\neq j}^{n} \frac{Gm_k \vec{r}_{jk}}{r_{jk}^3} \ . \tag{4}$$

Finally, we will put

$$r = \inf_{1 \le j < k \le n} r_{jk} \quad \text{and} \quad R = \sup_{1 \le j < k \le n} r_{jk} \ . \tag{5}$$

3.2 The Singularities of the N-Body Problem.

These singularities are well-known when all masses of interest are finite [3-5]. They are of two types: the collisions and the infinite expansions in a bounded interval of time.

3.2.1 The Collisions. If there is a collision at the time $t = t_c$ all bodies go to a definite position when $t \to t_c$:

$$t \to t_c \Longrightarrow \vec{r}_j \to \vec{r}_{j,c} \quad ; \quad j = \{1, \ldots, n\} \tag{6}$$

Since it is a collision, at least two of these $\vec{r}_{j,c}$ vectors are equal and the corresponding velocities $\vec{V}_j$ go to infinity at most as $(t_c - t)^{-1/3}$ while the corresponding acceleration $\vec{r}''_j$ goes to infinity at most as $(t_c - t)^{-4/3}$ (hence, $\vec{r}_j - \vec{r}_{j,c} = 0\ (t_c - t)^{2/3}$) .

Assume that $m_1, m_2, \ldots, m_L$ enter into collision, then:

$$\left.\begin{array}{l} \left[\sum\limits_{1\le j<k\le L} m_j m_k r_{jk}^2\right] (t_c - t)^{-4/3} \\ \text{and} \\ \left[\sum\limits_{1\le j<k\le L} m_j m_k r_{jk}^{-1}\right] (t_c - t)^{2/3} \end{array}\right\} \tag{7}$$

have a positive and bounded limit when $t \to t_c$ and the configuration of the L masses goes to a "central configuration" (in which the accelerations $\vec{r}''_j$ are proportional to the corresponding radius-vectors $(\vec{r}_j - \vec{r}_{j,c})$).

3.2.2 The Infinite Expansions in a Bounded Interval of Time. A motion of this type has been given by MacGehee [6]. It requires at least four bodies.

An infinite expansion at the singular instant t_s is characterized by:

$$t \to t_s \quad \text{implies} \quad R \to \infty \ . \tag{8}$$

It also requires:

$$r \to 0 \quad ; \quad r = 0(t_s - t)^{2/3} \quad ; \quad R^2 r \to 0 \ , \tag{9}$$

and the system has strong oscillations. Indeed there exist at least two mutual distances r_{ij} such that

$$\lim_{t \to t_s} \inf r_{ij} = 0 \quad ; \quad \lim_{t \to t_s} \sup r_{ij} = \infty \tag{10}$$

and

$$\left.\begin{aligned} &\int^{t_s} (t_s - t)dt/r_{ij}^2(t) = + \infty , \\ &\lim \sup [r_{ij}/R] \geq \frac{1}{n-3} \end{aligned}\right\} \tag{11}$$

3.2.3 Singularitities of Systems having Infinitesimal Masses. We extend the solution after the collision of two infinitesimal bodies. This problem, the regularization of collisions of infinitesimal bodies, can be answered as follows:

A) Either nothing particular happens and the two masses go on. (This possibility is given by the Siegel-regularization and also by the Easton-regularization). This solution reduces the n-body problem to sub-problems with at most one infinitesimal mass.

B) Or the deviation angle at collision is arbitrary (This possibility is obtained by continuity with the case of small masses). In this second case the collisions of two infinitesimal masses are not regular and the solution cannot be defined uniquely after such a collision. Before the collision we have again the reduction of the n-body problem into sub-problems with at most one infinitesimal mass.

C) Finally, in the particular case of rectilinear motions, it is usual to consider "elastic collisions" of infinitesimal masses. The motion is then uniquely defined after the collision of two infinitesimal masses.

Thus we face two kinds of problems:

P_1) The n-body problem with one infinitesimal mass and (n-1) finite masses.

P_2) The rectilinear n-body problem with elastic collision of infinitesimal masses.

Note that the motion of the finite masses is independent of the motion of the infinitesimal masses. The problem is thus decomposed into two sub-problems. The singularities of these motions are as follows:

S_1) Infinite expansions in a bounded interval of time. These expansions generalize those of 3.2.2 and they require an infinite expansion of the sub-system of finite masses. Indeed, if that sub-system remains bounded until t_s, all $\vec{r}_j$ have a definite limit $\vec{r}_{js}$ when $t \to t_s$ and:

$$\vec{r}_j - \vec{r}_{js} = O(t_s - t)^{2/3} \quad . \tag{12}$$

S_2) The generalization of the collisions of 3.2.1 If m_L is infinitesimal the following can be added to condition (7):

$$\left.\begin{array}{l} [\sum_{j=1}^{L-1} m_j r_{jL}^2] (t_c - t)^{-4/3} \\ \text{and} \\ [\sum_{j=1}^{L-1} m_j r_{jL}^{-1}](t_c - t)^{2/3} \end{array}\right\} \tag{13}$$

have a positive and bounded limit when $t \to t_c$ and the configuration of the colliding masses goes to a "central configuration."

S_3) The collisions of several infinitesimal masses in rectilinear motion.

S_4) A completely different type of collision is shown on Figures 1 and 2.

Let us consider three equal masses A, B and C (Figure 1) falling toward their center of mass with a triangular Lagrangian motion of zero energy:

$$\left.\begin{array}{ll} \vec{r}_A = \vec{A}(t_c - t)^{2/3} & ||\vec{A}|| = ||\vec{B}|| = ||\vec{C}|| = [\frac{3\sqrt{3}}{2} Gm_A]^{1/3} \\ \vec{r}_B = \vec{B}(t_c - t)^{2/3} & (\vec{A},\vec{B}) = (\vec{A},\vec{C}) = (\vec{B},\vec{C}) = 120° \\ \vec{r}_C = \vec{C}(t_c - t)^{2/3} & m_A = m_B = m_C \end{array}\right\} \tag{14}$$

It is possible that an infinitesimal mass D has the motion shown in Figures 1 and 2 with an infinite number of loops around

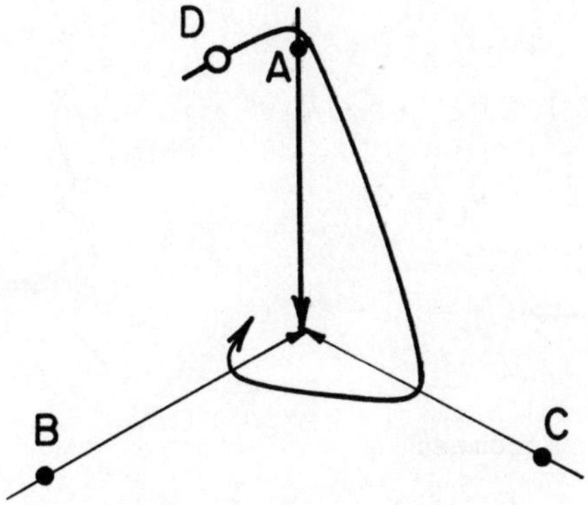

Fig. 1. The S_4 Type of Collision.

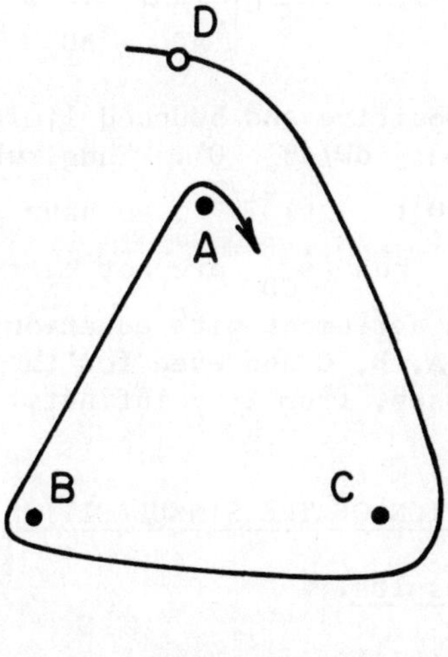

Fig. 2. The Collision of Figure 1 in Expanding Coordinates $\vec{s}_j$ of (15).

A, B, C during the fall. The existence of such motions is easy to demonstrate, but they are, of course, unstable and extremely sensitive to the initial conditions. The motion of D can easily be studied through the following transformation using the coordinates:

$$\left.\begin{array}{l} \vec{s}_j = \vec{r}_j (t_c - t)^{-2/3} \quad ; \quad j = A, B, C, D \\ \\ (\text{Hence } \vec{s}_A = \vec{A} \ ; \ \vec{s}_B = \vec{B} \ ; \ \vec{s}_C = \vec{C}) \\ \\ \theta = \operatorname{Log}(\frac{1}{t_c - t}) \quad ; \quad t \to t_{c-} \Longrightarrow \theta \to \infty \end{array}\right\} \tag{15}$$

The equation of motion of D becomes:

$$\frac{d^2\vec{s}_D}{d\theta^2} - \frac{2}{9}\vec{s}_D + \frac{1}{3}\frac{d\vec{s}_D}{d\theta} + Gm_A\left[\frac{\vec{s}_{DA}}{s_{DA}^3} + \frac{\vec{s}_{DD}}{s_{DB}^3} + \frac{\vec{s}_{DC}}{s_{DC}^3}\right] \tag{16}$$

The quantity W is given by

$$W = \left\{\frac{1}{2}\left[\frac{d\vec{s}_D}{d\theta}\right]^2 - \frac{s_D^2}{9} - G\left[\frac{m_A}{s_{AD}} + \frac{m_B}{s_{BD}} + \frac{m_C}{s_{CD}}\right]\right\} \cdot \exp\left\{-\frac{2\theta}{3}\right\}. \tag{17}$$

It has a positive and bounded limit when $\theta \to \infty$ and $t \to t_{c-}$ (Note that $dW/d\theta > 0$). Thus, while $\vec{V}_A, \vec{V}_B, \vec{V}_C = O(t_c - t)^{-1/3}$, we have $\vec{V}_D = O(t_c - t)^{-2/3}$ when s_{AD}, s_{BD} and s_{CD} are not extremely small. That kind of collision is in agreement with equation (12). It also exists for unequal masses A, B, C and even for the collision of any number of positive masses, from 2 to infinity.

4. REGULARIZATION OF THE SINGULARITIES OF THE N-BODY PROBLEM.

4.1 Present Results.

The binary collisions have been regularized by Sundman [7]. They are both Siegel and Easton-regularizable except for rectilinear motions for which the Easton-regularization is impossible, since all neighboring motions have also binary collisions. The particular case of two infinitesimal masses has already been discussed in Section 3.2.3.

The infinite expansion in a bounded interval of time is an

essential singularity never regularizable (neither for Siegel nor for Easton). It is also the case for the final type of collision (S_4).

It remains to study the multiple collisions of the two similar cases 3.2.1 and 3.2.3 (S_2), and of the case of infinitesimal masses in rectilinear motion (S_3).

4.2 Two and Three-Dimensional Motions; Attempt to Regularize Triple Collisions.

As already noted in 3.2.3, the problem of interest has at most one infinitesimal body and the three colliding masses approach a central configuration as approaching collision.

For 3 bodies there are two types of central configurations: the collinear (Fig. 3) and the triangular (Fig. 4) central configurations. The triple collisions have been studied extensively [2-5,8-15]. The impossibility of the Easton-regularization is shown in the following.

4.2.1 Case of Collinear Central Configurations. The simplest Eulerian motion is the rectilinear motion of zero-energy:

$$\left.\begin{aligned} \vec{r}_A &= \vec{A}(t_c - t)^{2/3} \\ \vec{r}_B &= \vec{B}(t_c - t)^{2/3} \\ \vec{r}_C &= \vec{C}(t_c - t)^{2/3} \quad , \end{aligned}\right\} \tag{18}$$

where t_c is the collision time, and the 3 constant and collinear vectors $\vec{A}$, $\vec{B}$, $\vec{C}$ are related to the usual equations of motion (4):

$$-\frac{2}{9}\vec{A} = Gm_B \frac{(\vec{B} - \vec{A})}{\|(\vec{B} - \vec{A})\|^3} + Gm_C \frac{(\vec{C} - \vec{A})}{\|(\vec{C} - \vec{A})\|^3} \tag{19}$$

and two other similar equations (implying $m_A\vec{A} + m_B\vec{B} + m_C\vec{C} = 0$) .

There are also elliptic, parabolic and hyperbolic Eulerian 3-body motions (Fig. 5). The three-masses have homothetic Keplerian motions and remain on a rotating straight line. If the mass B is between A and C , the equation of equilibrium is:

$$m_B\, r_{AC}^2(r_{BC}^3 - r_{AB}^3) = m_A\, r_{BC}^2(r_{AB}^3 - r_{AC}^3) + m_C\, r_{AB}^2(r_{AC}^3 - r_{BC}^3) \; . \tag{20}$$

This relation is in agreement with (19) and since $r_{AC} = r_{AB} + r_{BC}$ it can also be written as

$$\left.\begin{array}{l}\dfrac{r_{BC}}{r_{AB}} = x \quad ; \quad \dfrac{r_{AC}}{r_{AB}} = 1 + x \quad ; \\ (m_A + m_B)x^5 + (3m_A + 2m_B)x^4 + (3m_A + m_B)x^3 - (m_B + 3m_C)x^2 \\ - (2m_B + 3m_C)x - (m_B + m_C) = 0 \end{array}\right\} \tag{21}$$

Note that in the collinear central configurations the position of the center of mass can be obtained in terms of the positions of the three masses. Indeed, the quintic equation of (21) is equivalent to

$$r_A/r_{AB} = (x^5 + 3x^4 + 3x^3)/(x^4 + 2x^3 + x^2 + 2x + 1) \tag{22}$$

Let us consider now the 3-body motions in the vicinity of these Eulerian motions (with the same center of masses). The radius vectors are $\vec{r}_A + \vec{\delta}_A$; $\vec{r}_B + \vec{\delta}_B$; $\vec{r}_C + \vec{\delta}_C$, where $\vec{r}_A$, $\vec{r}_B$, $\vec{r}_C$ represent the Eulerian motion and $\vec{\delta}_A$, $\vec{\delta}_B$, $\vec{\delta}_C$ are the infinitesimal deviations.

Let us use the radial, circumferential and out-of-plane axes of Figure 6 with origins at A , B , C and investigate the neighboring motions that verify to the first order:

$$\left.\begin{array}{l}\dfrac{x_A}{m_B m_C r_{BC}} = - \dfrac{x_B}{m_A m_C r_{AC}} = \dfrac{x_C}{m_A m_B r_{AB}} = X \\[2ex] \dfrac{y_A}{m_B m_C r_{BC}} = - \dfrac{y_B}{m_A m_C r_{AC}} = \dfrac{y_C}{m_A m_B r_{AB}} = Y \\[2ex] \dfrac{z_A}{m_B m_C r_{BC}} = - \dfrac{z_B}{m_A m_C r_{AC}} = \dfrac{z_C}{m_A m_B r_{AB}} = Z \end{array}\right\} \tag{23}$$

The signs assume that B is between A and C . Note that $m_A\vec{\delta}_A + m_B\vec{\delta}_B + m_C\vec{\delta}_C = 0$.

Motions satisfying (23) are three-body motions (to the first order) provided that [16]:

$$\left.\begin{array}{l}(1 + e \cos v)[d^2X/dv^2 - 2dY/dv] = (2K + 3)X \\ (1 + e \cos v)[d^2Y/dv^2 + 2dX/dv] = -KY \\ (1 + e \cos v)[d^2Z/dv^2 + Z] = -KZ \quad , \end{array}\right\} \tag{24}$$

where

Fig. 3. A Collinear (or Euler) Central Configuration.

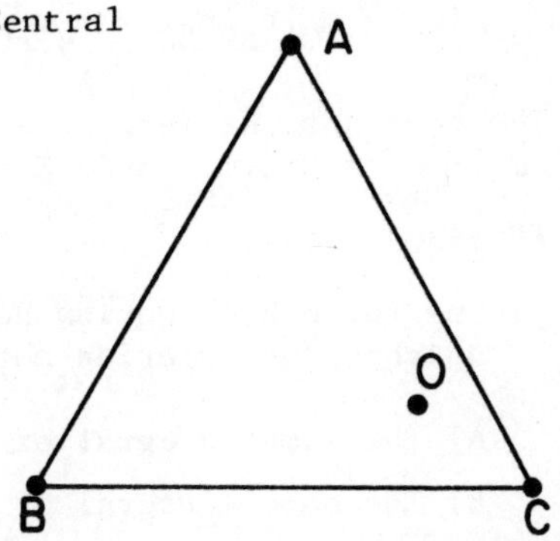

Fig. 4. A Triangular (or Lagrange) Central Configuration. The Triangle A, B, C is Equilateral.

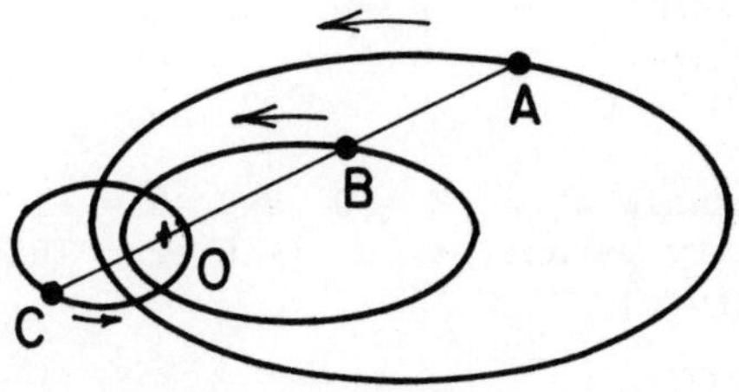

Fig. 5. An Elliptic Eulerian Motion.

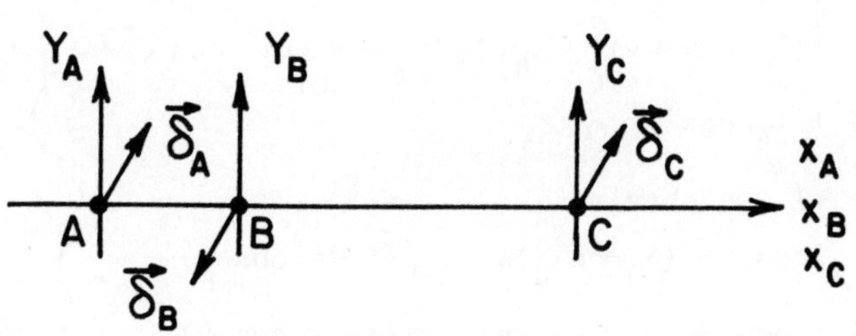

Fig. 6. Motion in the Vicinity of an Eulerian Motion.

$$\left.\begin{array}{l} \left.\begin{array}{l} v = \text{true anomaly} \\ e = \text{eccentricity} \end{array}\right\} \text{of the Eulerian motion of interest} \\ K = \dfrac{m_C r_{AB}(r_{AC}^3 - r_{BC}^3) + m_A r_{BC}(r_{AC}^3 - r_{AB}^3)}{(m_A + m_C) r_{AB}^2 r_{BC}^2 + m_B r_{AC}^2 (r_{AB}^2 + r_{BC}^2)} \; . \end{array}\right\} \quad (25)$$

The constant K is in the range $0 < K \leq 7$ since $K = 0$ when $m_A = m_C = 0 < m_B$ and $K = 7$ when $m_A = m_C > m_B = 0$. Note that:

I) The three-body motion described by (23) and (24) and the neighboring Eulerian motion have, to the first order:

A) The same integral of energy,

B) The same integral of angular momentum,

C) The value of the moment of inertia, $\Sigma m_j r_j^2$ is the same at all times.

They also satisfy the equation:

$$m_A \vec{r}_A \times \vec{\delta}_A + m_B \vec{r}_B \times \vec{\delta}_B + m_C \vec{r}_C \times \vec{\delta}_C = 0 \; .$$

II) The system (24) is integrable when $K = 0$ and $K = -1$. When $K = 0$, the auxiliary parameter u is used. This is defined by the quadrature:

$$\frac{du}{dv} = \frac{1}{(1 + e \cos v)^2} \; . \quad (26)$$

Six arbitrary constants λ_1 to λ_6 are also introduced, giving

$$\left.\begin{array}{l} X = [(\lambda_1 + 3u\lambda_2) e \sin v - \lambda_3 \cos v](1 + e \cos v) - 2\lambda_2 \\ Y = (\lambda_1 + 3u\lambda_2)(1 + e \cos v)^2 + \lambda_3 (2 \sin v + e \sin v \cos v) + \lambda_4 \\ Z = \lambda_5 \sin v + \lambda_6 \cos v \end{array}\right\} \quad (27)$$

When $K = -1$ we obtain:

$$\left.\begin{array}{l} X = [(\lambda_1 + u\lambda_2)\cos v + (\lambda_3 + u\lambda_4)\sin v](1 + e \cos v) \\ Y = [(\lambda_3 + u\lambda_4)\cos v - (\lambda_1 + u\lambda_2)\sin v](1 + e \cos v) \\ Z = (\lambda_5 + u\lambda_6)(1 + e \cos v) \end{array}\right\} \quad (28)$$

III) Another case of easy integrability is the case $e = 0$.

IV) We see that:

A) For a given 3-body motion in the vicinity of Eulerian motions the above note I shows that the corresponding proper Eulerian motion is unique and note II shows that these neighboring motions have, for 3 given masses, twelve arbitrary parameters. There are 6 from the Eulerian motions and 6 from the integration of Equation (24). Since, for 3 given masses, the general 3-body problem has also twelve arbitrary parameters, we find that <u>the neighboring motions given by (23) - (25) are the general neighboring motions of the Eulerian solutions</u>.

Let us return now to the problem mentioned at the beginning of 4.2. We will demonstrate the impossibility of the Easton-regularization of triple collisions of the collinear type (Figure 7) and we will examine the neighboring trajectories by two successive steps.

I) When approaching collision the three masses approach a collinear central configuration and also a rectilinear motion of zero energy as given in (18).

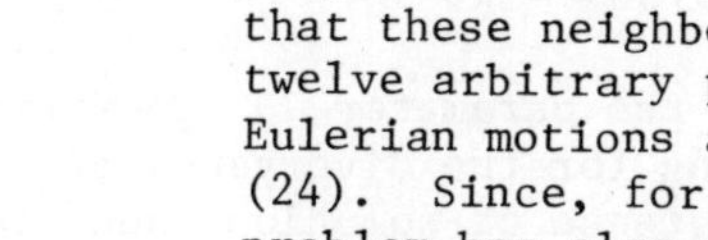

Fig. 7. A Triple Collision of Collinear Type.

II) There are parabolic Eulerian motions (Figure 8) in any vicinity of the rectilinear Eulerian motion (18). Hence it is sufficient to study the vicinity of these parabolic solutions and to prove that when the eccentricity $e = 1$ the solutions of (24) diverge.

The out-of-plane component Z cannot give divergence. Consider the non-negative function F

$$F = KZ^2 + (1 + e\cos v)[Z^2 + (dZ/dv)^2] \qquad (29)$$

and its derivative,

$$dF/dv = -e\sin v[Z^2 + (dZ/dv)^2] \qquad (30)$$

In the parabolic case $e = +1$; v goes from $-\pi$ to $+\pi$, F is maximum when $v = 0$ and thus Z remains bounded (in fact, it goes to 0 when $v \to \pm\pi$) . Thus, the divergence can only result from X and Y . For

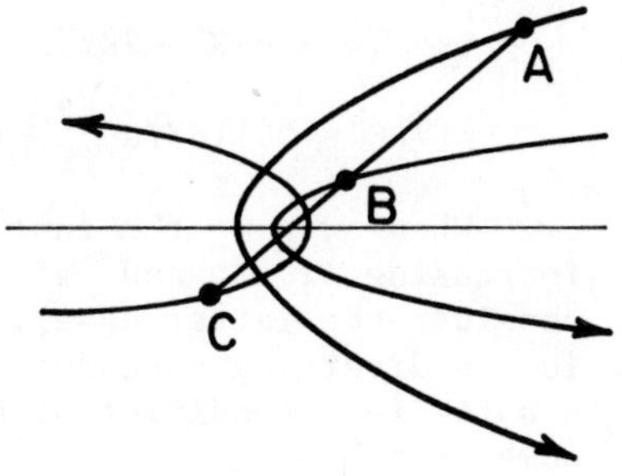

Fig. 8. A Parabolic Eulerian Motion.

$$e = 1 \quad \text{and} \quad \tan\frac{v}{2} = \theta \tag{31}$$

the two first equations of (24) become

$$\left.\begin{aligned} &(1+\theta^2)[d^2X/d\theta^2] + 2\theta[dX/d\theta] - 4dY/d\theta = (4K+6)X \\ &(1+\theta^2)[d^2Y/d\theta^2] + 2\theta[dY/d\theta] + 4dX/d\theta = -2KY\ . \end{aligned}\right\} \tag{32}$$

When v goes from $-\pi$ to $+\pi$ the parameter θ goes from $-\infty$ to $+\infty$, and since we are looking for the divergence of solutions, we want to verify the existence of solutions such that:

$$\left.\begin{aligned} &\theta \to -\infty \Rightarrow X \text{ and } Y \to 0 \\ &\theta \to +\infty \Rightarrow X \text{ and/or } Y \to \infty\ . \end{aligned}\right\} \tag{33}$$

The study of asymptotic properties when $\theta \to \pm\infty$ leads to two possibilities. Either X is the leading parameter:

$$\left.\begin{aligned} &X \sim \lambda|\theta|^n\ ; \quad Y \sim \frac{X}{\theta}\cdot\frac{4}{n - n^2 - 2K}\ ; \\ &\text{with } n^2 + n = 4K+6\ , \text{ that is } n = \tfrac{1}{2}[-1 \pm \sqrt{16K+25}]\ , \\ &\text{or } Y \text{ is the leading parameter:} \\ &Y \sim \mu|\theta|^m\ ; \quad X \sim \frac{Y}{\theta}\cdot\frac{4}{m^2 - m - 4K - 6} \\ &\text{with } m^2 + m = -2K\ , \text{ that is } m = \tfrac{1}{2}[-1 \pm \sqrt{1-8K}]\ . \end{aligned}\right\} \tag{34}$$

The only diverging case is the first with $2n = -1 + \sqrt{16K+25}$, that is, $n > 2$, in the three other cases the exponent is negative and both X and Y go to zero.

We can use as a test the following function,

$$\left.\begin{aligned} &L = (4K+6)X^2 - 2KY^2 - (1+\theta^2)[(dX/d\theta)^2 + (dY/d\theta)^2] \\ &dL/d\theta = 2\theta[(dX/d\theta)^2 + (dY/d\theta)^2]\ . \end{aligned}\right\} \tag{35}$$

When $\theta \geq 0$ the function L is non-decreasing, and it is increasing if X and Y are not identically zero (we will not consider the latter case). On the other hand, L goes to $+\infty$ in the diverging case and to zero in the converging case. Hence, a sufficient condition of divergence for $\theta \to +\infty$ is $L \geq 0$ for some $\theta \geq 0$.

The system (32) has some other properties:

A) It is linear and the solutions can be linearly combined.

B) It has the following symmetries:

If at $\theta = 0$

$$Y = 0 \ ; \quad (dX/d\theta) = 0 \Rightarrow X(\theta) \text{ is even; } Y(\theta) \text{ is odd.} \tag{36}$$

If at $\theta = 0$

$$X = 0 \ ; \quad (dY/d\theta) = 0 \Rightarrow X(\theta) \text{ is odd; } Y(\theta) \text{ is even.} \tag{37}$$

For instance, the solution for which at $\theta = 0$,

$$X_o = 1 \ ; \ Y_o = 0 \ ; \ X'_o = 0 \ ; \ Y'_o = 0 \tag{38}$$

is of type (36) and is diverging on both sides ($L_o = 4K + 6 > 0$).

In order to obtain a solution converging in the past and diverging in the future it is sufficient to obtain a diverging solution with the symmetry property of Equation (37), since a proper linear composition of this solution with the solution (38) will lead to the desired result.

We are thus led to the symmetrical solutions of type (37), for instance, the solutions starting at:

$$X_o = 0 \ , \ Y_o = 1 \ , \ X'_o = 0 \ , \ Y'_o = 0 \tag{39}$$

or at:

$$X_o = 0 \ , \ Y_o = 0 \ , \ X'_o = 1 \ , \ Y'_o = 0 \ . \tag{40}$$

These solutions have a negative initial L ($L_o = -2K$ and $L_o = -1$) and they have been integrated numerically by Mr. J. P. Peltier for various values of K , see [16]. They are always diverging since when θ goes to $+\infty$, the parameter X goes to $-\infty$ in the case (39) and to $+\infty$ in the case (40).

Thus triple collisions of the collinear type are never Easton-regularizable in the plane and in the three-dimensional n-body problem.

<u>4.2.2 Case of Triangular Central Configurations</u>. This study is similar to that of the previous section and leads to the same result: in the plane and in the three-dimensional n-body problem

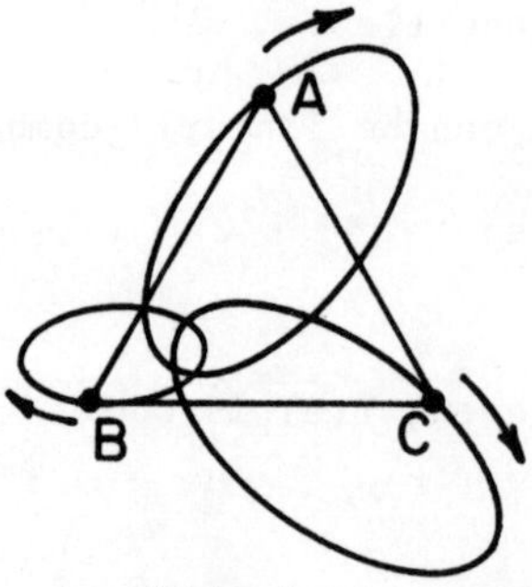

Fig. 9. An Elliptic Lagrangian Three-Body Motion.

triple collisions of triangular type are never Easton-regularizable.

As the Eulerian three-body motions, the Lagrangian motions (Figure 9) depend on six arbitrary parameters when the three masses are given. These three masses have Keplerian motions along three conics of the same eccentricity and the configuration of the three bodies remain forever an equilateral triangle.

Let us consider now the three-body motions in the vicinity of these Lagrangian motions (with the same center of masses).

It is assumed that:

A) The rotation of three bodies A, B and C is in the positive direction.

B) The succession A→B→C is in the positive direction.

C) The components of the discrepancies $\vec{\delta}_A$, $\vec{\delta}_B$, $\vec{\delta}_C$ between the Lagrangian motion and the neighboring motion will be measured in the rotating set of axes Oxyz (Figure 10), with $\vec{O}x$ parallel to $\vec{BC}$; $\vec{O}y$ the normal-in-plane component and $\vec{O}z$ the normal out-of-plane component.

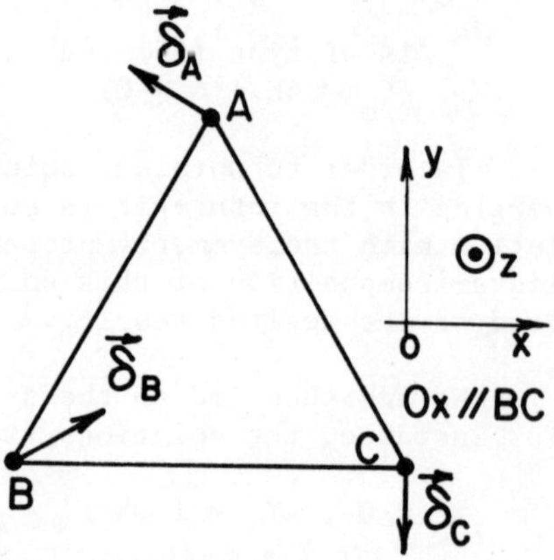

Fig. 10. The Rotating Set of Axes Oxyz.

The integration of the out-of-plane components z_A , z_B , z_C is simple (to the first order):

$$z_k = \frac{\lambda_{5k}\cos v + \lambda_{6k}\sin v}{1 + e\cos v} \quad ; \quad k = A,B,C \tag{41}$$

where

$$\left.\begin{array}{l} \lambda_{5k} , \lambda_{6k} \text{ are constants of integration ,} \\ e \text{ is the eccentricity} \\ v \text{ is the true anomaly} \end{array}\right\} \tag{42}$$

of the Lagrangian motion of interest. For the in-plane components x and y there are solutions satisfying to the first order:

$$m_A(x_A + iy_A) = m_B j(x_B + iy_B) = m_C j^2(x_C + iy_C) \quad , \tag{43}$$

where

$$i = \sqrt{-1} \quad ; \quad j = (i\sqrt{3} - 1)/2 \quad ; \quad j^3 = 1 \quad . \tag{44}$$

The differential equations of these solutions are:

$$\left. \begin{aligned} 2(1+e\cos v)[d^2X/dv^2 - 2dY/dv] &= 3X(1+N) \\ 2(1+e\cos v)[d^2Y/dv^2 + 2dX/dv] &= 3Y(1-N) \quad , \end{aligned} \right\} \tag{45}$$

where

I) The variables X and Y are real and such that:

$(X + iY)$ is proportional to

$$\left[(x_A + iy_A)(1 + e\cos v)\,[m_A + m_B j + m_C j^2]^{-1/2}\right] \tag{46}$$

(with an arbitrary <u>real</u> coefficient of proportionality).

II) The constant N is given by:

$$N = [m_A^2 + m_B^2 + m_C^2 - m_A m_B - m_A m_C - m_B m_C]^{1/2} \cdot [m_A\, m_B\, m_C]^{-1} \tag{47}$$

Hence

$$0 \leq N < 1 \quad . \tag{48}$$

Note that:

A) The solutions of (45) for $N = 1$ are the solutions of (27) of the system (24) for $K = 0$.

B) If m_B and/or $m_C = 0$ it implies $x_A + iy_A = 0$ (because of (43)) and then in (46) either the factor $j(x_B + iy_B)$ or the factor $j^2(x_C + iy_C)$ must be substituted for $x_A + iy_A$.

C) If $m_A = m_B = m_C$ we obtain $m_A + m_B j + m_C j^2 = 0$ and also $N = 0$. It is then sufficient to take $X + iY = (x_A + iy_A)(1 + e\cos v)$.

D) As for the vicinity of Eulerian motions, the solutions given by (41) - (47) are (to the first order) <u>the general</u>

<u>solutions of motions in the vicinity of Lagrangian motions</u> and we only need to choose the proper neighboring Lagrangian motion.

E) A remarkable property of the system (45) is its decomposability:

E_1) Decomposability when $N = 0$:

Let us put:

$$\begin{cases} p = X\cos v - Y\sin v \\ q = X\sin v + Y\cos v \end{cases} \tag{49}$$

The system (45) becomes separable:

$$\begin{cases} 2(1 + e\cos v)[p + d^2p/dv^2] = 3p \\ 2(1 + e\cos v)[q + d^2q/dv^2] = 3q \end{cases} \tag{50}$$

E_2) Decomposability when $N \neq 0$:

Let us consider the following second order system:

$$\begin{cases} 4N(1+e\cos v)[dX/dv] = [e^2\sin 2v - 2Ne\sin v]X + [e^2\cos 2v \\ \qquad + 4N(1+e\cos v) - 3N^2 - Q]Y \\ 4N(1+e\cos v)[dY/dv] = [e^2\cos 2v - 4N(1+e\cos v) - 3N^2 + Q]X \\ \qquad - [e^2\sin 2v + 2Ne\sin v]Y \quad . \end{cases} \tag{51}$$

The solutions of (51) are particular solutions of (45) when the constant Q satisfies:

$$Q = \pm [e^4 + 2N^2e^2 + 9N^4 - 8N^2]^{1/2} \tag{52}$$

With the two possible values of Q the two linear systems (51) give four independent solutions of (45) and the reduction is thus complete. When $Q = 0$ the complete reduction can only be obtained by continuity.

F) The system (45) and the two systems (51) have the symmetry properties of the systems (24) and (32).

If at $v = 0$

$$Y = 0 \quad ; \quad (dX/dv) = 0 \rightarrow X(v) \text{ is even } ; \ Y(v) \text{ is odd } . \tag{53}$$

If at $v = 0$

$$X = 0 \; ; \quad (dY/dv) = 0 \Rightarrow X(v) \text{ is odd } ; \; Y(v) \text{ is even } . \tag{54}$$

Let us now return to our original problem. We want to demonstrate the impossibility of the Easton-regularization of the collisions of triangular type (Figure 11) and we will examine the neighboring trajectories by two successive steps.

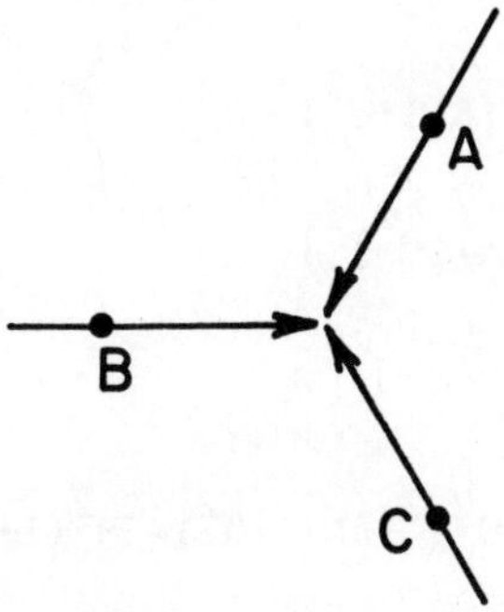

Fig. 11. A Triple Collision of Triangular Type.

I) When approaching collision the three bodies approach an equilateral triangle configuration and also a motion of the type of (18). The three vectors A, B, C are now on an equilateral triangle:

$$\left.\begin{aligned} \vec{r}_A - \vec{r}_{collision} &= \vec{A}(t_c - t)^{2/3} \\ \vec{r}_B - \vec{r}_{collision} &= \vec{B}(t_c - t)^{2/3} \\ \vec{r}_C - \vec{r}_{collision} &= \vec{C}(t_c - t)^{2/3} . \end{aligned}\right\} \tag{55}$$

II) In any vicinity of the motions of type (55) there are parabolic Lagrangian motions with a triple close approach and returning to infinity.

Hence, in order to prove the impossibility of the Easton-regularization of the triple collisions of triangular type, it is sufficient to study the vicinity of the parabolic Lagrangian solutions and to prove that when $e = 1$ the solutions of (45) diverge. These are solutions with X and $Y \to 0$ when $v \to -\pi$ and with X and/or $Y \to \infty$ when $v \to +\pi$. It is not expected to have a divergence for the out-of-plane components z_k given in (41). Let us put:

$$\left.\begin{aligned} e = 1 \quad &; \quad \tan \frac{v}{2} = \theta \\ X' = dX/d\theta \quad &; \quad X'' = d^2X/d\theta^2 \quad \text{etc.....} \end{aligned}\right\} \tag{56}$$

The system (45) becomes:

$$\left.\begin{aligned}(1+\theta^2)X'' + 2\theta X' - 4Y' &= 3X(1+N)\\ (1+\theta^2)Y'' + 2\theta Y' + 4X' &= 3Y(1-N)\end{aligned}\right\} \qquad (57)$$

and, with $Q = \pm(3N^2 - 1)$, the two systems (51) become:

$$\left.\begin{aligned}N(1+\theta^2)^2X' &= X[\theta - \theta^3 - N(\theta+\theta^3)] + 2Y[N(1+\theta^2) - \theta^2]\\ 2N(1+\theta^2)^2Y' &= X[(1-\theta^2)^2 - 4N(1+\theta^2) - 3N^2(1+\theta^2)^2]\\ &\quad + 2Y[\theta^3 - \theta - N(\theta^3+\theta)]\end{aligned}\right\} \qquad (58)$$

and

$$\left.\begin{aligned}2N(1+\theta^2)^2X' &= 2X[\theta-\theta^3-N(\theta+\theta^3)] + Y[(1-\theta^2)^2 + 4N(1+\theta^2) - 3N^2(1+\theta^2)^2]\\ N(1+\theta^2)^2Y' &= -2X[\theta^2 + N(1+\theta^2)] + Y[\theta^3 - \theta - N(\theta^3+\theta)] \ .\end{aligned}\right\} \qquad (59)$$

The study of the asymptotic properties when $\theta \to \pm\infty$ leads to the following.

In the system (58) Y is the leading parameter:

$$\left.\begin{aligned}&Y \sim \mu|\theta|^m \ ; \quad X \sim \frac{Y}{\theta}\cdot\frac{2 - 2N - 2mN}{3N^2 - 1}\\ &\text{where}\\ &m = \tfrac{1}{2}[-1 \pm \sqrt{13 - 12N}] \ .\end{aligned}\right\} \qquad (60)$$

In the system (59) X is the leading parameter:

$$\left.\begin{aligned}&X \sim \lambda|\theta|^n \ ; \quad Y \sim \frac{X}{\theta}\cdot\frac{2 + 2N + 2Nn}{1 - 3N^2}\\ &\text{where}\\ &n = \tfrac{1}{2}[-1 \pm \sqrt{13 + 12N}] \ .\end{aligned}\right\} \qquad (61)$$

In both cases positive and negative exponents appear, therefore, diverging and converging solutions are obtained.

Let us consider, for instance, system (59). For a given N it has only two independent symmetrical solutions:

(A) One solution of type (53); i.e., $X(\theta)$ even and $Y(\theta)$ odd, with conditions at $\theta = 0$:

$$X_o = 1 \ ; \ Y_o = 0 \ . \qquad (62)$$

(B) One solution of type (54); i.e., $X(\theta)$ odd and $Y(\theta)$ even, with conditions at $\theta = 0$:

$$X_o = 0 \quad ; \quad Y_o = 1 \quad . \tag{63}$$

For our demonstration it is sufficient that these two solutions diverge when $\theta \to +\infty$, indeed, a proper linear composition of these symmetrical solutions will lead to the solution we are looking for, with X and $Y \to 0$ when $\theta \to -\infty$ and with X and/or $Y \to \infty$ when $\theta \to +\infty$.

The divergence of the two solutions (62) and (63) of the system (59) can be shown in the following way. Let us put:

$$r = \frac{X}{Y} \quad ; \quad \frac{dr}{d\theta} = \frac{X'Y - Y'X}{Y^2} \tag{64}$$

Hence (59) implies:

$$\frac{dr}{d\theta} = \frac{(1 + 2r\theta - \theta^2)^2}{2N(1 + \theta^2)^2} + 2\,\frac{1 + r^2}{1 + \theta^2} - \frac{3N}{2} \,, \tag{65}$$

the system is thus reduced to the first order. Eqn. (61) shows that

$$\lim_{\theta\to\infty} [\frac{r}{\theta}] = \frac{1 - 3N^2}{2 + 2N + 2Nn} \tag{66}$$

In a case of divergence n is positive $(2n = -1 + \sqrt{13 + 12N})$, hence

$$\lim_{\theta\to\infty} (\frac{r}{\theta}) = \frac{1 - 3N^2}{2 + N + N\sqrt{13 + 12N}} = \frac{2 + N - N\sqrt{13 + 12N}}{4 + 4N} = R_D \tag{67}$$

and, in a case of convergence:

$$\lim_{\theta\to\infty} (\frac{r}{\theta}) = \frac{2 + N + N\sqrt{13 + 12N}}{4 + 4N} = R_C \quad . \tag{68}$$

Thus we have to integrate (65) from $\theta = 0$ to $\theta = +\infty$, from the initial conditions (62) and (63) (i.e., $r_o = \infty$ and $r_o = 0$) and we must verify that in both cases $\frac{r}{\theta}$ goes to R_D and not to R_C .

When $N = 1$ the integration of (65) gives:

$$\left.\begin{aligned} r_o = 0 &\Rightarrow r = \theta\ ;\quad \frac{r}{\theta} \to 1 = R_C \\ r_o = \infty &\Rightarrow r = \frac{-[5+5\theta^2+5\theta^4+\theta^6]}{10\theta+10\theta^3+4\theta^5}\ ;\quad \frac{r}{\theta} \to -\frac{1}{4} = R_D\ . \end{aligned}\right\} \tag{69}$$

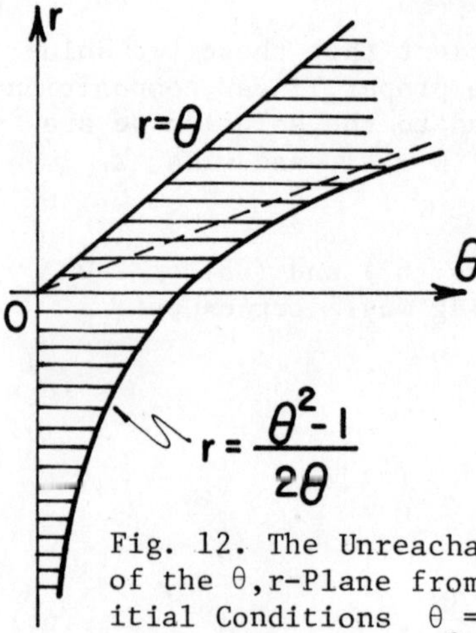

Fig. 12. The Unreachable Zone of the θ,r-Plane from the Initial Conditions $\theta_o = r_o = 0$ or $\theta_o = 0$; $r_o = \infty$.

When $0 < N < 1$ the ratio R_C remains in the range: $\frac{1}{2} < R_C < 1$ and it can be shown that R_c is unreachable from the initial conditions, either $r_o = 0$ or $r_o = \infty$ at $\theta_o = 0$. Let us consider the shadowed region of the θ,r plane (Figure 12).

A) This region contains all asymptotic directions with a slope between $\frac{1}{2}$ and 1 , i.e., corresponding to R_C .

B) From equation (65) it follows that the region is unreachable either from the initial condition $\theta_o = r_o = 0$ or from $\theta_o = 0$, $r_o = \infty$.

B_1) At $\theta = r = 0$ and along the straight line $r = \theta$ the slope given by (65) is larger than one:

$$\frac{dr}{d\theta} = \frac{1}{2N} + 2 - \frac{3N}{2} > 1\ . \tag{70}$$

B_2) For small values of θ the solution starting at $\theta = 0$, $r = \infty$, is:

$$r = \frac{-1}{2\theta} + \frac{1-N}{2}\theta + 0(\theta^3) < \frac{\theta^2-1}{2\theta}\ . \tag{71}$$

B_3) Finally, along the lower limit curve the slope given by (65) is smaller than that of the limit curve:

$$\frac{dr}{d\theta} = \frac{1}{2\theta^2} + \frac{1}{2} - \frac{3N}{2} < \frac{d}{d\theta}\left(\frac{\theta^2-1}{2\theta}\right) = \frac{1}{2\theta^2} + \frac{1}{2}\ . \tag{72}$$

Hence, when $0 < N < 1$, the symmetric solutions of the system (59) are diverging and the triple collisions of triangular type are not Easton-regularizable.

Remarks:

A) A similar demonstration [16] gives the same results for the symmetric solutions of the system (58).

B) These demonstrations are not valid when $N = 0$ and when $N = \sqrt{1/3}$. Fortunately, these cases are simple [16] and lead to the same result.

C) A side-result of these demonstrations is the following:

Let us consider a three-body system with three given positive masses, a zero energy and a given angular momentum different from zero.

The final evolutions are of 7 possible different types: 3 hyperbolic elliptic evolutions (according to the escaping mass) and 4 tri-parabolic evolutions (according to the final configuration).

The original evolutions are also of the same 7 possible types. The new result is the demonstration of the existence of all $7 \times 7 = 49$ possible original-final evolutions for all ratios of positive masses. For instance, from a tri-parabolic equilateral original evolution to any hyperbolic-elliptic final evolution, etc. .... This result can be extended to 3-body systems with one infinitesimal mass with only a minor modification: the three hyperbolic-elliptic evolutions become one hyperbolic-parabolic and two parabolic-elliptic evolutions.

D) Another side-result concerns the stability of elliptic Lagrangian motions. This stability has been studied numerically in the restricted case ($m_3 = 0$) by Danby and Bennett [17,18].

These results can be extended into three directions:

I) The stability of the system under interest is that of the system (45) that is related to the masses through the only parameter N of (47). Hence, Figure 13 can be extended to the general three-body problem with the following relation obtained from (47):

$$N^2 = 1 - 3x + 3x^2 \ . \tag{73}$$

It leads to the abscissa of the points A and B :

$$\left.\begin{aligned} N_A &= \sqrt{11/12} \quad ; \quad x_A = 0.0286... \\ N_B &= \sqrt{8/9} \quad ; \quad x_B = 0.0385... \end{aligned}\right\} \tag{74}$$

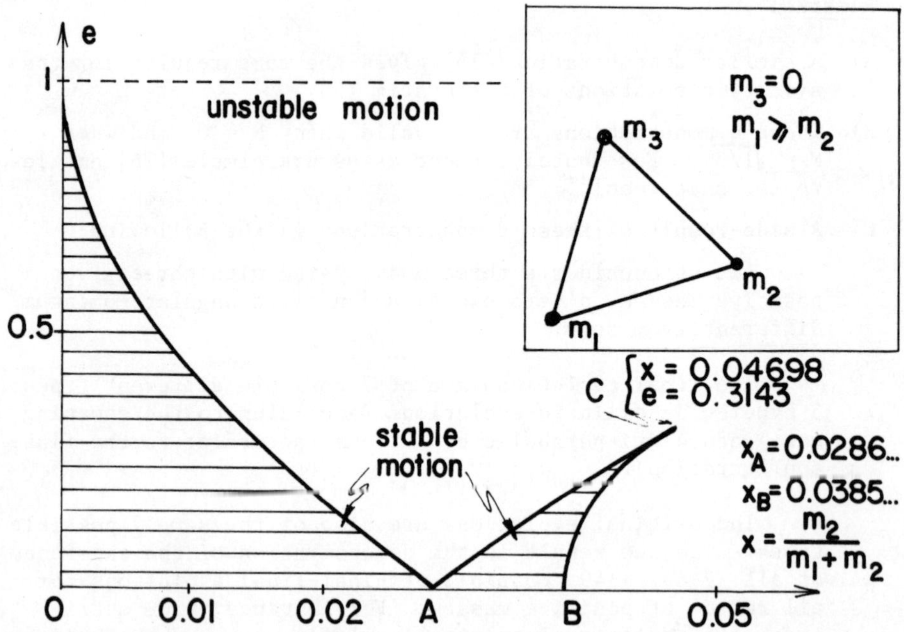

Fig. 13. Stability Zones of Elliptic Lagrangian Motions (Restricted Case $m_3 = 0$), in Terms of the Mass Ratio x and the Eccentricity e.

Similarly, the Bennett study [18], that leads to the instability of Eulerian motions, can also be extended (with the same result) to the general three-body Eulerian motions.

II) The minimum of $1 - 3x + 3x^2$ is $\frac{1}{4}$, hence the above extension is valid only when $N \geq \frac{1}{2}$. Fortunately, it is easy to demonstrate the instability of the system (45) when $N < \frac{1}{2}$ and $e < 1$.

III) The system (45) is decomposable into the two systems given by (51) and (52); it implies that the curve BC of the Figure 13 corresponds to $Q = 0$. Its equation is thus:

$$\left.\begin{aligned} e^2 &= \sqrt{8(N^2 - N^4)} - N^2 , \\ &\text{where} \\ N^2 &= 1 - 3x + 3x^2 . \end{aligned}\right\} \quad (75)$$

Thus the triple collisions of two or three dimensional n-body motions are never Easton-regularizable and this conclusion can be extended to multiple collisions. Thus the only Easton-regularizable singularity of these motions are the binary collisions (with a special mention of the binary collisions of infinitesimal masses as discussed in 3.2.3).

The close triple approaches lead generally to the escape of a very small binary in a direction opposite to the escape of the third body [9]. For close triple approaches of collinear type the small escaping binary contains generally the central mass of the collinear configuration; for close triple approaches of triangular type the small escaping binary contains generally the largest mass [15].

4.3 Regularization of the Singularities of the Rectilinear N-Body Motions.

The rectilinear n-body problem has a particular property: all its solutions have collisions and the Easton-regularization of binary collisions is impossible (the neighboring solutions are also colliding). Fortunately, the binary collisions are Siegel-regularizable and it leads to the usual picture of rectilinear motions with their invariable succession of bodies and, therefore, the study of the Easton-regularization can be carried out.

The collision of two infinitesimal masses will be of the type "elastic collision", which is also the limit of the case of small masses. We are thus led to the study of triple collisions among rectilinear motions.

4.3.1 Triple Collisions of Positive Masses. The rectilinear three-body problem has been the subject of many studies [2, 16, 19-22]. A useful image of the motion is shown on Figure 14 where $\tan\beta = \left[\frac{(m_A+m_B+m_C)m_B}{m_A m_C}\right]^{1/2}$. The representative point P moves in the x, y-plane, with:

$$\left.\begin{aligned}
x_A &= -x\left[\frac{m_C}{m_A+m_B}\right]^{1/2} - Y\left[\frac{(m_A+m_B+m_C)m_A}{m_A(m_A+m_B)}\right]^{1/2}\\
x_B &= -x\left[\frac{m_C}{m_A+m_B}\right]^{1/2} + Y\left[\frac{(m_A+m_B+m_C)m_B}{m_B(m_A+m_B)}\right]^{1/2}\\
x_C &= x\left[\frac{m_A+m_B}{m_C}\right]^{1/2} .
\end{aligned}\right\} \qquad (76)$$

It may be verified that:

$$\left.\begin{aligned}
&m_A x_A + m_B x_B + m_C x_C = 0\\
&(m_A+m_B+m_C)(x^2+y^2) = m_A x_A^2 + m_B x_B^2 + m_C x_C^2 .
\end{aligned}\right\} \qquad (77)$$

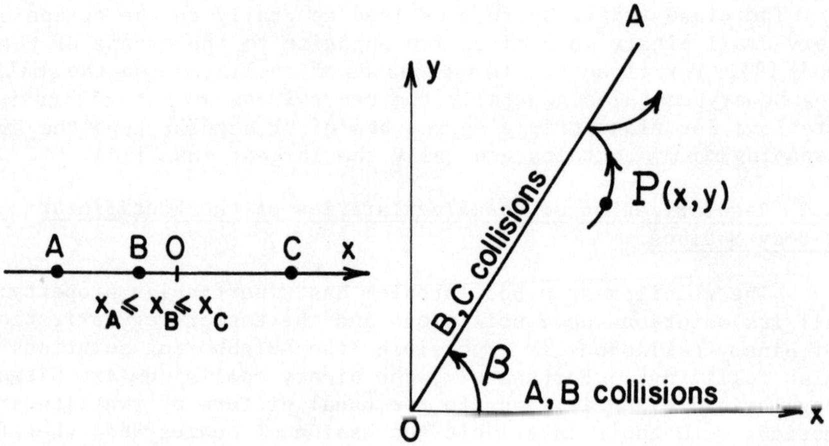

Fig. 14. Analysis of a Three-Body Rectilinear Motion.

Since $x_A \leq x_B \leq x_C$, the representative point P moves in the angle $x\hat{O}A$ and

$$(\text{measure } x\hat{O}A) = \beta \; ; \; \tan\beta = \left[\frac{(m_A + m_B + m_C)m_B}{m_A m_C}\right]^{1/2} \tag{78}$$

The two sides Ox and OA correspond to the A,B and B,C collisions.

The motion of P corresponds to the three-body rectilinear motion of interest if we assume that P moves in the force field F(x,y) defined by:

$$\left.\begin{aligned} F(x,y) &= G\left[\frac{m_A m_B}{x_B - x_A} + \frac{m_A m_C}{x_C - x_A} + \frac{m_B m_C}{x_C - x_B}\right] \\ M\frac{dx^2}{dt^2} &= +\frac{\partial F}{\partial x} \; ; \; M\frac{d^2y}{dt^2} = +\frac{\partial F}{\partial y} \; ; \; M = m_A + m_B + m_C . \end{aligned}\right\} \tag{79}$$

The energy integral is

$$h = \frac{1}{2}(m_A x_A'^2 + m_B x_B'^2 + m_C x_C'^2) - F = \frac{M}{2}(x'^2 + y'^2) - F(x,y) . \tag{80}$$

Figures 15-17 show the representation of various solutions. An important property is related to the scale transformation of

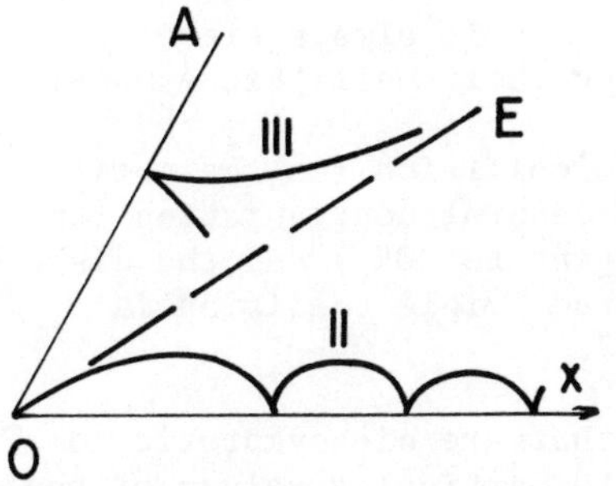

Fig. 15. Case h = 0. The Eulerian Motion Along OE.
II. A Solution with a Triple Collision.
III. A Solution with a Tri-Parabolic Escape.

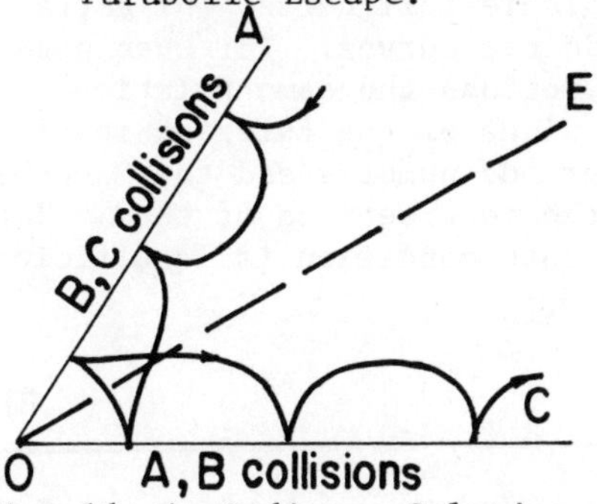

Fig. 16. An Ordinary Solution of Exchange Type. If h=0 the Representative Curve Tends Toward a Cycloid at Large Distances.

the n-body problem:

Let us transform a representative curve by a homothetic transformation of center 0 and ratio k. The new curve corresponds to another solution with the energy integral h/k and the velocities are divided by $\sqrt{k}$. That property is applied in Figure 17 (where h = 0). It shows that if a tri-parabolic arrival leads to a hyperbolic-elliptic escape, then neighboring tri-parabolic arrivals can lead to hyperbolic-elliptic escapes with arbitrarily small escaping binaries (at arbitrarily large velocity of escape).

This phenomenon leads to the MacGehee necessary condition [2] of Easton-regularizability of rectilinear triple collisions: "Motions with tri-parabolic arrival must lead to tri-parabolic escapes."

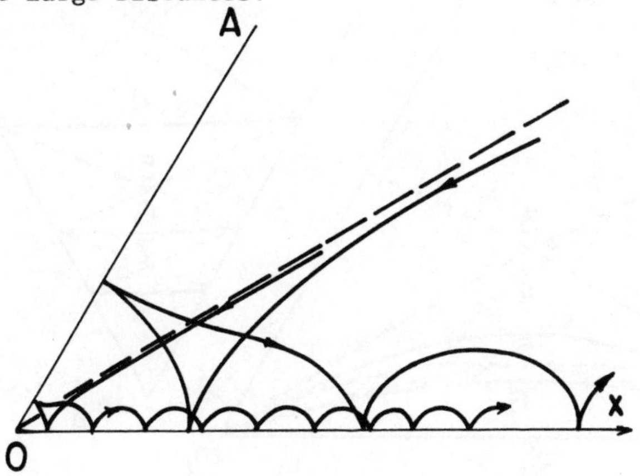

Fig. 17. From a Tri-Parabolic Arrival to a Hyperbolic-Elliptic Escape.

A) When $h = 0$ the final evolution is always either a tri-parabolic escape or a hyperbolic-elliptic escape.

B) Motions approaching a triple collision (as the motion II in Figure 15) approach a central configuration (the representative curve is tangent to OE) and the discussion in the vicinity of the triple collision is similar to that of Figure 17.

We must then look for motions that are bi-asymptotic to Eulerian motions. They correspond to particular values of the ratios of masses that have been computed by Simo [21]. They are represented in the "mass diagram" and in the "inverse mass diagram" of Figure 18.

Note that the bi-asymptotic motions with an even number of binary collisions correspond to a single full curve, while the odd numbers correspond to the two dotted curves. For even numbers and with the reversibility of motions the same solution (Figure 19) corresponds to the two sides of the Eulerian straight line OE . This is not the case for odd numbers and the MacGehee condition requires mass ratios at the intersection of the two dotted curves, hence $m_A = m_C$. The MacGehee condition is insufficient for the Easton-regularization.

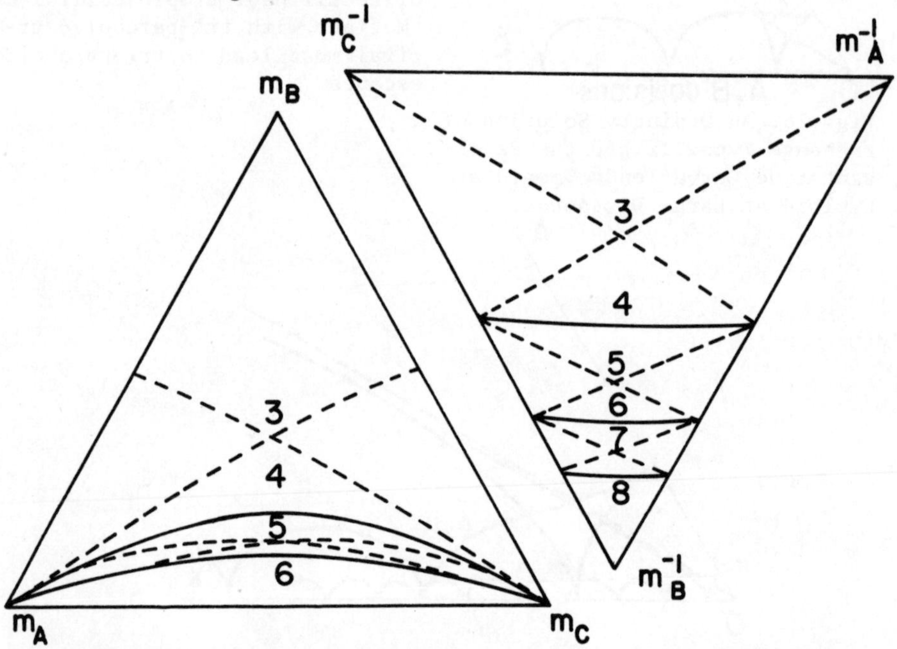

Fig. 18. Mass Diagram and Inverse Mass Diagram given by the MacGehee Condition. The Integers 3, 4, 5 . . . give the Number of Binary Collisions of Bi-Asymptotic Motion of Interest.

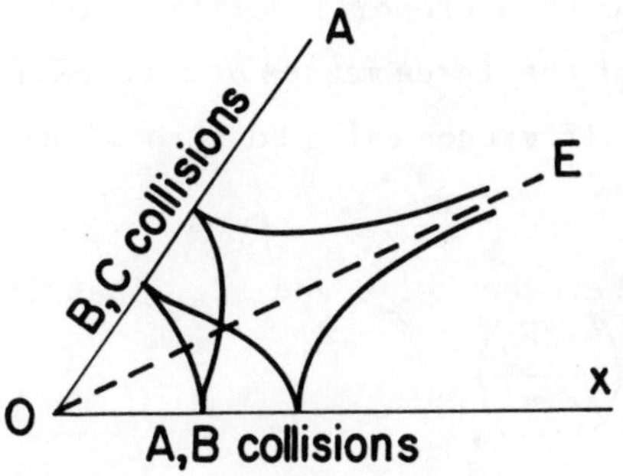

Fig. 19. A Bi-Asymptotic Motion with Four Collisions.

Let us examine the solutions in the vicinity of the Eulerian straight line OE , and let us call $\vec{i}$ the unit vector of OE and $\vec{j}$ the normal unit vector (Figure 20). The radius vector $\vec{OP}$ can be developed in terms of time t , the four parameters ε_1 , ε_2 , ε_3 , t_c and the constant A :

$$\vec{OP} = \vec{i}[A\tau^{2/3} + \varepsilon_1\tau^{4/3}] + \vec{j}[\varepsilon_2\tau^{F_2} + \varepsilon_3\tau^{F_3}] + \text{second order terms} , \tag{81}$$

where

$$\tau = |t - t_c| .$$

The case $\varepsilon_1 = \varepsilon_2 = \varepsilon_3 = 0$ is the Eulerian case, A is thus a function of the masses and K is defined in (25): (82)

$$F_2 = \frac{3 - \sqrt{16K + 25}}{6} < -\frac{1}{3}$$

$$F_3 = \frac{3 + \sqrt{16K + 25}}{6} > \frac{4}{3} \tag{83}$$

If we consider a very close triple approach we can use a development as (81) for the arrival and another one for the departure (with ε_{1-} , ε_{2-} , ε_{3-} , t_{c-} and ε_{1+} , ε_{2+} , ε_{3+} , t_{c+}) and the question of Easton-regularizability is that of continuity of the relation between ε_{1-} , ε_{2-} , ε_{3-} , t_{c-} and ε_{1+} , ε_{2+} , ε_{3+} , t_{c+} in the vicinity of the triple collision, i.e., in the vicinity of $\varepsilon_{2-} = 0$.

ε_{1-} and ε_{1+} are equal and proportional to the energy integral h .

Fig. 20. Analysis of Solutions in the Vicinity of OE .

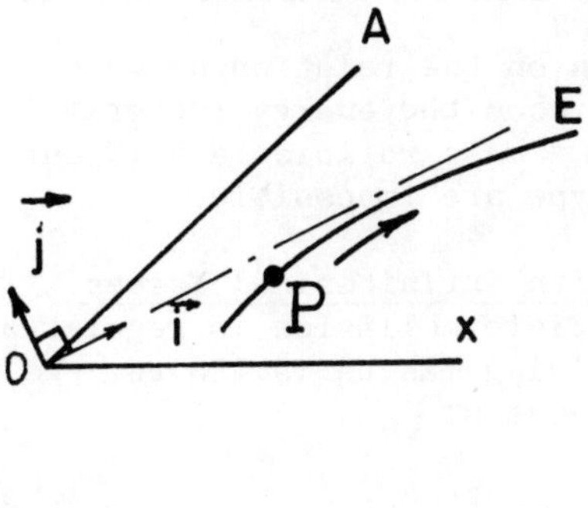

ε_{2-} and ε_{2+} are proportional with a proportionality coefficient depending only of their sign and of the three masses of interest. t_{c-} and t_{c+} are very close, with a difference going to zero with ε_{2-}

Unfortunately, the relation between ε_{3-} and ε_{3+} has the form:

$$\varepsilon_{3+} + \frac{\varepsilon_{2-}}{\varepsilon_{2+}} \varepsilon_{3-} = B\, \varepsilon_1 \cdot |\varepsilon_{2-}|^{\left(\frac{4-3F_3}{3F_3-1}\right)} \qquad (84)$$

The coefficient B only depends on the three masses and the sign of ε_{2-} . On the other hand, the exponent $\left(\frac{4-3F_3}{3F_3-1}\right)$ being negative, ε_3 may have arbitrary large values in any vicinity of a triple collision with the exception of the case $B = 0$.

The condition $B = 0$ corresponds to the following final condition of Easton-regularizability [16]: "Bi-asymptotic motions must correspond to two Eulerian motions with the same instant of collision."

According to the numerical computations of Mrs. Richa [16] that "Richa condition" is never satisfied for the MacGehee and Simo mass ratios. The Eulerian motion asymptotic in the future has always an instant of collision later than that of the Eulerian motion asymptotic in the past.

Thus triple collisions of positive masses are never Easton-regularizable (in the particular meaning adopted for rectilinear motions) and this conclusion can be extended to triple collisions between three masses, one of which is infinitesimal.

However, the Simo results of Figure 18 have interest.

A) They give the case of regularizable triple collision <u>when we restrict the analysis to zero energy motions</u>.

B) They give some information on the relation between original and final evolutions when the energy integral is zero. For instance, with a mass ratio on a full curve, the motions of exchange type are impossible.

<u>4.3.2 Multiple Collisions with Infinitesimal Masses</u>. The previous section shows that a multiple collision is regularizable only if, at most, one of the colliding masses is finite. We are thus led to the two following sub-cases:

4.3.2.1 Multiple Collisions of Infinitesimal Masses. This case is very simple. It is of the type "rectilinear billiard games" without even the rotations of the balls, and the curves of the representative Figures 14-17 and 19-20 become straight lines with ordinary mirror reflections along Ox and OA .

The Easton-regularization of the triple collisions requires the study of the successive reflections in the vicinity of 0 : the triple collision is Easton-regularizable only if the angle $x\hat{O}A = \beta$ is equal to π/q where q is an integer. That condition is sufficient when $q \geq 3$ but for $q = 2$ it requires only $[\inf(m_A, m_C)/m_B] = 0$ and simple analysis shows that in this case the Easton-regularization requires both $\frac{m_A}{m_B} = 0$ and $\frac{m_C}{m_B} = 0$.

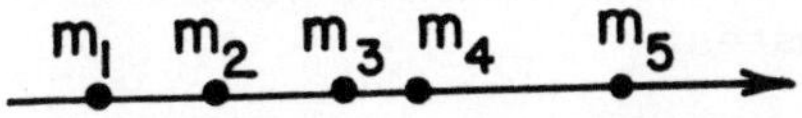

Fig. 21. Multiple Collision of Infinitesimal Masses in Rectilinear Motions.

In a case of multiple collision (Figure 21) there remains one more condition of Easton-regularizability: a mass must not be between two infinitely larger masses. If not all neighboring motions have triple or multiple collisions, they are not "regular", and the solution of interest cannot be extended by continuity, which remark is also valid when there are finite masses. Thus, if we call m_1 , m_2 , . . . m_k the k infinitesimal masses that collide, their rectilinear motion is Easton-regularizable if and only if:

A) For all p in the range $[2,k-1]$:

$$\beta_p = \frac{\pi}{q_p} \text{ , with } q_p \text{ integer,} \tag{85}$$

that is

$$\frac{[m_{(p-1)} + m_p + m_{(p+1)}]m_p}{m_{(p-1)} \quad m_{(p+1)}} = \tan^2 \frac{\pi}{q_p} \ . \tag{86}$$

B) If $q_p = 2$ we must have both

$$\frac{m_{(p-1)}}{m_p} = 0 \quad ; \quad \frac{m_{(p+1)}}{m_p} = 0 \ . \tag{87}$$

C) For all triplets a , b , c such that $1 \leq a < b < c \leq k$ we must not have at the same time:

$$\frac{m_b}{m_a} = 0 \quad ; \quad \frac{m_b}{m_c} = 0 \quad . \tag{88}$$

When k = 3 the representative motion is that of a light beam inside an angle $\frac{\pi}{q}$, i.e., in the smallest symmetry element of a regular q-polygon. Similarly, if k = 4 , $q_2 \geq 3$, $q_3 \geq 3$ the representative motion is that of a light beam inside a trihedron. This trihedron being the smallest symmetry element of a regular polyhedron. This can be extended to k-tuple collisions and to regular polyhedrons in a (k-1)-dimensional space.

4.3.2.2 Multiple Collisions with One Finite Mass and Several Infinitesimal Masses in Rectilinear Motion. It is obvious that the two sides of the finite mass are independent; we have thus Easton-regularizable triple collisions when the finite mass is between the two infinitesimal masses.

Let us now consider a given side of the finite mass. In a triple collision case the phenomenon leading to the MacGehee condition can also happen here (formation of a very small binary and escape of the third body, the furthest infinitesimal mass, with a velocity as large as desired). In order to avoid that effect we must use one of the Simo ratios that can be seen in the inverse mass diagram of Figure 18 at the ends of curves 4 , 6 , 8 , etc.

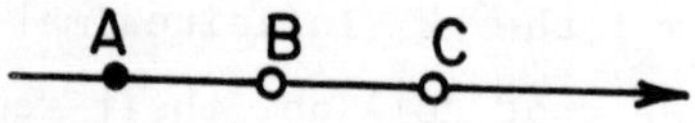

Fig. 22. Triple Collision of a Finite Mass A and Two Infinitesimal Masses B and C .

For instance, if A is the finite mass (Figure 22) we must have:

$$\frac{m_B}{m_A} = 0 \tag{89}$$

and

$$\frac{m_C}{m_B} = \begin{array}{ll} \text{either} & 1 \\ \text{or} & 2.3014 \\ \text{or} & 3.7501 \\ \text{or} & \text{etc.} \end{array} \tag{90}$$

Note that here the motions of m_B and m_C are two-body motions separated by elastic collisions, and these can be computed with a pocket calculator.

Let us put:

$$x_A = 0 \quad ; \quad dx_B/dt = V_B \quad ; \quad dx_C/dt = V_C \tag{91}$$

Between two successive B,C collisions the two following differences are constant:

$$h_B = \frac{V_B^2}{2} - \frac{Gm_A}{x_B} \quad ; \quad h_C = \frac{V_C^2}{2} - \frac{Gm_A}{x_C} \tag{92}$$

and the energy integral is:

$$h = m_B h_B + m_C h_C \quad . \tag{93}$$

The final question is: "Are the conditions (89) and (90) sufficient for the Easton-regularizability of triple collisions?" The answer is obviously yes if $\frac{m_C}{m_B} = 1$. The BC collisions represent just an exchange of B and C ; the two initial two-body motions are carried out with only successive reversal of the subscripts B and C .

For the other values of $\frac{m_C}{m_B}$ the answer depends on the energy integral h given in (93):

If $h = 0$ the answer is yes and solutions approaching infinity near the triple collision have the same value of h_B and also of h_C before and after the triple close approach (with $m_B h_B + m_C h_C = h = 0$ and $h_B < 0 < h_C$).

If $h \neq 0$ the answer is no, since we still have before and after the triple close approach:

$$\left.\begin{array}{ll} m_B h_{B-} + m_C h_{C-} = h & h_{B-} < h_{C-} \\ m_B h_{B+} + m_C h_{C+} = h & h_{B+} \leq h_{C+} \end{array}\right\} \tag{94}$$

but the relation between h_{B-} and h_{B+} depends on the first collision (either A,B or B,C) and for trajectories approaching infinity near the triple collision we obtain:

$$\left.\begin{aligned} h_{B+} &= h_{B-} - k\frac{|h|}{m_B} \quad \text{in one case, and} \\ h_{B+} &= \inf\left\{h_{B-} + k\frac{|h|}{m_B}\,;\ \frac{2h}{m_B+m_C} - k\frac{|h|}{m_B} - h_{B-}\right\} \text{ in the other case.} \end{aligned}\right\} \quad (95)$$

The positive scalar k is a function of the ratio $\frac{m_C}{m_B}$. For instance, for the second Simo ratio $\frac{m_C}{m_B} = 2.3014$ it is $k = 0.28069$.

One may notice that the "Richa condition" of the end of Section 4.3.1 is just satisfied at the limit when $\frac{m_B}{m_A} = 0$; $\frac{m_B}{m_C} = 1$ while it is not satisfied for the other values of the ratio m_C/m_B given in (90).

For multiple collisions on a given side of the finite mass the rules of this section and of the previous one lead to an even simpler result. The only Easton-regularizable multiple collisions are those for which all infinitesimal masses are equal. In a case of unequal m_B and m_C the presence of a fourth mass m_D can destroy the necessary condition $m_B h_B + m_C h_C = 0$ through a C,D collision. Thus for triple or multiple collisions with one finite mass and several infinitesimal masses in rectilinear motion the rules of Easton-regularizability are simple:

On each side of the finite mass we must have either two unequal masses with a Simo ratio and zero energy or an arbitrary number of equal masses. Note that this number can also be zero or one.

ACKNOWLEDGEMENTS.

This study is based on the thesis of Mrs. Claude Richa [16] and I thank very much this young lady who completed her work under the hard conditions of present-day Lebanon.

CONCLUSIONS.

The Easton-regularization of the singularities of the n-body problem has been discussed, and it was shown that, aside from the binary collisions, the singularities are generally not regularizable. The only positive new results are given in the last two sections and concern multiple collisions of the rectilinear n-body problem when all colliding masses but perhaps one are infinitesimal.

It should be pointed out that the main value of Easton's work is the introduction of a completely new idea and an exciting new point of view concerning collisions. His point of view originated several important new contributions to the field which could not have been produced without Easton's fundamental contribution.

It would be of interest to find the transformations that regularize regularizable singularities, and also to extend the notion of Easton-regularization through the analysis of neighboring solutions with the same values of the integral of motion. They will avoid the dispersion of the other neighboring solutions and will thus perhaps lead to new Easton-regularizable cases and to the corresponding regularizing transformations that are so useful for numerical integrations.

REFERENCES.

[1] Easton, R., Journal of Differential Equations, 10, p. 92, 1971.

[2] MacGehee, R., "Triple Collisions in Newtonian Gravitational Systems," Dynamical Systems, Theory and Applications, J. Moser, ed., pp. 550-572, 1975.

[3] Saari, D. G., "Singularities and Collisions of Newtonian Gravitational Systems," Archiv. for Rational Mechanics, 49, No. 4, pp. 311-320, 1973.

[4] Marchal, C., "Qualitative Methods and Results in Celestial Mechanics," (survey paper) 26th International Astronautical Congress (Lisbonne, September 1975); also, ONERA, TP n° 1975-77, 1975.

[5] Pollard, H., "Gravitational Systems," Journal of Mathematics and Mechanics, 17, pp. 601-612, 1967.

[6] Mather, J. N. and MacGehee, R., "Solutions for the Collinear Four-Body Problem which becomes Unbounded in Finite Time," National Science Foundation, Grant No. GP4313X and GP38955, 1974.

[7] Sundman, K. F., Acta. Mathematics, 36, p. 105, 1912.

[8] Siegel, C. L., Ann. Math. 2, 42, p. 127, 1941.

[9] Szebehely, V., Astronomical Journal, 179, pp. 981 and 1449, 1974.

[10] Marchal, C. and Losco, L., Astronomy and Astrophysics, 84, pp. 1-6, 1980.

[11] Waldvogel, J., Celestial Mechanics, 11, p. 429, 1975.

[12] Waldvogel, J., Celestial Mechanics, 14, p. 287, 1976.

[13] Waldvogel, J., Bulletin de l'Academie Royale de Belgique 5, 63, p. 34, 1977.

[14] Losco, L., Celestial Mechanics, 15, p. 477, 1977.

[15] Marchal, C., Astronautica Acta., 7, pp. 123-126, 1980.

[16] Richa, C., "Cas de régularisation des singularités du problème des n-corps," Thèse de 3e cycle à l'Université Pierre et Marie Curie (Paris 6), 15 Octobre 1980.

[17] Danby, J. M. A., "Stability of Triangular Points in the Elliptic Restricted Problem of Three Bodies," Astronomical Journal, 69, p. 165, 1964.

[18] Bennett, A., "Characteristic Exponents of the Five Equilibrium Solutions in the Elliptically Restricted Problem," Icarus, 4, p. 177, 1965.

[19] Nahon, F., "Trajectoires rectiliqnes du problème des 3 corps lorsque la constante des forces vives est nulle, Celestial Mechanics, 18, pp. 169-175, 1973.

[20] Irigoyen, M. and Nahon, F., "An Integrable Case of the Rectilinear Problem of Three Bodies," Astronomy and Astrophysics, 17, pp. 286-295, 1972.

[21] Simo, C., "Masses for which Triple Collision is Regularizable," Celestial Mechanics, 21, pp. 25-36, 1980.

[22] Waldvogel, J., "The Rectilinear Restricted Problem of Three Bodies," Celestial Mechanics, 8, pp. 189-198, 1973.

LES GEODESIQUES DE L'ELLIPSOIDE A TROIS AXES INEGAUX, D'APRES JACOBI, WHITTAKER, ARNOLD

F. NAHON

Université de Paris 6

ABSTRACT. The problem of determining the geodesics of the ellipsoïd with three inequal axes presents three interesting features :

1) It can be solved by quadratures, by the use of confocal coordinates (Jacobi, 1858) ;

2) so it admits a second integral, the geometrical interpretation of which was shown by Whittaker ;

3) for a critical value of this integral, all the orbits are doubly asymptotic to the only one which is periodic (the section of the ellipsoïd which passes by the "ombilic"). This description was given by Arnold.

We give the proofs missing in Whittaker's and Arnold's papers and throw light on the whole problem by considering the limiting case of the billard-ball problem.

1. INTRODUCTION

La détermination des géodésiques de l'ellipsoïde à trois axes inégaux a été ramenée aux quadratures par Jacobi en 1858. C'est le premier exemple historique d'intégration de l'équation aux dérivées partielles de Hamilton-Jacobi par séparation des variables pour un choix convenable des coordonnées.

Puisque le problème est intégrable, c'est qu'il admet une "deuxième intégrale" en plus de l'intégrale de l'énergie cinétique. Elle est susceptible d'une interprétation géométrique signalée par Whittaker.

V. Szebehely (ed.), Applications of Modern Dynamics to Celestial Mechanics and Astrodynamics, 237–247.

Les géodésiques se groupent en familles qui ont chacune leur enveloppe. Chaque famille est composée ou bien de trajectoires partout denses, ou bien de trajectoires fermées, dans la région intérieure à l'enveloppe. Un cas particulièrement remarquable est celui où l'enveloppe se réduit à deux des quatre ombilics de l'ellipsoïde. Toutes les géodésiques issues d'un ombilic se rassemblent au point ombilical opposé. Une seule est fermée, c'est l'ellipse section droite de l'ellipsoïde qui contient les quatre ombilics. Toutes les autres sont doublement asymptotiques à cette ellipse. Telle est la description donnée par Arnold.

Le but de cette leçon est de présenter, avec le schéma des démonstrations, tous ces résultats épars dans la littérature.

I - LES COORDONNEES DE JACOBI

Soit à déterminer les géodésiques d'une surface S de $\mathbb{R}_3$.

Si S est définie par son equation implicite $f(x_1, x_2, x_3) = 0$ on doit intégrer le système

$$\begin{aligned} \ddot{x}_1 &= Rf'_{x_1} \\ \ddot{x}_2 &= Rf'_{x_2} \\ \ddot{x}_3 &= Rf'_{x_3} \\ f(x_1 x_2 x_3) &= 0 \end{aligned} \tag{1}$$

où R est la mesure algébrique de la réaction $\vec{R} = R\vec{N}$ et $\vec{N}$ le vecteur normal à S, gradient de f.

Si S est définie par les équations paramétriques $x_i = \overline{x_i}(X,Y)$ d'où le ds^2 :

$$ds^2 = EdX^2 + GdY^2 \tag{2}$$

et le Lagrangien :

$$L = \frac{1}{2}(E\dot{X}^2 + G\dot{Y}^2) \tag{3}$$

le théorème de Maupertuis montre que les trajectoires sont les extrêmales de

$$J = \int\sqrt{EdX^2 + GdY^2} \tag{4}$$

Soit pour S l'ellipsoïde

$$\begin{gathered} \frac{x_1^2}{a_1} + \frac{x_2^2}{a_2} + \frac{x_3^2}{a_3} = 1 \\ a_1 < a_2 < a_3 \end{gathered} \tag{5}$$

Jacobi introduit comme courbes coordonnées les courbes :

$$C_1 = S \cap K(Y), \qquad C_2 = S \cap H(X) \tag{6}$$

où H(X) et K(Y) sont les deux quadriques homofocales de l'ellipsoïde S définies par :

$$H(X) : \frac{x_1^2}{a_1 - X} + \frac{x_2^2}{a_2 - X} + \frac{x_3^2}{a_3 - X} = 1$$
$$a_1 < X < a_2 \tag{7}$$

$$K(Y) : \frac{x_1^2}{a_1 - Y} + \frac{x_2^2}{a_2 - Y} + \frac{x_3^2}{a_3 - Y} = 1$$
$$a_2 < Y < a_3 \tag{8}$$

Ce choix des coordonnées sur S a les propriétés suivantes :

a) On peut obtenir l'expression des $x_i(X,Y)$

$$x_i^2 = \frac{a_i(a_i - X)(a_i - Y)}{(a_i - a_j)(a_i - a_k)} \tag{9}$$

On obtient une représentation de l'octant $x_1 > 0$, $x_2 > 0$, $x_3 > 0$ de S sur l'intérieur du rectangle R : $a_1 < X < a_2$, $a_2 < Y < a_3$. On passe aux 7 autres images sur l'ellipsoïde par symétries par rapport aux faces du trièdre $x_1 > 0$, $x_2 > 0$, $x_3 > 0$.

Le passage se fait par continuité lorsqu'on a au préalable déterminé les formes limites de H(X) pour $X \to a_1$, $X \to a_2$ et de K(Y) pour $Y \to a_2$, $Y \to a_3$

b) Le ds^2 de la surface (S) a la forme remarquable :

$$ds^2 = (Y - X)[m(X)dX^2 + n(Y)dY^2] \tag{10}$$

donc le changement de variables

$$dt = (Y - X)d\tau \tag{11}$$

met le Lagrangien sous la forme

$$L = m(X)\frac{X'^2}{2} + n(Y)\frac{Y'^2}{2} + h(Y - X) \tag{12}$$

qui est séparable (cas de Liouville-Stäckel) :

$$\begin{cases} m(X)X'^2 = 2h(k - X) \\ n(Y)Y'^2 = 2h(Y - k) \end{cases} \quad (13)$$

A chaque valeur de k correspond une famille $\mathcal{F}_k$ de trajectoires, dont l'équation est obtenue par quadratures

$$\tau = \phi(X,k) = \psi(Y,k) \quad (14)$$

II - LA DEUXIEME INTEGRALE ET L'INTERPRETATION DE WHITTAKER

Soient t_1, t_2 les vecteurs unitaires tangents aux courbes C_1, C_2 et $\theta = (t_1, \vec{V})$ où $\vec{V} = d\vec{M}/dt$ est le vecteur vitesse d'une géodésique.
Pour une surface de $ds^2 = EdX^2 + GdY^2$, on sait que :

$$tg^2\theta = \frac{GdY^2}{EdX^2}$$

donc pour les géodésiques de l'ellipsoïde :

$$tg^2\theta = \frac{n(Y)dY^2}{m(X)dX^2} = \frac{Y - k}{k - X}$$

La famille $\mathcal{F}_k$ est donc composée des géodésiques pour lesquelles la direction de la géodésique qui passe au pont (X,Y) est donnée par :

$$tg^2\theta = \frac{Y - k}{k - X} \quad (15)$$

L'interprétation de Whittaker est la suivante :
Soit (P) le plan tangent en M à l'ellipsoïde, $\vec{V}$ le vecteur vitesse en M.

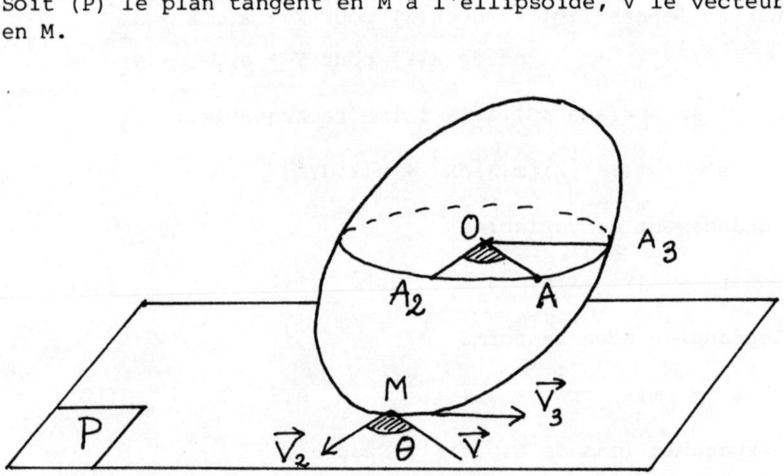

Par le centre O, mener le plan P' parallèle à P qui coupe l'ellipsoïde suivant une ellipse E et le demi-diamètre $\vec{OA}$ parallèle à $\vec{V}$.

Soit $d = |\vec{OA}|$ et p = distance de O au plan tangent en P. Lorsque M parcourt la géodésique, d et p varient mais :

$$pd = p_o d_o \tag{16}$$

Tel est le résultat donné par Whittaker sans démonstration ; on peut l'établir en utilisant les équations extrinsèques des géodésiques.

Montrons maintenant l'équivalence des relations (15) et (16). Soient $\vec{OA}_1$ et $\vec{OA}_2$ les deux demi-diamètres de l'ellipse E, parallèles aux vecteurs tangents aux courbes coordonnées t_1 et t_2. On peut établir que $\vec{OA}_1$ et $\vec{OA}_2$ sont les axes de l'ellipse E et que

$$OA_1^2 = X, \qquad OA_2^2 = Y$$

de sorte que l'équation de l'ellipse E est :

$$\frac{x^2}{X} + \frac{y^2}{Y} = 1$$

On obtient donc une interprétation géodésique des coordonnées X, Y ; de plus, avec les définitions données pour d et θ, on a :

$$x = d\cos\theta, \qquad y = d\sin\theta$$

d'où :

$$d^2[\cos^2\theta Y + \sin^2\theta X] = XY \tag{17}$$

Le théorème d'Apollonius, sur le volume du parallélépipède construit sur trois demi-diamètres conjugués, nous apprend d'autre part que

$$p^2 XY = \text{cste} = a_1 a_2 a_3 \tag{18}$$

En multipliant membre à membre (17) et (18), on obtient

$$p^2 d^2 (X\sin^2\theta + Y\cos^2\theta) = a_1 a_2 a_3$$

d'où l'équivalence annoncée de (15) et (16).

III - LE CAS PARTICULIER $k = a_2$ (DESCRIPTION D'ARNOLD)

(A) <u>Représentation de l'ellipsoïde sur le rectangle R</u> :

$a_1 < X < a_2 \qquad a_2 < Y < a_3$

On établira les propositions suivantes, en utilisant les formules (9) de Jacobi.

a) Soit Ω le point de l'arc AC de l'ellipsoïde $X = Y = a_2$

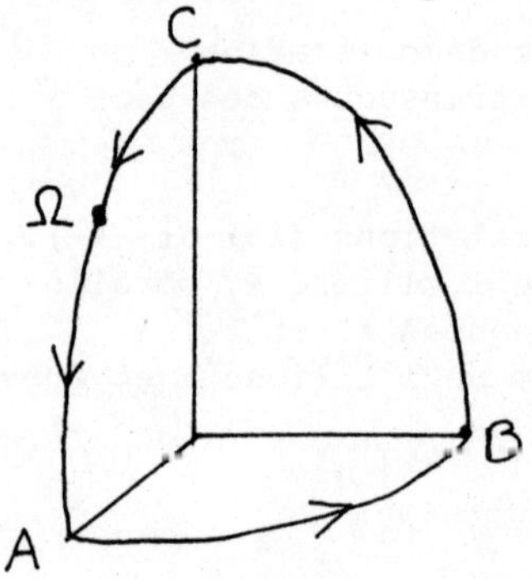

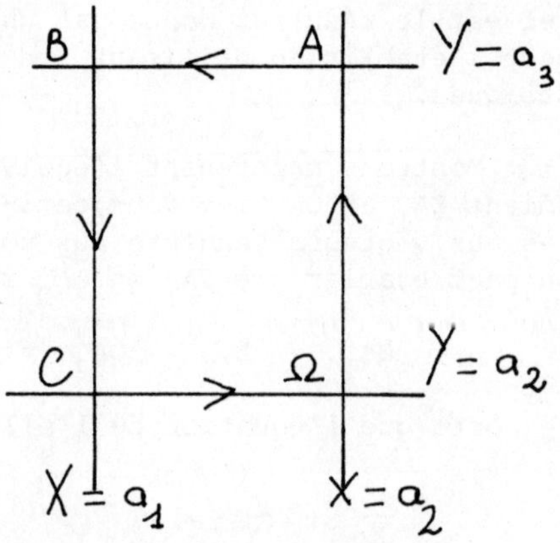

Au contour ABCΩA sur l'ellipsoïde correspond le contour (désigné par les mêmes lettres) du rectangle R.

On remarque que l'image de l'arc CA de l'ellipsoïde est la ligne brisée CΩA du rectangle.

On remarque aussi que le plan mené par O parallèle au plan tangent en Ω de l'ellipsoïde (S) coupe (S) suivant un cercle : en effet, les demi-diamètres sont égaux : $X = Y = a_2$. On dit que Ω est un <u>ombilic</u>.

Les trois autres ombilics de (S) sont : Ω' symétrique de Ω par rapport au centre O, $\overline{\Omega}$ symétrique de Ω par rapport au plan OAB et $\overline{\Omega}'$ symétrique de $\overline{\Omega}$ par rapport au centre O.

b) Une trajectoire qui traverse l'arc AB sur l'ellipsoïde est l'image d'une trajectoire tangente à AB dans le rectangle ; de même pour une trajectoire qui traverse l'arc BC sur l'ellipsoïde.

Une trajectoire qui passe par B sur l'ellipsoïde a pour image une trajectoire qui a un point de rebroussement en B sur le rectangle.

Une trajectoire qui franchit le point Ω sur l'ellipsoïde a pour image dans le rectangle une ligne brisée composée de deux trajectoires l'une qui arrive en Ω avec une pente M_o^+ ; l'autre qui part de Ω avec une pente M_1^- ;

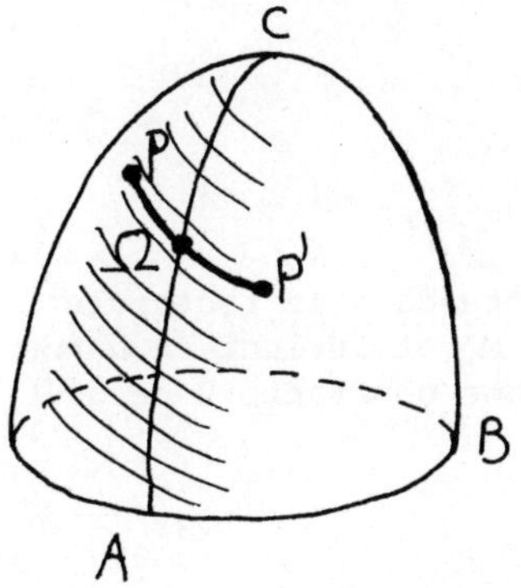

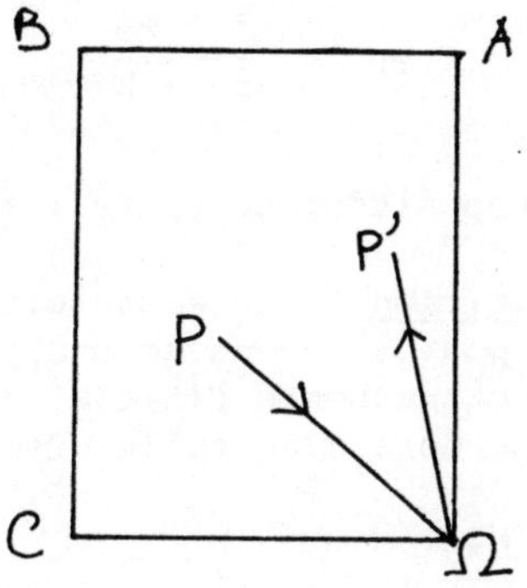

ces deux trajectoires sont symétriques par rapport à la bissectrice de l'angle $C\Omega A$ du rectangle :

$$M_1^- \cdot M_o^+ = +1 \qquad (19)$$

où M_o^+ et M_o^- sont dans le rectangle R les pentes respectives de ΩP et $\Omega P'$.

(B) <u>Etude des trajectoires images des géodésiques de la famille $(k = a_2)$ dans le rectangle</u>.

Le système différentiel (13) s'écrit en choisissant 2h = 1 donc ds = dt

$$\left(\frac{dX}{d\tau}\right)^2 = (a_2 - X)^2 f(X)$$

$$\left(\frac{dY}{d\tau}\right)^2 = (Y - a_2)^2 f(Y) \qquad (20)$$

$$f(u) = \frac{(u - a_1)(a_3 - u)}{u}$$

L'arc s se calcule par $ds = dt = (Y - X)a_2$, c'est-à-dire

$$ds = (Y - a_2)d\tau + (a_2 - X)d\tau$$

d'où :

$$ds = \frac{dX}{\sqrt{f(X)}} + \frac{dY}{\sqrt{f(Y)}} \qquad (21)$$

Le système différentiel (20) associe à tout point $M_o(X_o, Y_o)$ quatre

demi-trajectoires obtenues pour τ variant de 0 à $+\infty$ ou deux trajectoires complètes obtenues pour τ variant de $-\infty$ à $+\infty$.
Soit :

$$M(X,Y) = \frac{Y - a_2}{a_2 - X} \qquad (22)$$

On a la proposition suivante :

Proposition 1 : Une trajectoire complète dans le rectangle part de Ω pour $\tau \to -\infty$ avec une pente finie M^- et revient en Ω pour $\tau \to +\infty$ avec une pente finie M^+. Il existe une constante μ ne dépendant que de $a_1 a_2 a_3$ telle que

$$M^- . M^+ = \mu^2 \qquad (23)$$

(C) Trajectoires correspondantes sur l'ellipsoïde :

L'image d'une trajectoire complète du rectangle est une trajectoire qui part de Ω , franchit les arcs AB puis BC ou bien BC puis AB, donc aboutit au point Ω' symétrique de Ω par rapport au centre 0 de S.

La formule (20) nous donne la longueur de l'arc $\Omega\Omega'$; c'est :

$$\ell(\Omega\Omega') = \int_{a_2}^{a_3} \frac{dy}{f(y)} + \int_{a_1}^{a_2} \frac{dx}{f(x)} = \int_{a_1}^{a_3} \frac{du}{f(u)}$$

c'est-à-dire :

$$\ell(\Omega\Omega') = \int_{a_1}^{a_3} \frac{u}{(u-a_1)(a_3-u)}\, du \qquad (24)$$

Cette longueur ne dépend pas de la géodésique ; elle ne dépend pas non plus de a_2 et on peut la retrouver en remarquant qu'elle donne en particulier le demi-périmètre de l'ellipse d'axes (OA,OC) c'est-à-dire $\sqrt{a_1}$, $\sqrt{a_3}$ qui est une géodésique remarquable de la famille.

Résumons-là par :

Proposition 2 : Toutes les géodésiques de la famille $(k = a_2)$ partent d'un ombilic et aboutissent à l'ombilic symétrique par rapport au centre 0 de (S). Soit $\Omega\Omega'$ le couple d'ombilics choisis : la longueur $\Omega\Omega'$ est la même pour toutes les trajectoires et elle ne dépend pas de a_2, mais seulement de a_1, a_3.

Prolongement d'une géodésique $\Omega\Omega'$

Cette géodésique est définie sur l'ellipsoïde par l'angle $\theta_0 = (\vec{\sigma},\vec{V}_0)$ où $\vec{\sigma}$ est le vecteur unitaire de la tangente à l'ellipse $C\Omega A$, orientée de Ω vers A, et $\vec{V}_0$ le vecteur vitesse initiale de la géodésique considérée.

Soit θ'_0 l'angle correspondant au premier retour de cette géodésique par Ω sur l'ellipsoïde. On a la formule :

$$\operatorname{tg} \frac{\theta'_0}{2} = \mu^2 \operatorname{tg} \frac{\theta_0}{2}$$

Démonstration :

1) A l'arc $\Omega\Omega'$ de l'ellipsoïde correspond dans le rectangle la trajectoire complète issue de Ω avec la pente M_0^- et arrivant en Ω avec la pente M_0^+ telles que :

$$M_0^- \cdot M_0^+ = \mu^2$$

2) Cette géodésique se prolonge sur l'ellipsoïde par l'arc $\Omega'\Omega$ qui a pour image dans le rectangle la trajectoire complète issue de Ω avec la pente M_1^- telle que :

$$M_0^+ \cdot M_1^- = 1 \quad \text{(formule 19)}$$

et qui retourne en Ω avec la pente M_1^+ telle que $M_1^- \cdot M_1^+ = \mu^2$

3) Elle repart sur l'ellipsoïde de Ω avec la trajectoire qui a pour image dans le rectangle la trajectoire issue de Ω avec la pente M_2^- telle que : $M_1^+ \cdot M_2^- = 1$. On peut donc calculer :

$$\frac{M_2^-}{M_0^-} = \frac{1}{\mu^4} \qquad (26)$$

Or à une géodésique de l'ellipsoïde partant de Ω avec l'angle θ correspond la trajectoire du rectangle issue de Ω avec la pente M^- telle que :

$$M^- = \operatorname{cotg}^2 \frac{\theta}{2} \qquad (27)$$

La formule (26) est donc équivalente à la formule (25).

Conséquence :

Proposition 3 : Une géodésique quelconque issue de Ω sur l'ellipsoïde est asymptotique pour $t \to +\infty$ à la géodésique section de l'ellipsoïde par le plan $x_2 = 0$, parcourue dans le sens $A\Omega C$, et pour $t \to -\infty$, elle est asymptotique à la même géodésique parcourue en sens contraire.

Cette géodésique, la seule périodique, est donc doublement asymptotique à toutes les géodésiques de la famille $(k = a_2)$.

Démonstration :

La proposition (3) est démontrée par la formule (25) si on a établi que $\mu > 1$.

C'est ce que l'on peut vérifier dans un cas particulier remarquable, qui donne en même temps la visualisation des trois propositions de ce chapitre III. C'est le cas où $a_1 = 0$. L'ellipsoïde s'aplatit suivant l'ellipse (CB)

$$x_1 = 0 \qquad \frac{x_2^2}{a_2} + \frac{x_3^2}{a_3} = 1 \tag{29}$$

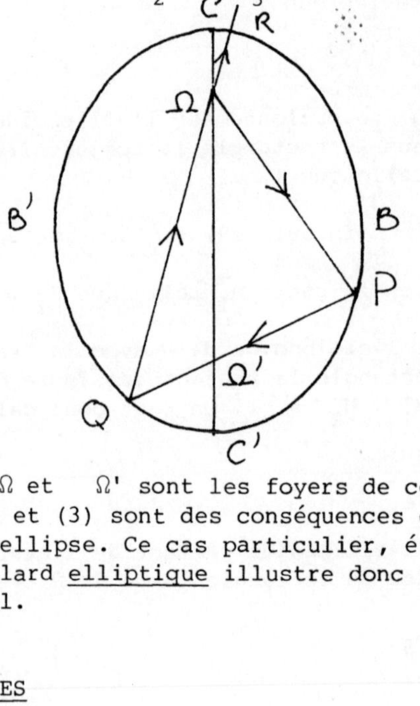

Les ombilics Ω et Ω' sont les foyers de cette ellipse ; les propositions (2) et (3) sont des conséquences élémentaires des propriétés de l'ellipse. Ce cas particulier, étudié par Birkhoff sous le nom de billard elliptique illustre donc toutes les circonstances du cas général.

REFERENCES

Jacobi, (1958), Vorlesungen über Dynamik, (leçons 26, 27, 28).

Whittaker, Analytical Dynamics of particles and rigid bodies, p. 393 de l'édition de 1970.

Arnold, Méthodes mathématiques de la mécanique classique, p. 262.

Birkhoff, Dynamical systems, p. 169.

On pourra comparer la présentation donnée dans cet exposé avec celle de Eisenhart :

Luther Pfahler Eisenhart, A treatise on differential geometry of curves and surfaces.

D'autre part, Whittaker attire l'attention sur l'existence, liée à celle d'une "deuxième intégrale", d'une transformation infinitésimale qui transforme l'une dans l'autre les géodésiques d'une même famille. Ce point de vue est abordé de façon plus générale dans :

Sophus Lie, Untersuchungen über geodätische curven, Math. Annalen, Vol. 20, 1882.

COORDONNÉES SYMÉTRIQUES SUR LA VARIÉTÉ DE COLLISION TRIPLE DU PROBLÈME PLAN DES TROIS CORPS.

Jörg Waldvogel

École Polytéchnique Fédérale, Zurich, Suisse

RÉSUMÉ. Ce travail concerne la variété des solutions du problème plan des trois corps avec le moment cinétique $p_\varphi=0$ et l'énergie $h=0$. Cette variété est équivalente à une sous-variété invariante N (correspondant aux solutions avec $p_\varphi=0$) de la variété de collision triple introduite par McGehee. On décrit la variété N de 4 dimensions en utilisant 6 paramètres inspirés par la réduction symétrique de Murnaghan et la régularisation de Lemaître. La topologie de N est celle d'un tore à 4 dimensions $(S^2 x S^2)$ moins 12 points. Le flot sur N est du type gradient avec une fonction de Lyapounov stationnaire en 40 points d'équilibre. Les 6 paramètres sont en plus très bien adaptés à l'intégration numérique du flot.

1. COORDONNÉES SYMÉTRIQUES.

Soient $m_j>0$ $(j=1,2,3)$ les trois masses d'un problème plan des trois corps et soient $x_j \in \mathbb{C}$ leurs positions par rapport au centre de gravité O (nous adoptons la notation complexe pour les vecteurs dans le plan).

Introduisons les différences

$$A_j=X_\ell-X_k=a_j e^{i\varphi_j} \tag{1}$$

où $i=\sqrt{-1}$; les indices (j,k,ℓ) prennent comme valeurs les 3 permutations cycliques de (1,2,3) dans le travail

V. Szebehely (ed.), Applications of Modern Dynamics to Celestial Mechanics and Astrodynamics, 249–266.

entier. Le modul $a_j = |A_j|$ est le côté du triangle opposé à la masse m_j, et $\varphi_j = \arg(A_j)$ est l'angle entre l'axe réel et A_j (voir fig. 1). Introdusions aussi les moments

$$P_j = m_j \frac{dX_j}{dt}$$

conjugués aux X_j (t est le temps).

Les intégrales premières du moment linéaire sont

$$\Sigma m_j X_j = 0, \qquad \Sigma P_j = 0, \tag{2}$$

et l'hamiltonien du problème plan des trois corps est donné par

$$H = T - U, \tag{3}$$

où

$$T = \frac{1}{2} \Sigma \frac{1}{m_j} |P_j|^2, \quad U = \Sigma \frac{m_k m_\ell}{a_j} . \tag{4}$$

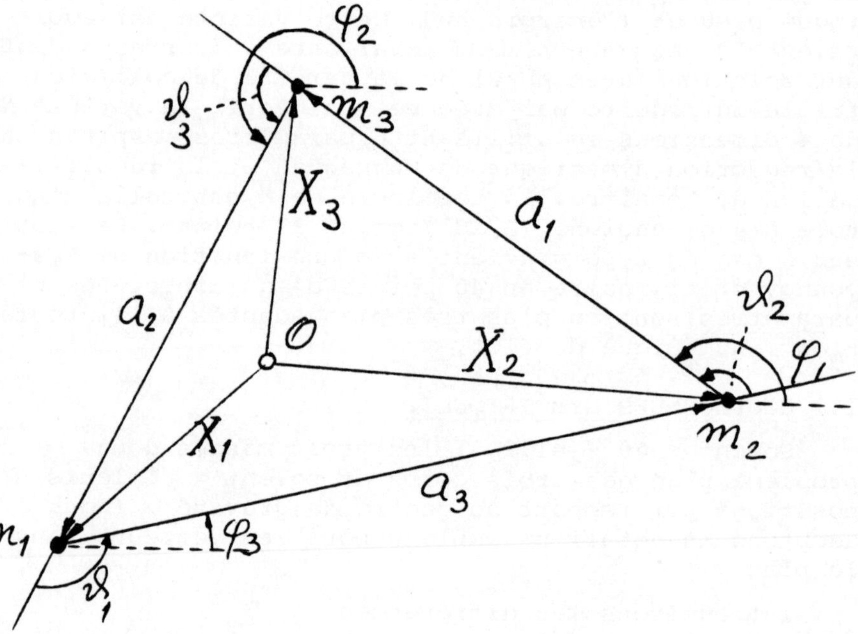

Figure 1

Dans ce système de 4 degrés de liberté nous introduisons les 4 coordonnées symétriques de Murnaghan (1936), c'est-à-dire les 3 côtes a_1, a_2, a_3 du triangle et l'angle

$$\varphi = \frac{1}{3}(\varphi_1+\varphi_2+\varphi_3). \tag{5}$$

En suivant van Kampen et Wintner (1937) nous éliminons les noeuds en introduisant les moments p_1, p_2, p_3, $p_\varphi \in \mathbb{R}$ conjugués aux coordonnées a_1, a_2, a_3, $\varphi \in \mathbb{R}$. On déduit la définition des moments conjugués de la fonction génératrice

$$G = a_1p_1+a_2p_2+a_3p_3+\varphi p_\varphi \tag{6}$$

au moyen des relations

$$P_j = 2\frac{\partial G}{\partial X_j} \qquad (j=1,2,3),$$

qui donnent explicitement

$$P_j=p_k e^{i\varphi_k}-p_\ell e^{i\varphi_\ell}+\frac{p_\varphi}{3} i\left(\frac{1}{a_k} e^{i\varphi_k}-\frac{1}{a_\ell} e^{i\varphi_\ell}\right). \tag{7}$$

Pour manipuler les nouvelles variables nous avons besoin de quelques relations géométriques du triangle. Les angles extérieurs

$$\vartheta_j = \varphi_\ell - \varphi_k \quad (\text{mod } 2\pi) \tag{8}$$

s'expriment par les côtes au moyen de la relation

$$\cos\vartheta_j = \frac{a_j^2-a_k^2-a_\ell^2}{2a_ka_\ell} . \tag{9}$$

En utilisant l'expression pour l'aire S du triangle,

$$S=\sqrt{\sigma(\sigma-a_1)(\sigma-a_2)(\sigma-a_3)}, \qquad \sigma = \frac{1}{2}(a_1+a_2+a_3) \tag{10}$$

on obtient

$$\sin\vartheta_j = \frac{2S}{a_ka_\ell} , \tag{11}$$

$$\tan\frac{\vartheta_j}{2} = \frac{4S}{a_j^2-(a_k-a_\ell)^2} \tag{12}$$

Pour retrouver les coordonnées cartésiennes, les coordonnées symétriques étant données, on commence avec l'équ. (12) qui fournit les angles ϑ_j mod 2π. Ensuite les φ_j et les X_j sont données par

$$\varphi_j = \varphi + \frac{1}{3}(\vartheta_\ell - \vartheta_k) \tag{13}$$

$$mX_j = m_\ell a_k e^{i\varphi_k} - m_k a_\ell e^{i\varphi_\ell}, \qquad m = \sum m_j, \tag{14}$$

conséquences de (5), (8) ou (1), (2). Enfin, l'équ. (7) fournit les moments P_j.

En utilisant les relations (7), (14) on obtient l'identité

$$\sum_{j=1}^{3} \bar{X}_j P_j = \sum_{j=1}^{3} a_j p_j + i p_\varphi \tag{15}$$

qui aide à interpréter les nouvelles coordonnées.

Pour exprimer l'énergie cinétique T par les coordonnées symétriques nous calculons

$$\frac{1}{2}|P_j|^2 = \frac{1}{2}(p_k^2 + p_\ell^2) - p_k p_\ell \cos\vartheta_j$$

$$+ \frac{p_\varphi}{3}\left(\frac{p_k}{a_\ell} - \frac{p}{a_k}\right)\sin\vartheta_j + \frac{p_\varphi^2}{9}\left(\frac{1}{2a_k^2} + \frac{1}{2a_\ell^2} - \frac{\cos\vartheta_j}{a_k a_\ell}\right);$$

au moyen de (9) et (11) nous obtenons

$$T = \frac{1}{2}\sum \frac{1}{m_j}\left(p_k^2 + p_\ell^2 + p_k p_\ell \frac{a_k^2 + a_\ell^2 - a_j^2}{a_k a_\ell}\right)$$

$$+ \frac{2p_\varphi}{3}\sum \frac{1}{m_j}\left(\frac{p_k}{a_\ell} - \frac{p_\ell}{a_k}\right)\frac{S}{a_k a_\ell} + \frac{p_\varphi^2}{9}\sum \frac{1}{m_j}\frac{a_k^2 + a_\ell^2 - \frac{1}{2}a_j^2}{a_k^2 a_\ell^2} \tag{16}$$

Avec l'hamiltonian (3) les équations du mouvement sont données par

$$\frac{da_j}{dt} = \frac{\partial H}{\partial p_j}, \quad \frac{dp_j}{dt} = -\frac{\partial H}{\partial a_j} \quad (j=1,2,3) \tag{17}$$

$$\frac{d\varphi}{dt} = \frac{\partial H}{\partial p_\varphi}, \quad \frac{dp_\varphi}{dt} = -\frac{\partial H}{\partial \varphi} = 0. \tag{18}$$

L'angle φ est une variable cachée; par conséquent la dernière équation (17) signifie que p_φ est une intégrale première: d'après l'identité (15)

$$p_\varphi = \mathrm{Im} \sum_{j=1}^{3} \bar{X}_j P_j = \text{const} \tag{19}$$

est le moment cinétique du système des trois corps par rapport au centre de gravité O.

Ainsi l'élimination des noeuds est terminée: les équations (17) forment un système de 6 équations différentielles pour les variables a_j, p_j $(j=1,2,3)$, et équ. (18),

$$\frac{d\varphi}{dt} = \frac{2}{3}\sum \frac{S}{m_j a_k a_\ell}\left(\frac{p_k}{a_\ell} - \frac{p_\ell}{a_k}\right) + \frac{2}{9} p_\varphi \sum \frac{a_k^2 + a_\ell^2 - \frac{1}{2}a_j^2}{m_j a_k^2 a_\ell^2}, \tag{20}$$

sert à déterminer φ. L'intégrale première de l'énergie est

$$H = T - U = h = \text{const}. \tag{21}$$

2. RÉGULARISATION

Nous introduisons les variables α_j de Lemaître (1954, 1964), définies par

$$a_1 = \alpha_2^2 + \alpha_3^2, \quad a_2 = \alpha_3^2 + \alpha_1^2, \quad a_3 = \alpha_1^2 + \alpha_2^2 \tag{22}$$

ou

$$a_j = \alpha_k^2 + \alpha_\ell^2 .$$

Le carré α_j^2 est la distance (sur a_k ou a_ℓ) entre m_j et le point de contact du cercle inscrit au triangle (voir fig.2)

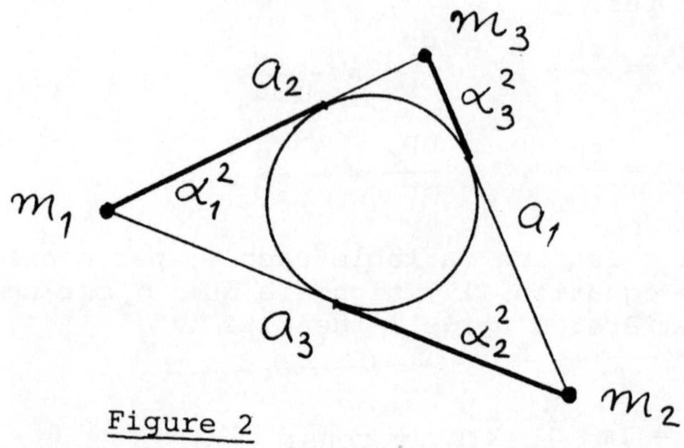

Figure 2

La transformation inverse est

$$\alpha_j = \pm\sqrt{\alpha^2 - a_j}, \qquad \alpha^2 := \alpha_1^2+\alpha_2^2+\alpha_3^2 = \sigma, \tag{23}$$

et l'aire signée s'écrit :

$$S = \alpha\ \alpha_1\alpha_2\alpha_3 \ . \tag{24}$$

Les équations (12) se transforment en

$$\tan\frac{\vartheta_j}{2} = \frac{\alpha\alpha_j}{\alpha_k\alpha_\ell} \ . \tag{25}$$

Chaque triplet $(\alpha_1,\alpha_2,\alpha_3)$ représente un triangle parce que les inégalités du triangle

$$a_j+a_k = a_\ell + 2\alpha_\ell^2 \geq a_\ell$$

sont satisfaites automatiquement. Les triangles de périmètre $2\alpha^2$ sont représentés par les points $(\alpha_1,\alpha_2,\alpha_3)$ de la sphère $\sum\alpha_j^2=\alpha^2$. L'orientation des triangles est donnée par le signe $\mathrm{sign}(\alpha_1\alpha_2\alpha_3)$. Sur la sphère un triangle non-dégénéré est représenté 4 fois ainsi que son image symétrique. Les triangles dégénérés qui satisfont $a_j+a_k = a_\ell$ correspondent à l'équateur $\alpha_\ell=0$. Une collision entre m_j et m_k

($a_\ell=0$, $a_j=a_k$) correspond à un des poles $\alpha_j=\alpha_k=0$, $\alpha_\ell=\pm\alpha$ (voir fig. 3).

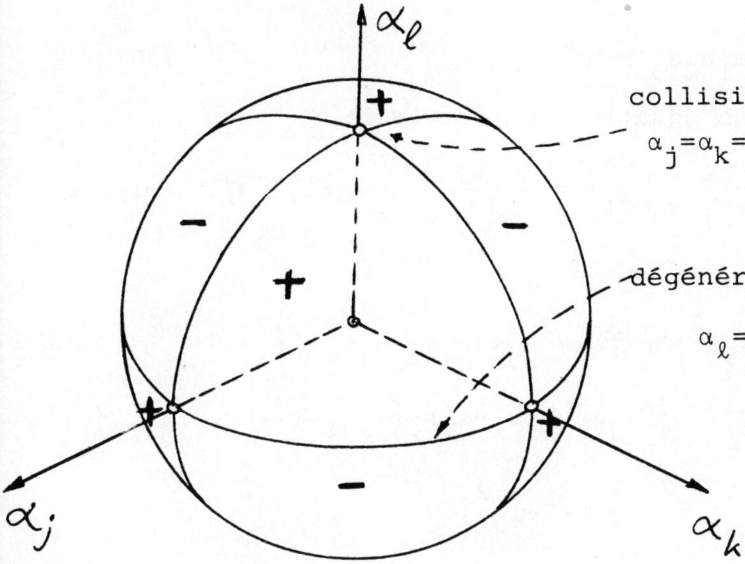

Figure 3

En suivant Waldvogel (1979 a) nous introduisons les variables π_j conjugués aux α_j. Avec la fonction génératrice

$$G = \alpha_1\pi_1+\alpha_2\pi_2+\alpha_3\pi_3 \tag{26}$$

on trouve

$$p_j = \frac{\partial G}{\partial a_j} = \frac{1}{4}\left(-\frac{\pi_j}{\alpha_j}+\frac{\pi_k}{\alpha_k}+\frac{\pi_\ell}{\alpha_\ell}\right). \tag{27}$$

Le théorème d'Euler sur les fonctions homogènes implique

$$\sum a_jp_j=\sum a_j\frac{\partial G}{\partial a_j} = \frac{1}{2}G = \frac{1}{2}\sum\alpha_j\pi_j. \tag{28}$$

Enfin, nous introduisons la nouvelle variable indépendante τ (le temps modifié) par la relation différentielle

$$dt = a_1a_2a_3d\tau \tag{29}$$

qui était suggérée par Sundman (1913), Lemaître (1954), Waldvogel (1972) et autres. La forme canonique des équations du mouvement peut être conservée si l'on ne considère que des solutions d'énergie h. En suivant Poincaré nous prenons comme nouvel hamiltonien la fonction

$$K = a_1 a_2 a_3 (H-h) \tag{30}$$

qui définit les équations du mouvement

$$\frac{d\alpha_j}{d\tau} = \frac{\partial K}{\partial \pi_j}, \quad \frac{d\pi_j}{d\tau} = -\frac{\partial K}{\partial \alpha_j} \qquad (j=1,2,3). \tag{31}$$

On trouve

$$\frac{1}{2}a_1 a_2 a_3 |p_j|^2 = \frac{a_j}{8}[\alpha^2\pi_j^2 + (\alpha_k\pi_\ell - \alpha_\ell\pi_k)^2]$$
$$+ \frac{2}{3}p_\varphi \cdot \frac{a_j}{a_k a_\ell} \cdot \frac{\alpha}{4}[\pi_j\alpha_k\alpha_\ell(\alpha_\ell^2-\alpha_k^2) + (\alpha_k\pi_\ell-\alpha_\ell\pi_k)\alpha_j(\alpha^2+\alpha_j^2)]$$
$$+ \frac{p_\varphi^2}{9}\frac{a_j}{a_k a_\ell}[2\alpha^2\alpha_j^2 + \frac{1}{2}(\alpha_\ell^2-\alpha_k^2)^2]$$

d'où

$$K = \frac{1}{8}\sum \frac{a_j}{m_j}[\alpha^2\pi_j^2 + (\alpha_k\pi_\ell-\alpha_\ell\pi_k)^2]$$
$$+ \frac{p_\varphi}{6}\sum \frac{a_j\alpha}{m_j a_k a_\ell}[\pi_j\alpha_k\alpha_\ell(\alpha_\ell^2-\alpha_k^2) + (\alpha_k\pi_\ell-\alpha_\ell\pi_k)\alpha_j(\alpha^2+\alpha_j^2)]$$
$$+ \frac{p_\varphi^2}{9}\sum \frac{a_j}{m_j a_k a_\ell}[2\alpha^2\alpha_j^2 + \frac{1}{2}(\alpha_\ell^2-\alpha_k^2)^2] \tag{32}$$
$$- \sum m_k m_\ell a_k a_\ell - h a_1 a_2 a_3;$$

les a_j en fonction des α_j sont donnés par la transformation (22).
Avec les vecteurs

$$\underline{\alpha} = (\alpha_1, \alpha_2, \alpha_3)^T, \quad \underline{\pi} = (\pi_1, \pi_2, \pi_3)^T$$

l'hamiltonien (32) s'écrit

$$K = K_o(\underline{\alpha}, \underline{\pi}, p_\varphi) - h a_1 a_2 a_3, \tag{33}$$

où la fonction $K_o(\underline{\alpha},\underline{\pi},p_\varphi)$ contient les quatre premiers termes de K; elle est homogène du degré 4 en $(\alpha_1,\alpha_2,\alpha_3,p_\varphi)$. Seulement les solutions de (31) avec K =0 correspondent à des solutions physiques; K = 0 est une relation invariante.

Dans le cas particulier $p_\varphi=0$ l'hamiltonien K est complètement régulier (en fait il est un polynôme); par conséquent toutes les collisions binaires sont régularisées. Dans une collision entre m_k et m_ℓ, K = 0 implique

$$\pi_k^2+\pi_\ell^2 = \frac{8m_k m_\ell}{m_k^{-1}+m_\ell^{-1}} .$$

Dans le cas $p_\varphi=0$ la fonction K_o se réduit à

$$K_o(\underline{\alpha},\underline{\pi},0) = \frac{1}{8}\,\underline{\pi}^T\,B(\underline{\alpha})\underline{\pi} - \sum m_k m_\ell a_k a_\ell, \tag{34}$$

où $B(\underline{\alpha})$ est la matrice symétrique

$$B(\underline{\alpha})=\begin{pmatrix} \frac{a_1}{m_1}\alpha^2+\frac{a_2}{m_2}\alpha_3^2+\frac{a_3}{m_3}\alpha_2^2 & -\frac{a_3}{m_3}\alpha_1\alpha_2 & -\frac{a_2}{m_2}\alpha_1\alpha_3 \\ -\frac{a_3}{m_3}\alpha_1\alpha_2 & \frac{a_2}{m_2}\alpha^2+\frac{a_3}{m_3}\alpha_1^2+\frac{a_1}{m_1}\alpha_3^2 & -\frac{a_1}{m_1}\alpha_2\alpha_3 \\ -\frac{a_2}{m_2}\alpha_1\alpha_3 & -\frac{a_1}{m_1}\alpha_2\alpha_3 & \frac{a_3}{m_3}\alpha^2+\frac{a_1}{m_1}\alpha_2^2+\frac{a_2}{m_2}\alpha_1^2 \end{pmatrix}. \tag{35}$$

Les équations du mouvement sont

$$\frac{d}{d\tau}\underline{\alpha} = \frac{1}{4}B(\underline{\alpha})\underline{\pi}, \qquad \frac{d}{d\tau}\underline{\pi} = -\frac{\partial K}{\partial\underline{\alpha}} \tag{36}$$

Par contre, le cas $p_\varphi\neq 0$ est plus compliqué: ni l'hamiltonien ni les $\pi_j(\tau)$ sont réguliers. Dans une collision binaire à $\tau=0$ on a $\pi_j(\tau)=O(\tau^{-1})$.

3. HOMOTHÉTIE.

En suivant McGehee (1974, 1978), Devaney (1980) nous introduisons une échelle variable au moyen du moment d'inertie

$$I = \sum_{j=1}^{3} m_j |x_j|^2 \tag{37}$$

qui peut être exprimé par les distances a_j :

$$mI = \sum m_j m_k a_\ell^2, \qquad m = \sum m_j \tag{38}$$

(voir Wintner (1941), § 322 bis). La racine carrée r du moment d'inertie (un multiple du rayon d'inertie) est introduite comme distance typique dans le système des trois corps:

$$r^2 = I. \tag{39}$$

Dans une collision triple $r \to 0$, et les variables dépendantes sont $O(r^\nu)$ avec des exposants $\nu > 0$. Nous introduisons des variables normalisées, indiquées par le symbole ($\tilde{\ }$), qui ont des limites finies lorsque $r \to 0$:

$$a_j = \tilde{a}_j r, \qquad p_j = \tilde{p}_j r^{-1/2}, \qquad p_\varphi = \tilde{p}_\varphi r^{1/2} \tag{40}$$

$$\alpha_j = \tilde{\alpha}_j r^{1/2}, \qquad \pi_j = \tilde{\pi}_j, \qquad d\tau = d\tilde{\tau} . r^{-3/2} . \tag{41}$$

La relation entre le temps physique t et $\tilde{\tau}$ est donnée par l'équation différentielle

$$\frac{dt}{d\tilde{\tau}} = r^{3/2} \tilde{a}_1 \tilde{a}_2 \tilde{a}_3 . \tag{42}$$

Equ. (39) implique l'existence d'une nouvelle relation invariante dans les variables normalisées:

$$\sum m_j m_k \tilde{a}_\ell^2 = m \tag{43}$$

ou explicitement

$$m_1 m_2 (\tilde{\alpha}_1^2 + \tilde{\alpha}_2^2)^2 + m_2 m_3 (\tilde{\alpha}_2^2 + \tilde{\alpha}_3^2)^2 + m_3 m_1 (\tilde{\alpha}_3^2 + \tilde{\alpha}_1^2)^2 = m. \tag{44}$$

La surface (44) d'ordre 4 dans l'espace $(\tilde{\alpha}_1, \tilde{\alpha}_2, \tilde{\alpha}_3)$ a la topologie d'une S^2.

A présent nous allons établier une équation différenteille pour r. Introduisons

$$v = \frac{dI}{dt} . \tag{45}$$

La différentiation de (37) par rapport au temps donne

$$v = 2\mathrm{Re}\sum m_j \bar{X}_j \frac{dX_j}{dt} = 2\ \mathrm{Re}\sum \bar{X}_j P_j, \tag{46}$$

et avec (15), (28) on obtient

$$v = \sum \alpha_j \pi_j = \underline{\alpha}^T \underline{\pi}. \tag{47}$$

Enfin, nous introduisons

$$v = \tilde{v} \cdot r^{1/2}, \qquad \tilde{v} = \sum \tilde{\alpha}_j \pi_j \ ; \tag{48}$$

en utilisant (39), (42), (48) nous obtenons de l'équation (45)

$$\frac{dr}{d\tilde{\tau}} = \frac{r}{2}\, \tilde{a}_1\, \tilde{a}_2\, \tilde{a}_3\, \tilde{v}. \tag{49}$$

Pour transformer les équations du mouvement (31) aux variables normalisées on exprime K par $\underline{\tilde{\alpha}}$, $\underline{\pi}$, $\tilde{p}_\varphi$,

$$K = r^2 K_0(\underline{\tilde{\alpha}}, \underline{\pi}, p_\varphi) - r^3 h\, \tilde{a}_1\, \tilde{a}_2\, \tilde{a}_3, \tag{50}$$

et on trouve les équations différentielles

$$\begin{aligned} \frac{d\underline{\tilde{\alpha}}}{d\tilde{\tau}} &= \frac{\partial}{\partial \underline{\pi}} K_0(\underline{\tilde{\alpha}}, \underline{\pi}, \tilde{p}_\varphi) - \frac{1}{4}\, \tilde{a}_1\, \tilde{a}_2\, \tilde{a}_3\, \tilde{v}\, \underline{\tilde{\alpha}} \\ \frac{d\underline{\pi}}{d\tilde{\tau}} &= - \frac{\partial}{\partial \tilde{\alpha}} K_0(\underline{\tilde{\alpha}}, \underline{\pi}, \tilde{p}_\varphi) + hr \frac{\partial}{\partial \underline{\tilde{\alpha}}} (\tilde{a}_1\, \tilde{a}_2\, \tilde{a}_3) \end{aligned} \tag{51}$$

qui admettent les intégrales premières

$$K_0(\underline{\tilde{\alpha}}, \underline{\pi}, \tilde{p}_\varphi) - rh\, \tilde{a}_1\, \tilde{a}_2\, \tilde{a}_3 = 0 \tag{52}$$

$$\sum m_j m_k \tilde{a}_\ell^2 = m.$$

On établit une équation différentielle pour $\tilde{v}$ en commençant avec l'identité bien connue

$$\frac{dv}{dt} = \frac{d^2 I}{dt^2} = 2U + 4h \tag{53}$$

(voir Wintner (1941), § 315). En utilisant les relations (4), (41), (42), (48), (49) on obtient

$$\frac{d\tilde{v}}{d\tilde{\tau}} + \frac{\tilde{v}^2}{4} \tilde{a}_1 \tilde{a}_2 \tilde{a}_3 = 2 \sum m_k m_\ell \tilde{a}_k \tilde{a}_\ell + 4rh \tilde{a}_1 \tilde{a}_2 \tilde{a}_3. \quad (54)$$

Une relation même plus importante s'obtient de l'inégalité de Sundman (1909),

$$I \ddot{I} - \frac{1}{4} \dot{I}^2 \geq 2hI + p_\varphi^2. \quad (55)$$

Avec les équations (39), (45), (40), (48) on transforme cette inégalité classique en

$$r \frac{dv}{dt} - \frac{1}{4} \tilde{v}^2 \geq 2hr + \tilde{p}_\varphi^2 .$$

D'autre part, la différentiation de l'équ. (48), en utilisant (42) et (49), donne

$$r \frac{dv}{dt} = \frac{1}{\tilde{a}_1 \tilde{a}_2 \tilde{a}_3} \frac{d\tilde{v}}{d\tilde{\tau}} + \frac{1}{4} \tilde{v}^2,$$

d'où

$$\frac{d\tilde{v}}{d\tilde{\tau}} \geq \tilde{a}_1 \tilde{a}_2 \tilde{a}_3 \, (2hr + \tilde{p}_\varphi^2). \quad (56)$$

4. LA VARIÉTÉ DE COLLISION TRIPLE.

Les équations différentielles (42), (49), (51) pour les variables t, r, $\underline{\tilde{\alpha}}$, $\underline{\pi}$ permettent en principe une description complète des solutions physiques du problème des trois corps (pour lesquelles $r>0$). D'autre part, la fonction $r(\tilde{\tau})=0$ est solution de l'équation (49). Considérons donc les équations (51) avec $r=0$. La variété des solutions de (51) avec $r=0$ est la variété de collision triple T introduite par McGehee (1972) pour l'exemple du problème des trois corps à une dimension. T est une limite de la variété de toutes les solutions du problème plan des trois corps. Les trajectoires sur T n'ont pas de signification physique directe: le temps t n'avance pas (équ. (42)); elles caractérisent le comportement à l'instant de collision triple.

Dans les équations (51), (52) les quantités r, h ne se présentent que sous forme de leur produit $\tilde{h}=rh$.

Par conséquent la variété *T* est équivalente à la variété des solutions d'énergie h=0 (Waldvogel, 1979). Pour construire un mouvement d'énergie nul correspondant à une trajectoire de *T* on ajoute l'équation (49) avec r>0.

Sur la variété *T* l'équ. (56) implique

$$\frac{d\tilde{v}}{d\tilde{\tau}} \geq \tilde{a}_1\tilde{a}_2\tilde{a}_3\tilde{p}_\varphi^2 \geq 0; \tag{57}$$

par conséquent le flot sur *T* est du type gradient, la quantité $\tilde{v}$ jouant le rôle d'une fonction de Lyapounov. Parceque la relation (57) se déduit de l'inégalité de Sundman (55) nous introduisons le terme "fonction de Sundman" pour $\tilde{v}$.

Maintenant nous allons discuter l'intégrale première du moment cinétique. D'aprés l'équ. (49),

$$p_\varphi = \tilde{p}_\varphi \cdot r^{1/2}, \tag{58}$$

le moment cinétique p_φ est nécessairement nul sur la variété *T* (où r=0). Mais en général $\tilde{p}_\varphi \neq$ const sur *T*, c'est-à-dire l'intégrale première du moment cinétique est perdue. D'autre part, pour r>0 les équations (58), (49) impliquent

$$\frac{d\tilde{p}_\varphi}{d\tilde{\tau}} = -\frac{1}{4}\,\tilde{a}_1\tilde{a}_2\tilde{a}_3 \cdot \tilde{p}_\varphi. \tag{59}$$

Par conséquent, $\tilde{p}_\varphi=0$ est une relation invariante même pour r=0. Introduisons donc la sous-variété invariante $N \subset T$, définie par r=0, $\tilde{p}_\varphi=0$. *N*, dénotée la variété non tournante, est donnée par les équations différentielles (51) avec $r=p_\varphi=0$,

$$\frac{d\underline{\tilde{\alpha}}}{d\tilde{\tau}} = \frac{1}{4}\,B(\underline{\tilde{\alpha}})\underline{\pi} - \frac{1}{4}\tilde{a}_1\tilde{a}_2\tilde{a}_3\tilde{v}\underline{\tilde{\alpha}}, \qquad \tilde{v}=\underline{\tilde{\alpha}}^T\underline{\pi}$$

$$\frac{d\underline{\pi}}{d\tilde{\tau}} = -\frac{\partial}{\partial\underline{\tilde{\alpha}}}\,K_0(\underline{\tilde{\alpha}},\underline{\pi},0), \tag{60}$$

où

$$K_0(\underline{\tilde{\alpha}},\underline{\pi},0) = \frac{1}{8}\,\underline{\pi}^T B(\underline{\tilde{\alpha}})\underline{\pi} - \sum m_j m_k \tilde{a}_j\tilde{a}_k, \tag{34}$$

avec les 2 intégrales premières

$$\sum m_j m_k (\tilde{\alpha}_j^2+\tilde{\alpha}_k^2)^2 = m, \tag{61}$$

$$K_o(\underline{\tilde{\alpha}},\underline{\pi},0)=0. \tag{62}$$

Le flot sur N est du type gradient: la fonction de Sundman $\tilde{v}$ ne décroît jamais. Toutes les collision binaires sont régularisées sur N.

Le type topologique de la variété N se déduit des intégrales premières (61), (62). L'intégrale de la normalisation, (61), représente une surface de degré 4 dans l'espace des configurations $\underline{\tilde{\alpha}}$. Pour $m_j>0$ elle est du type topologique de la sphère S^2.

Dans les figures 4 et 5 nous présentons cette surface dans une projection axonométrique où on a tracé les niveaux U = const de la fonction des forces.

L'intégrale première (62) est une forme quadratique en $\underline{\pi}$ avec la matrice symétrique $B(\underline{\tilde{\alpha}})$. En général $B(\underline{\tilde{\alpha}})$ est définie positive; donc l'équ. (62) représente un ellipsoïde. Les seules exceptions sont les 6 poles $\tilde{\alpha}_j=\tilde{\alpha}_k=0$ où l'ellipsoïde dégénère en un cylindre. La variété N a donc le type topologiques $S^2 \times S^2$ moins 12 points. Ces 12 points, identifiés en pairs, correspondent aux 6 évasions possibles (3 binaires, $t\to\pm\infty$). La topologie de T sans régularisation a été discutée par Simò (1980).

5. LES POINTS D'ÉQUILIBRE.

Dans ce chapitre nous discutons quelques résultats sur les points critiques (points d'équilibre) du flot sur N. Les points critiques sont les solutions stationnaires $\underline{\tilde{\alpha}}(\tilde{\tau})=\underline{\tilde{\alpha}}$=const, $\underline{\pi}(\tilde{\tau})=\underline{\pi}$=const des équ. (60) qui se réduisent à un système d'équations algébriques. Si la configuration $\underline{\tilde{\alpha}}$ d'un point d'équilibre est connue le moment $\underline{\pi}$ s'obtient à partir du système linéaire

$$B(\underline{\tilde{\alpha}})\underline{\pi} = \tilde{a}_1\tilde{a}_2\tilde{a}_3\tilde{v}\underline{\tilde{\alpha}}, \tag{63}$$

où $\tilde{v}$ est donné par

$$\tilde{v} = \pm\sqrt{8\tilde{U}}, \qquad \tilde{U} = \sum \frac{m_k m_\ell}{\tilde{a}_j}, \tag{64}$$

Figure 4.

$m_1=m_2=m_3=1$

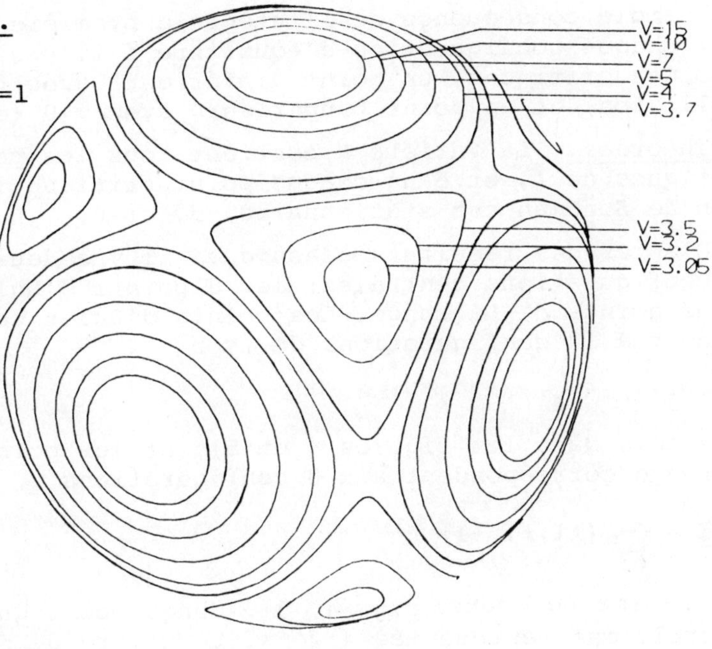

Figure 5.
$m_1=1$, $m_2=10$,
$m_3=100$

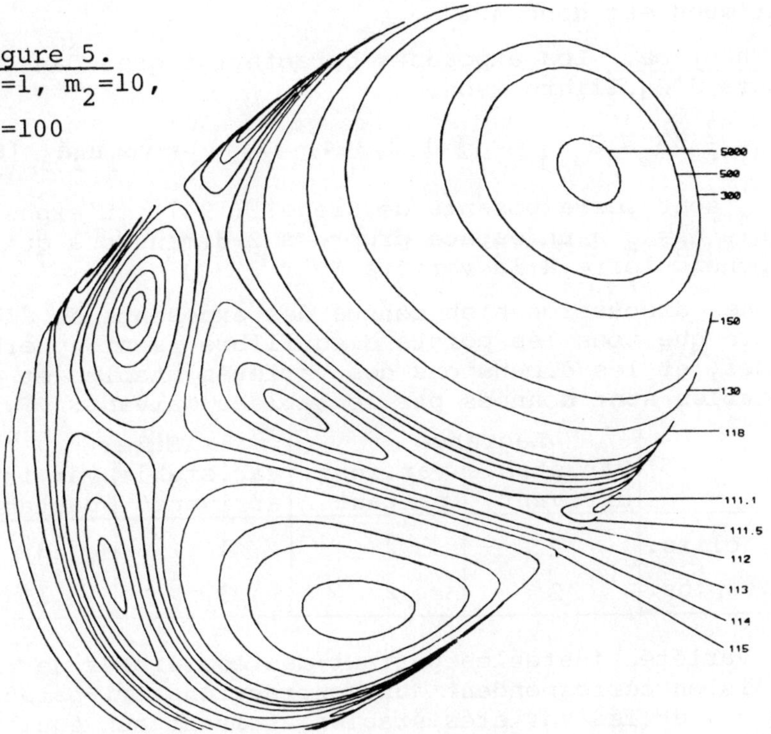

une simple conséquence de l'intégrale première (62). Pour chaque configuration d'équilibre $\underline{\tilde{\alpha}}$ il y a donc 2 points critiques, un point "inférieur" avec $\tilde{v}<0$ (collision) et un point "supérieur" avec $\tilde{v}>0$ (explosion).

Théorème. La variété N contient tous les points critiques de T, et dans chaque point critique la fonction de Sundman est stationnaire, $d\tilde{v}/d\tilde{\tau}=0$.

D'aprés des résultats classiques il y a deux types de configurations centrales: les 3 points d'Euler et les 2 points de Lagrange. Les points d'Euler correspondent à 12 configurations du type

$$\pm\tilde{\alpha}_j, \quad \pm\tilde{\alpha}_k, \quad \tilde{\alpha}_\ell=0$$

(les cols dans les figures 4 et 5), et les points de Lagrange correspondent aux 8 configurations

$$\underline{\tilde{\alpha}} = \frac{\tilde{\alpha}}{\sqrt{3}}(\pm 1,\pm 1,\pm 1)$$

(les minima de U dans fig. 4, 5). Chaque configuration centrale est représentée 4 fois. Le nombre de points critiques est donc 40.

Théorème. Les exposants caractéristiques dans les points d'équilibre sont

$$\lambda_j = \frac{3}{4}\tilde{v}\tilde{a}_1\tilde{a}_2\tilde{a}_3\sigma_j \quad (j=1,2,3,4) \quad \lambda_5=\lambda_6=-\tilde{v}\tilde{a}_1\tilde{a}_2\tilde{a}_3 \quad (65)$$

où σ_j sont les exposants de Siegel (1941). L'exposant double $\lambda_5=\lambda_6$ a un espace propre à 2 dimensions qui est perpendiculaire à la variété N.

Une discussion bien connue des exposants de Siegel montre que tous les points d'équilibre sont hyperboliques, et les dimensions des variétés stables et instables sont données par le tableau suivant.

	Lagrange		Euler	
	var.stable arrivant	var.inst. partant	var.stable arrivant	var.inst. partant
$\tilde{v}<0$,collis.	2	2	1	3
$\tilde{v}>0$,explos.	2	2	3	1

Les variétés instables partant des équilibres de collision correspondent aux contractions tri-paraboliques, et les variétés stables arrivant aux équilibres d'explosion correspondent aux évasions tri-paraboliques.

Le comportement du flot sur *N* dépend fortement des connections entre les points d'équilibre. Pour terminer nous mentionnons deux connections qui ont été trouvées numériquement dans le cas $m_1=m_2=m_3$: Lagrange, $\tilde{v}<0 \rightarrow$ Euler, $\tilde{v}>0$ par Waldvogel (1976) et Lagrange, $\tilde{v}>0 \rightarrow$ Euler, $\tilde{v}>0$ (solution isoscèle) par Marchal et Losco (1980).

REMERCIEMENTS.

L'auteur exprime sa sincère gratitude à l'Université de Paris VI pour le support généreux de ce travail. Je remercie vivement le Professor Fernand Nahon des discussions stimulantes, des suggestions valables concernant ce travail et de son hospitalité chaleureuse.

RÉFÉRENCES.

Devaney, R.L. (1980): Triple Collision in the Planar Isosceles Three-Body Problem. Inventiones Math. 60, 249-267.

Lemaître, G. (1954): Régularisation dans le problème des trois corps. Bull. Classe des Sci., Acad.Roy. Belg. 40, 759-767.

Lemaître, G. (1964): The Three-Body Problem. NASA CR-110.

Marchal, C. and Losco, L. (1980): Analysis of the Neighborhood of Triple Collisions and Tri-Parabolic Escapes in the Three-Body Problem. Astron. Astroph. 84, 1-6.

McGehee, R. (1974): Triple Collision in the Collinear Three-Body Problem. Inventiones Math. 27, 191-227.

McGehee, R. (1978): Singularities in Classical Celestial Mechanics. Proc. Internat. Congress of Mathematicians, Helsinki, 827-834.

Murnaghan, F.D. (1936): A Symmetric Reduction of the Planar Three-Body Problem. Amer. J. Math. 58,829-832.

Siegel, C.L. (1941): Der Dreierstoss. Ann. Math. 42, 127-168.

Simò, C. (1980): Masses for which Triple Collision is Regularizable. Celest. Mech. 21, 25-36.

Sundman, K.F. (1909): Nouvelles recherches sur le problème des trois corps. Acta Soc. Sci. Fenn. 35, 1-27.

Sundman, K.F. (1913): Mémoire sur le problème des trois corps. Acta Math. 36, 105-179.

van Kampen, E.R. and Wintner, A. (1937): On a Symmetric Reduction of the Problem of Three Bodies. Amer. J. Math. 59, 153-166, 269.

Waldvogel, J. (1972): A New Regularization of the Planar Problem of Three Bodies. Celest. Mech. 6, 221-231.

Waldvogel, J. (1976): The Three-Body Problem Near Triple Collision. Celest. Mech. 14, 287-300.

Waldvogel, J. (1979): Stable and Unstable Manifolds in Planar Triple Collision. V. Szebehely (ed.), Instabilities in Dynamical Systems, Reidel, 263-271.

Waldvogel, J. (1979a): La variété de collision triple. C.R. Acad. Sc. Paris 288, Sér. A., 635-637.

Wintner, A. (1941): The Analytical Foundations of Celestial Mechanics. Princeton University Press.

ABSTRACT. The planar problem of three bodies is described by means of Murnaghan's symmetric variables (the sides a_j of the triangle and an ignorable angle), which directly allow for the elimination of the nodes. Then Lemaître's regularized variables $\alpha_j=\sqrt{\alpha^2-a_j}$, where $\alpha^2=\frac{1}{2}(a_1+a_2+a_3)$, as well as their canonically conjugated momenta are introduced. By finally applying McGehee's scaling transformation $\alpha_j=r^{1/2}\tilde{\alpha}_j$, where r^2 is the moment of inertia, a system of 7 differential equations (with 2 first integrals) for the 5-dimensional triple collision manifold T is obtained.
Moreover, the zero angular momentum solutions form a 4-dimensional invariant submanifold $N\subset T$ represented by 6 differential equations with polynomial right-hand sides. The manifold N is of the topological type $S^2 x S^2$ with 12 points removed, and it contains all 5 restpoints (each one in 8 copies). The flow on T is gradient-like with a Lyapounov function stationary in the 40 restpoints.
These variables are well suited for numerical studies of planar triple collision.

ERGODIC THEORY AND AREA PRESERVING MAPPINGS

Robert W. Easton

University of Colorado, Boulder, Colorado

ABSTRACT. Much work has been done recently in numerical (computer) studies of area preserving mappings of the plane. Often the orbit of a single point seems to fill a region in the plane having positive area. Such a region is called an "ergodic zone". In this paper we give an exposition of some mathematical techniques which can be used to show that under suitable hypotheses, the closure of an orbit of a single point has positive Lebesque measure. We then apply these techniques to show that a family of mappings called linked twist maps are ergodic.

INTRODUCTION.

The qualitative (topological) study of solutions of Hamiltonian systems of differential equations has a long history. This subject is important for applications and it is also beautiful in its mathematical development. I will concentrate in this paper on a very small part of the subject, the study of area preserving mappings of the plane. Such a mapping might arise for example, as the Poincaré map of a surface of section transverse to a periodic orbit of a Hamiltonian system of differential equations with two degrees of freedom. Many experiments have been performed and are continuing to be performed which use a computer to iterate the action of an area preserving map T, on a point x. The computer is programmed to apply T to an input consisting of a pair of numbers $x = (x_1, x_2)$, and to plot the first $N+1$ points $x, T(x), T^2(x), \ldots, T^n(x)$ of the orbit of x. Usually $N \leq 10^7$ and three different results occur. The first is that the plot consists of P points where P is much smaller than N. An acceptable explanation for this result is that x is a periodic point of period P so that

V. Szebehely (ed.), Applications of Modern Dynamics to Celestial Mechanics and Astrodynamics, 267–276.

$T^P(x) = x$. The second result is that the picture consists of many points which seem to lie on a simple closed curve. This result is explained by the celebrated KAM theorem, (Arnold, 1978 and Moser, 1973) which states (roughly) that for many initial conditions x, the closure in R^2 of the set $\{T^n(x) : n = 1,2,\ldots\}$ is a smooth simple closed curve. The third result is the one which interests us and which does not as yet have a satisfactory mathematical explanation. The result is that the picture consists of many points which do not seem to be regularly distributed in any way. One explanation which is not entirely satisfying is that x is a periodic point of very long period. The persistent experimenter will increase N again and again and plot $\{T^n(x) : N_1 \leq n \leq N\}$ to see how the picture changes. Usually it either looks about the same or looks "worse". This suggests the following mathematical question: Does there exist an x such that the closure of the orbit of x has positive measure? The answer depends of course, on the transformation T. It is known that if there is a compact set C having positive measure and if T, restricted to C is ergodic, then for almost every x belonging to C the closure of the orbit of x is equal to C. Recently, techniques have been developed (Katok et al., 1977 and Pesin, 1977) which can be used to prove that a given transformation T is ergodic in a given region C. Our goal is to explain these techniques and to show how they can be applied. At present the theoretical background to which I refer is not complete and much work remains to be done. A question which is not answered at this time is the following: Does there exist a $C^{3+\varepsilon}$ perturbation T^* of a twist map T and a compact set C having positive measure such that T^* restricted to C is ergodic?

1. Examples of Area Preserving Mappings of the Plane.

Example (1). Twist maps. Here $\mathbb{C}$ will denote the complex plane. Let $A = \{z \in \mathbb{C} : 1 \leq |z| \leq 2\}$. A twist mapping of the annulus A is a mapping $T: A \to A$ given by the equation $\tau(z) = e^{i\theta(r)}z$ where $r = |z|$ and $\theta:[1,2] \to R^1$ is a smooth increasing function. Twist mappings preserve Lebesque measure on $\mathbb{C}$. Given $z \in A$ the orbit of z lies on the circle of radius $r = |z|$. If $\theta(r)/2\pi$ is irrational, the orbit of z is dense in this circle. If $\theta(r)/2\pi$ is rational the orbit of z is periodic. Thus the orbit structure of a twist map has a very simple description. However, the orbit structure of a map which is close (in the C^r topology) to a twist mapping is still not well understood. It follows from the celebrated invariant curve theorem (Katok, 1977) that if $T^* : A \to \mathbb{C}$ is an area preserving map which is sufficiently close in the C^5 topology to T, then there is a set $X \subset A$ of positive measure such that for each $x \in X$ the closure of the T^* orbit of x is an invariant simple closed curve which is close in the C^5 topology to the circle $|z| = |x|$. This result is still true in the $C^{3+\varepsilon}$ topology, and is false in the $C^{3-\varepsilon}$ topology (Herman, 1981). Little is known about the orbit structure of T^* in $A-X$. However, the measure of X approaches the measure of A as $T^* \to T$ in the C^5 topology.

Example (2). Linked twist maps. Let $\{A_k\}$ be a collection of annuli contained in a smooth 2-manifold M. Suppose that ω is a two-form on M and that T_k is a C^r diffeomorphism which preserves ω. Further suppose that T_k is the identity map on $M-A_k$, and is topologically conjugate to a twist map on A_k. The composition map $T = T_n o T_{n-1} o \ldots T_1$ is called a <u>linked twist map</u>. Linked twist maps were first studied by Easton (1975). The simplest examples of linked twist maps occur on the torus T^2. Under suitable conditions linked twist maps on T^2 were shown to be ergodic on the union of the annuli and even Bernouillian (Burton and Easton, 1979 and Wojtkowski, 1979). Consequently, the orbit of almost every point is dense in the union of the annuli. This work was extended to cover a wide class of linked twist maps by Przytycki (1981). The annuli are required to cross transversely, and thus linked twist maps are not close to twist maps even in the C^1 topology. A proof of ergodicity is outlined in section 2.

Example (3). Hyperbolic twist maps. Perhaps the simplest way to perturb a twist map is to combine it with a linear hyperbolic map. Such maps have the form

$$f: \begin{pmatrix} x \\ y \end{pmatrix} \to \begin{pmatrix} a\cos(q)\, x + a \sin(q)\, y \\ -a^{-1}\sin(q)\, x + a^{-1}\sin(q)\, y \end{pmatrix}$$

where $a > 1$ is a real number and q is a function of x and y, say $q = (x^2 + y^2)^b$ where b is a positive constant. Such maps were studied numerically by Easton (1979). Simo (1981) has discovered that such maps occur in the problem of Hénon-Heiles.

The origin is a hyperbolic fixed point of f. For $a = b = 1.5$ the stable and unstable manifolds of the origin according to numerical evidence cross transversally giving rise to the complicated orbit structure associated with transverse homoclinic points. Little is known about the orbit structure of hyperbolic twist maps, but I conjecture that the closure C of the orbit of a transverse homoclinic point has positive measure in this case and that the hyperbolic twist map restricted to C is ergodic.

2. Ergodicity of Linked Twist Maps.

The intention in this section is to present the ideas that enter the proof that linked twist maps are ergodic. For more detail the reader should consult Burton and Easton (1979), Przytycki (1981) and Wojtkowski (1979). For simplicity we will consider linked twist mappings on the union of two annuli contained in the torus. This study of linked twist maps involves some of the difficulties which are associated with studying so-called "ergodic zones" in celestial mechanics.

Let $T^2 = R \times R/Z \times Z$ where R^1 and Z respectively denote the groups of real numbers and integers. $R^1 \times R^1/Z \times Z$ denotes the quotient group formed from the direct product $R^1 \times R^1$ of R^1 and the subgroup $Z \times Z$. Let $q : R^1 \times R^1 \to T^2$ be the quotient projection. Define

$$A_1 = q([0,1/2] \times R^1)$$

$$A_2 = q(R^1 \times [0,1/2]).$$

Then A_1 and A_2 are annuli in the torus T^2. Suppose that for $k = 1,2; \alpha_k : R^1 \to R^1$ is a C^2 function with the following properties

(a) $\alpha_k(0) = 0$ and $\alpha_k(t) = n$ for $t \in [1/2,1]$ where $n > 1$ is an integer

(b) $\alpha_k(t+1) = \alpha_k(t) + n$ for all t

(c) $\alpha_k'(t) > 0$ whenever $0 < t < 1/2$ and $\alpha'(0) = \alpha'(1/2) = 0$.

Define $\tau_1 : R^2 \to R^2$ by $\tau_1(x,y) = (x+\alpha_1(y),y)$,

$\tau_2 : R^2 \to R^2$ by $\tau_2(x,y) = (x,y+\alpha_2(x))$, and define

$\tau = \tau_2 \circ \tau_1$. Thus $\tau(x,y) = (x+\alpha_1(y),y+\alpha_2(x+\alpha_1(y)))$. Define

$T : L \to L$ by $T = qT_2qT_1$ where $L = A_1 \cup A_2$. Then T is a linked twist map which preserves the measure μ on L defined by $\mu(E) = m(q^{-1}(E) \cap [0,1] \times [0,1])$ where m denotes Lebesque measure on R^2. We picture L as the shaded region in the unit square as shown in Figure 1. $q : [0,1) \times [0,1) \to T^2$ is 1-1, onto and we will use q^{-1} to determine coordinates on T^2.

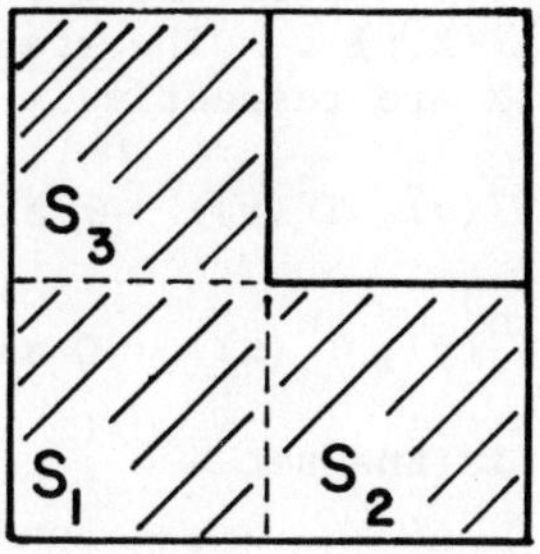

Figure 1

Partition L into three sets S_1, S_2, S_3 as follows: Let

$S_1 = A_1 \cap A_2$, $S_2 = A_1 - S_1$, $S_3 = A_2 - S_1$.

We need to recall some well known results from ergodic theory and stable manifold theory which can be used to prove that a smooth measure preserving diffeomorphism of a smooth manifold is ergodic.

Let (X,d) be a compact metric space and let μ be a Borel measure on X with $\mu(X) = 1$ and μ positive on open sets. We assume throughout this section that $T: X \to X$ is a homeomorphism of X which preserves μ in the sense that $\mu(T(E)) = \mu(T^{-1}(E)) = \mu(E)$ for every Borel set E. For $f \in L^1(X,\mu)$ and $n \geq 0$ define

$$A_n^+(f)(x) = 1/n \sum_{j=0}^{n-1} f(T^j(x)),$$

$$A_n^-(f)(x) = 1/n \sum_{j=0}^{-n+1} f(T^j(x)).$$

Consider A_n^+ and A_n^- as transformations of $L^1(X,\mu)$. We have the following Theorem: (Birkhoff-Kinchin):

a) $P^{\pm}(f) = \lim_{n\to\infty} A_n^{\pm}(f)$ exists in $L^1(X,\mu)$.

b) $P^+(f)(x) = P^-(f)(x)$ for almost every $x \in X$

c) $\|P^{\pm}(f)\|_1 = \|f\|_1$ where $\| \ \|_1$ denotes the L^1 norm on $L^1(X,\mu)$.

For a proof of the theorem see Billingsley (1965).

Definitions: T is ergodic provided $P^+(f)$ is constant almost everywhere for each $f \in L^1(X,\mu)$. The stable and the unstable manifolds of a point $x \in X$ are respectively the sets

$$W^s(x) = \{y \in X : d(T^n(y), T^n(x)) \to 0 \text{ as } n \to \infty\}$$

$$W^u(x) = \{y \in X : d(T^n(y), T^n(x)) \to 0 \text{ as } n \to -\infty\}.$$

The orbit of a point x is the set

$$O(x) = \{T^n(x) : n \text{ is an integer}\}.$$

T is topologically transitive if for each pair of non-empty open sets U and V in X there exist an integer n such that $T^n(U) \cap V \neq \phi$. $f \in L^1(X,R^1)$ is locally constant almost everywhere if there exists a set $Q(f)$ with $\mu(Q(f)) = 0$ such that for each $x \in X - Q(f)$, f is constant almost everywhere on some neighborhood of x.

Proposition: If T is topologically transitive and if for each continuous function f, P^+f is locally constant almost everywhere, then T is ergodic.

Proof: Since $C^0(X,R^1)$ is dense in $L^1(X,R^1)$ and P^+ is continuous and since the constant functions form a closed subspace of $L^1(X,R^1)$ it is sufficient to show that P^+f is constant almost everywhere for $f \in C^0(X,R^1)$. By hypothesis there exists a set $Q(f)$ of measure zero such that for $x,y \in X - Q(f)$ there exist neighborhoods $V(x)$ and $V(y)$ of x and y such that P^+f is constant almost everywhere on $V(x)$ and on $V(y)$. Also by hypothesis there exists n such that $T^n(V(x)) \cap V(y) \neq \phi$. Because P^+f is constant on orbits it follows that P^+f is equal to the same constant on $V(x)$, and on $T^n(V(x)) \cap V(y)$ and hence on $V(y)$. It follows that P^+f is constant almost everywhere on X. This completes the proof.

A procedure for showing that P^+f is locally constant almost everywhere involves examining the "foliations" or partitions of X by the stable and unstable manifolds of points. This procedure is well known to ergodic theoretists and is described briefly by Weiss (1975).

Proposition: If f is a continuous (hence uniformly continuous) function on X then $P^+(f)$ is constant on the stable manifold of x provided $P^+(f)(x)$ exists. Similarly $P^-(f)$ is constant on the unstable manifold of x provided $P^-(f)(x)$ exists.

Proof: Let $y \in W^s(x)$. Given $\varepsilon>0$ choose $\delta>0$ such that $d(x_1,x_2)<\delta$ implies that $|f(x_1)-f(x_2)|<\varepsilon$. Choose $m>0$ such that $d(T^k(y),T^k(x))<\delta$ whenever $k \geq m$. For $n>m$,

$$A_n^+(f)(y) - A_n^+(f)(x) = 1/n \sum_{j=0}^{m} f(T^j(y)) - f(T^j(x))$$

$$+ 1/n \sum_{j=m+1}^{n-1} f(T^j(y)) - f(T^j(x)).$$

Hence $|A_n^+(f)(y) - A_n^+(f)(x)| \leq \frac{1}{n}\left|\sum_{j=0}^{m} f(T^j(y)) - f(T^j(x))\right| + \varepsilon/n.$

Therefore

$|A_n^+(f)(y) - A_n^+(f)(x)| \to 0$ as $n \to \infty$ and it follows that $P_n^+(f)(y)$ is defined and is equal to $P_n^+(f)(x)$.

<u>Definitions:</u> Let $x \in X$ and $F \subset X$. Define $\mathrm{Net}(x,F) = \bigcup\{W^u(y) : y \in W^s(x),\ y \notin F\}$. If f is a continuous real valued function on X define $F(f) = \{x \in X : \text{either } (P^+f)(x) \neq (P^-f)(x) \text{ or } (P^+f)(x) \text{ is undefined}\}$. The following result follows from the definition:

<u>Proposition</u>: P^-f is constant on $\mathrm{Net}(x,F(f))$.

<u>Definition</u>: A transformation T of X has <u>thick nets</u> provided that given $F \subset X$ with $\mu(F)=0$, for almost every $x \in X$ there exists a neighborhood U of x such that $\mu(\mathrm{Net}(x,F) \cap U) = \mu(U)$.

It follows immediately from this definition that if a transformation has thick nets then P^+f is locally constant almost everywhere for each continuous function f on X.

In order to verify that a transformation has thick nets one must leave the abstract development above and assume that X is a smooth Riemanian manifold and that T is a C^2 smooth diffeomorphism of X which preserves a smooth measure μ. The reader is referred to the work of Pesin (1977) which can be used to show that such a transformation has thick nets provided that for almost every x, $\chi^+(x,v) \neq 0$ where $\chi^+(x,v)$ is the Lyapunov characteristic exponent associated with x and a vector $v \neq 0$ tangent to X at x.

By definition $\chi^+(x,v) = \lim_{n\to\infty} \sup(\frac{1}{n} \log \| dT^n v\|)$.

Consequently, to show that a transformation is ergodic one must show that it is topologically transitive and that the Lyapunov characteristic exponents associated with almost every point are different from zero.

The steps in doing this for a linked twist map on the torus are briefly described below. For more detail see Burton and Easton (1979).

Step 1: There is a general result from ergodic theory which we call the <u>Frequency Lemma</u>: Suppose that $T: X\to X$ is a Borel measure preserving homeomorphism of a metric space X. Suppose E is a Borel set with $0<\mu(E)<\infty$. Let $R(E) = \{x:T^n(x)\in E$ for infinitely many $n\geq 0\}$.

Let $R_\nu(E) = \{x \in R(E) : (P^+\chi_E)(x) \geq \nu\}$.

Then $\mu(\bigcup_{\nu>0} R_\nu(E)) = \mu(R(E))$.

Thus for almost every point x whose orbit intersects E infinitely often we have the result that the orbit of x intersects E with some positive frequency ν. For a proof of this Lemma see Burton and Easton (1979). Now consider $E = T_1^{-1}(A_1)\cap A_2$. For $q = (x,y) \in E$ the differential of the transformation T at q is the matrix

$$dT(q) = \begin{pmatrix} 1 & a \\ b & 1+ab \end{pmatrix}$$

where $a = \alpha_1'(y)$ and $b = \alpha_2'(x+\alpha_1(y))$. From the definition of E it follows that $a \neq 0$ and $b \neq 0$ provided that q does not belong to the boundary of L. The matrix $dT(q)$ is hyperbolic. One can show that the orbit of almost every point $q \in L$ intersects E infinitely often and therefore intersects E with some positive frequency. From this one can show that almost every point $q \in L$ has all its Lyapunov characteristic exponents different from zero. Hence from Pesin's work T has thick nets.

Step 2: To show that a linked twist map T of the torus is topologically transitive it is convenient to study the map τ of the plane which covers T. The following result can be established Devaney (1979).

Theorem: For almost every $x \in q^{-1}(L)$, there exists a Lipschitz continuous function $\gamma : R^1 \to R^1$ such that

(1) $W^u(x) = \{(t,\gamma(t)) : t \in R^1\}$

(2) $\gamma(t) \to \infty$ as $t \to \infty$ and $\gamma(t) \to -\infty$ as $t \to -\infty$.

Similarly there exists a Lipschitz function $\tilde{\gamma}$ such that

$W^s(x) = \{(\tilde{\gamma}(t),t) : t \in R^1\}$.

From the theorem we conclude that for almost every pair of points x,y we have some point z belonging to $W^u(x) \cap W^s(y)$. The proof that T is topologically transitive goes as follows: Let U and V be two open sets in L. Let $\underline{U} = q^{-1}(U)$ and $\underline{V} = q^{-1}(V)$. By the theorem, for almost every pair of points $x \in \underline{U}$, and $y \in \underline{V}$, there exists a point $z \in W^u(x) \cap W^s(y)$. Since the transformation τ covers T we have $q(W^u(x)) = W^u(q(x))$.

Consequently, for almost every pair of points $a \in U$ and $b \in V$ there exists a point $c \in W^u(a) \cap W^s(b)$. By Poincaré's recurrence theorem almost every point d in L is recurrent in the sense that there exists a sequence of integers $\{n_k\}$ with $n_k \to \pm\infty$ such that $\tau^{n_k}(d) \to d$ as $k \to \pm\infty$. Hence we may suppose without loss of generality that $T^{n_k}(a) \to a$ as $k \to -\infty$ and $T^{n_k}(b) \to b$ as $k \to +\infty$. Since $c \in W^u(a)$, $\tau^{n_k}(c) \to \tau^{n_k}(a)$ as $k \to -\infty$ and therefore $\tau^{n_k}(c) \in U$ for some $n_k < 0$. Similarly, $\tau^{n_\ell}(c) \in V$ for some $n_\ell > 0$. Consequently, $\tau^{(n_\ell - n_k)}(U) \cap V \neq \phi$, proving that T is topologically transitive. This completes the outline of the proof that a toral linked twist map T is ergodic.

REFERENCES.

Arnold, V., (1978), Math. Methods in Classical Mech., Graduate Texts in Math #60, Springer-Verlag.

Burton R. and R. Easton, (1979), Ergodicity of Linked Twist Maps, Lecture Notes in Math. 819, Springer-Verlag.

Billingsley, P., (1965), Ergodic Theory and Information, John Wiley and Sons Inc.

Easton, R. (1979), Perturbed Twist Maps, Homoclinic Points and Ergodic Zones, Instabilities in Dynamical Systems, (ed. V. Szebehely), D. Reidel Publ. Co.

Easton, R., Chain Transitivity and the Domain of Influence of an Invariant Set, Lect. Notes in Math. No. 668, Springer-Verlag.

Devaney, D., (1979), Linked Twist Maps are Almost Anosov, Lect. Notes in Math. 819, Springer-Verlag.

Herman, M., (1981), to appear.

Katok, A., Y. Sinai, A. Stepin, (1977) Theory of Dynamical Systems and General Transformation Groups with an Invariant Measure, J. of Soviet Math., 7, pp. 974-1065.

Moser, J., (1973), Stable and Random Motions in Dynamical Systems, Annals of Math. Study 77, Princeton Univ. Press.

Pesin, R., (1977), Lyapunov Characteristic Exponents and Smooth Ergodic Theory, Russian Math. Surveys, 32, pp. 55-114.

Przytycki, F., (1981), Linked Twist Mappings: Ergodicity, IHES Pub. M/81/20.

Ruelle, D., (1978), Ergodic Theory of Differentiable Dynamical Systems, IHES. Pub. P/78/240.

Simo, C., (1981) See this volume, p. 357.

Weiss, B., (1975), The Geodesic Flow on Surfaces of Negative Curvature, Lect. Notes in Physics No. 38, Springer-Verlag.

Wojtkowski, M., (1979), Linked Twist Mappings have the K-Property, In Nonlinear Dynamics, Annals N. Y. Academy of Sciences, No. 357 (ed. R. Helleman).

EXPLODING DYNAMICAL SYSTEMS

Okan Gurel

IBM Cambridge Scientific Center
545 Technology Square, Cambridge, Mass. 02139 USA

ABSTRACT. In these lectures various generic properties of "exploding" systems in terms of their multiple solutions are presented. Dynamical models with varying "explosion complexities" are discussed and the relationship between the stabilities and explosions are drawn. It is shown that a "stability index" in the sense of Poincare may be introduced to identify the stability of the characteristic solutions and used as a tool of global analysis of exploding systems. Cardinality, stability and dimensionality issues of the systems are also considered, and illustrated with specific examples. Certain applications to celestial mechanics and astrodynamics are also conjectured and discussed.

1. INTRODUCTION

In the 1978 NATO Advanced Study Institute, bifurcation theory and its applications was the topic of presentation (1) which precedes the current lectures. There it was discussed that bifurcation phenomenon plays a significant role in nonlinear systems dynamics. Here we extend that notion to cover further behavior of systems experiencing bifurcations, and in fact going beyond that we explore how some dynamical systems explode, however remain in a finite domain to create some intuitionally unexpected conclusions.

Starting with the elements of the set of solutions which we encounter we study singular solutions, limit cycles, attractors, etc. The interesting solutions of a dynamical system are the ones which remain in a "bounded" subspace of the solution space. The <u>singular solution</u> where the right-hand sides of the differential

V. Szebehely (ed.), Applications of Modern Dynamics to Celestial Mechanics and Astrodynamics, 277–299.

(system) equations, $dx/dt=f(x)$ vanish, thus once at the singular solution the system remains there as $t \to +\infty$, is the most elaborated one in the literature. Similarly limit cycles have been incorporated into studies of physical systems as bounded solutions. Study of other bounded objects such as "attractors", "chaotic objects", etc. have increased recently, and examples of simple systems of low dimension, $N \geq 3$, have been constructed.

Poincare's observations on the singular solutions and subsequent classification as focus, center and saddle point, and should be extended to the new objects. Stability property of these "local" solutions as characterizing the "global" stability property of the system possessing such points has already been considered. Extension of the stability property to higher dimensional objects is a natural development in (nonlinear) systems.

The difficulty with such "characteristic" objects in higher dimensions, $N > 3$, is due to many factors. However, we recently realized the relationship between such "new" objects and known "old" physical phenomena, and thus we are forced to consider these objects seriously and to understand their behavior better. This requires modifying our already established thinking, if not inventing new ways to cope with such new discoveries.

2. AN INDEX FOR CHARACTERISTIC SOLUTIONS

In studying the characteristic solutions of a bifurcating dynamical system we may refer to an index in the spirit of Poincare. We define this index by referring to a dimension, here named f-dimension indicating the dependency of the dimension on functions characterizing the system. This index is based on characteristic flows. Recent discussions by Neimark (2) are also based on similar arguments. We first present definitions on characteristic solutions, then discuss f-dimension.

2.1. Characteristic Solutions, (3)

A dynamical system in E^n is characterized by

$$f_p : X \to X , \quad X \subset E^n$$

where the subscript p refers to the dependency of f on the parameter space.

Definition 1a. Singular point, x_o,

$$f_p = 0, \text{ thus } f_p = x_o \to x_o .$$

Linearized system: Lf_p at x_o

Characteristic equation: $|Lf_p - s| = 0$

Eigenvalues: $s_1, \ldots, s_n$

Half flows: $1/2\ w_s$ and $1/2\ w_u$

Here for each negative (positive) eigenvalue a pair of stable (unstable) half flows is denoted by $1/2\ w_{m1}^{1}$ $(1/2\ w_{m2}^{1})$.

Definition 1b. Singular point, x_o is embedded into n-dimensional solution space with n pair of (stable and unstable) w_{m1}^{1} or w_{m2}^{1},

$$(n-k)w_{m1}^{1} + kw_{m2}^{1}$$

k=0 implying stable singular solution.

Definition 2. Exploded point, $f_p : x_e \to x_e$ is embedded into n-dimensional solution space with only one stable (or unstable) flow w_{m1}^{1} (or w_{m2}^{1}) approaching(or leaving) it.

Definition 3. Limit cycle, $f_p : L \to L$, is a solution and it is embedded into the n-dimensional solution space $L \in X \subset E^n$, with two stable (or unstable) flows w_{m1}^{1} (or w_{m2}^{1}), approaching (or leaving) it, respectively.

2.2. f-dimension, (3)

Definition 3. The f-dimension (functional dimension) of solution x is one less than the combined dimension of the stable and unstable manifolds d, which is the number of approaching and leaving flows, thus

$$\boxed{\text{f-dim } x = d-1}$$

Remark:

* Singular point, f-dim $x_o = n-1$,
(Notice this is the dimension of the boundary of X)

Exploded point, f-dim $x_e = 1-1 = 0$,

* Limit cycle, f-dim $L = 2-1 = 1$, a
(A closed line)

Remark:

x_e does not satisfy $f_p=0$, thus it is not stationary but oscillating, however without any period

2.3. An Index, (3)

Definition 4. (A Stability Index). Stability index of a singular point is defined as the difference between the number of w_{m1}^{1}, i.e. the dimension of the stable manifold, (n-k), and that of w_{m2}^{1}, i.e., the dimension of the unstable manifold, k.

$$I_s = (n-k) - k = n-2k$$

3. CLASSIFICATION OF BIFURCATIONS

3.1. Classification of Bifurcating Solutions

For systems with increasing dimensions 1, 2, 3,... depending on the interactions of the characteristic flows one can classify various results as various combinations of Limit cycles (LC) exploded points (EP) and separatrices (SX). Table I summarizes this classification of solutions as they appear following bifurcations. The increasing dimension is indicated on the first column of the table. On the subsequent columns increasing interaction of individual pairs of half flows as characteristic flows are shown. The figures show that all the half flows are unstable, that is, a stability change has taken place, and a solid dot on each pair of half flows indicates a point lying on the bifurcated object. A single interaction leads to a limit cycle, or multiple interactions results in limit surfaces. If no interaction, either a set of exploded points or a separatrix is obtained.

3.2. X-Based Classifications of Solutions

A classification of solutions bifurcating from a generating solution, and based on the solution space X, may be given in two parts. These are cardinality classification, that is the number of solutions appearing, and stability classification, that is the distinction of various types of solutions based on the stability considerations only.

TABLE I: CLASSIFICATION OF BIFURCATING SOLUTIONS

Dimension	Number of interacting pairs of half flows Decomposed None= 0	Single One=1	Multiple Two=2
1	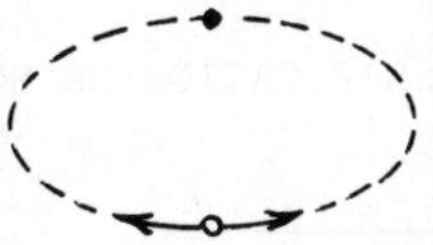		
2	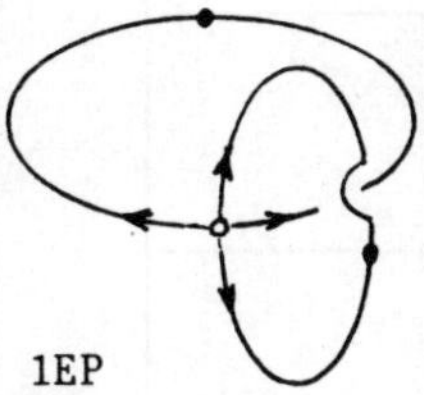 1EP 2EP	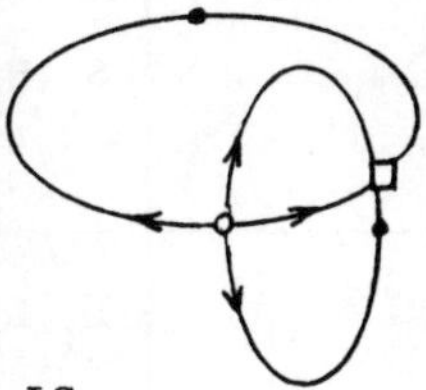 LC	
3	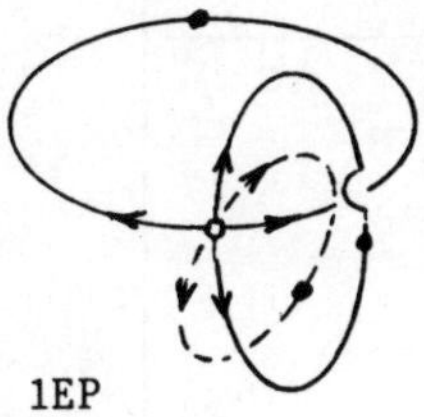 1EP 2EP 3EP	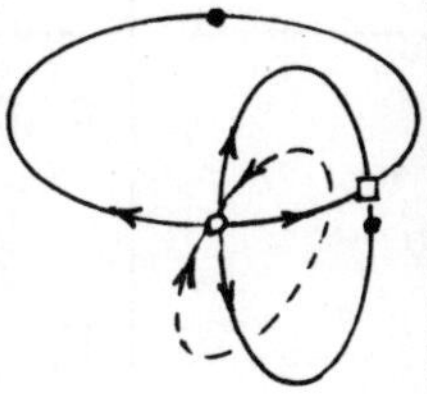 LC	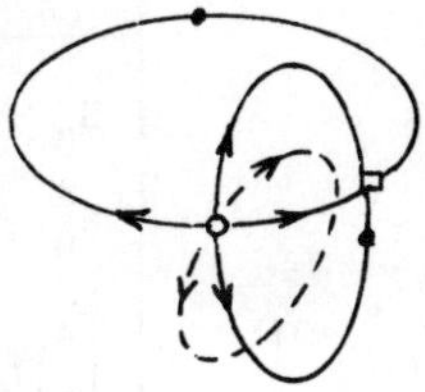 LC EP
		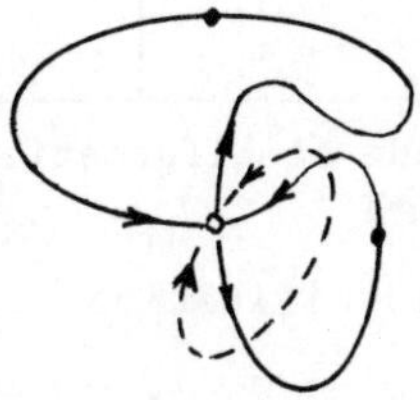 SEPRX.	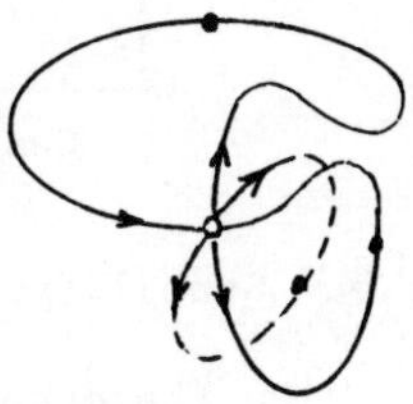 SEPRX. EP

3.2. a. Cardinality Classification

The cardinality classification is illustrated in Table II below. In this classification a single and a multiple number of various solutions are considered such as S, a singular point, L, a limit cycle, and O a limit object. In Figure 1, a singular solution S_o is shown to bifurcate into a) a limit cycle, L, or b) a pair of singular solutions S_1 and S_2.

TABLE II. X-BASED CARDINALITY CLASSIFICATION OF SOLUTIONS

Bifurcation from	Bifurcation to
S_o	$S_1, S_2, \ldots$
S_0 S_0', S_0''	$L_1, L_2, \ldots$
S_0 $S_0', S_0'', \ldots$ L_1 $L_1, L_2, \ldots$ O_1 $O_1, O_2, \ldots$	$O_1, O_2, \ldots$

Cascade of bifurcations

3.2. b. Stability Classification

On the other hand, a classification can be given for the stability variations in the bifurcation solutions. This is based on the stabilities spacifically at the generating solution. In each of the following cases there is a bifurcation into a finite domain.

1. Full Peeling

All of the pairs of half flows change their directions.

2. Partial Peeling

Some of the pairs of half flows change their direction.

3. Decomposed Partial Peeling

The generating singular point splits into multiple solutions and each one of the multiple solutions exhibit a different change of stability pattern of partial peeling. However, the combined change makes up the stability change in the original solution.

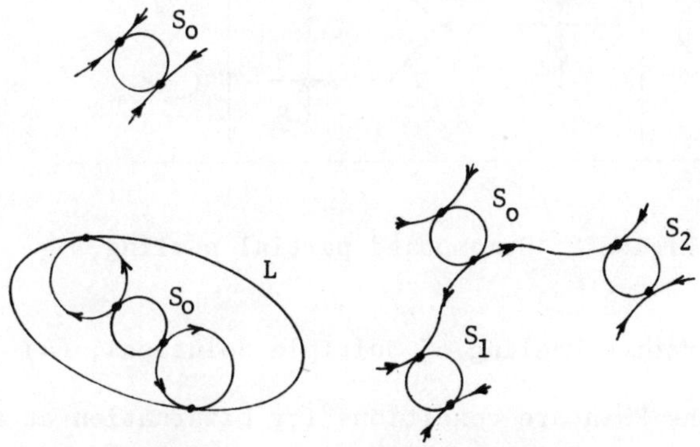

Figure 1. Bifurcating into a) a limit cycle, b) multiple singular solutions.

3.3. X-Based Classification of Bifurcations

Similar to the classification of solutions one can classify the bifurcation by considering the X-space. We have two forms of bifurcation into a finite domain. Also, we list decomposed partial peeling and cooperative peeling of multiple solutions as X-based classifications.

3.3. a. Partial Peeling to the Critical Limiting Set (CLS)

The set of solutions obtained after bifurcation belong to the critical limiting set (CLS) determining the stability of the system.

3.3. b. Partial Peeling to a Noncritical Limiting Set (NCLS)

The set of solutions obtained after bifurcation does not belong to the critical set of the entire system, thus there is a local bifurcation.

3.3. c. Decomposed Partial Peeling

In the solution space multiple solutions appear. Their stability changes are complementary. An example of decomposed partial peeling is illustrated in Figure 2.

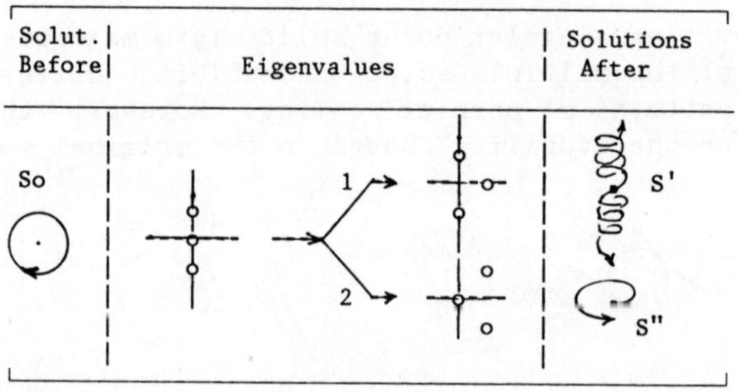

Figure 2. Decomposed partial peeling.

3.3. d. Cooperative Peeling of Multiple Solutions, (4)

One of the Poincare conditions for bifurcation at a singular point x^o is that the determinant evaluated at x^o vanishes at the parameter value, p^o. Thus denoting

$$\Delta(x,p) = |M(x,p)|$$

where $M(x,p)$ = nxn matrix, $M_c(x,p)$ = canonical nxn matrix corresponding to $M(x,p)$, at x^o

$$\Delta_c(x^o,p) = m_{11}(x^o,p) \ldots m_{nn}(x^o,p)$$

Critical limiting set CLS(X) contains the characteristic solutions such as singular points. Let us say that there are k singular points, thus

$M_c^{\,o}$, $M_c^{\,1}$, ...; $M_c^{\,k}$. Therefore the set of canonical matrices will form the cooperative peeling matrix shown in Figure 3.

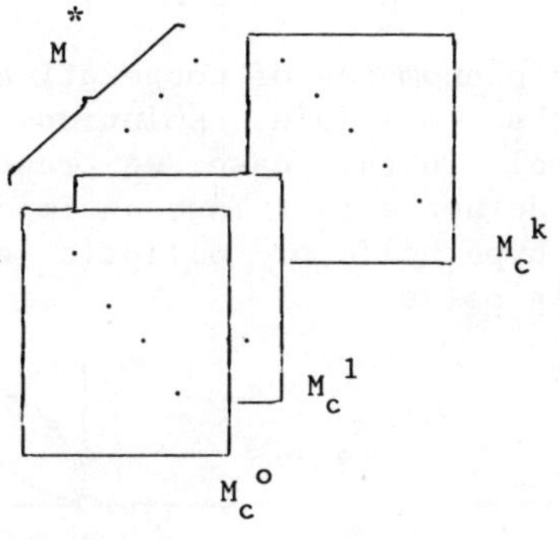

Figure 3. Cooperative peeling matrix.

This leads to the combined matrix which indicates the cooperative behavior. The elements of the combined matrix are

$$M^* = \begin{vmatrix} m_{11}^0 & m_{11}^1 & \cdots & m_{11}^k \\ m_{22}^0 & m_{22}^1 & \cdots & m_{22}^k \\ \cdot & \cdot & & \cdot \\ \cdot & \cdot & & \cdot \\ m_{nn}^0 & m_{nn}^1 & \cdots & m_{nn}^k \end{vmatrix}$$

In the case k is not equal to n, dummy rows or dummy columns of some constant value may be assigned to make the combined matrix a square matrix.

Cooperative Peeling (necessary) Condition I.
While the determinants of all the individual matrices do not vanish, the vanishing determinant of the combined matrix, M*, implies "cooperative peeling"

The two other conditions of Poincare, the stability change and change in the number of solutions, are combined in the following condition:

Cooperative Peeling (sufficient) Condition II.
For the combined peeling to take place, in addition to Condition I, it is necessary that CLS(X) is replaced by CLS'(X).

Here CLS and CLS' are the two different sets of solutions before and after bifurcation (peeling).

3.3. e. Independent Peeling of Multiple Solutions, (5)

Contrary to the phenomenon of cooperative peeling, it is also possible that the multiple solutions may individually (noncooperative) peel. In this case we treat each singular point as if other points do not exist. Thus an important characteristic of this is the hyperbolic or elliptic neighborhood of the individually peeling point.

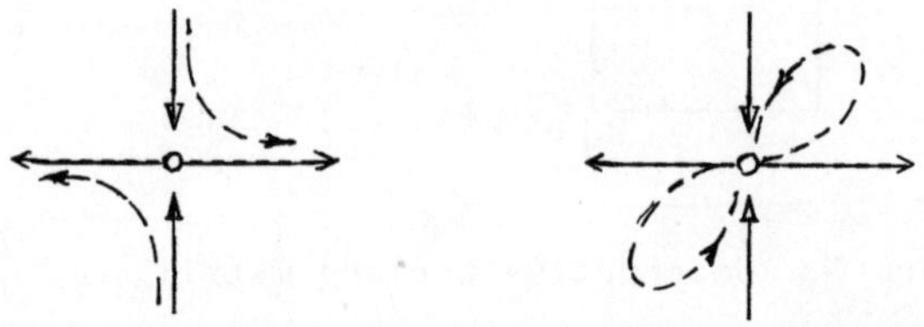

Hyperbolic neighborhood Elliptic neighborhood

Figure 4. Hyperbolic and elliptic neighborhoods.

In the case of hyperbolic point w_{m2} would not be balanced by a w_{m1}, in elliptic case w_{m2} is balanced by w_{m1}.

3.4. P-Based Classification of Bifurcations

Since a dynamic system depends on X as well as P, one can also classify bifurcations based on the parameter space P. Some types are discussed below.

3.4. a. Peeling into Noncritical Limiting Set

As some parameters vary there can be a bifurcation, however not leading to CLS but to a NCLS. These bifurcations should be recognized because they can create locally valid solutions, and globally undesirable situations.

3.4. b. Combined Peeling in Multi-parameter Systems

In this section, we discuss peeling of the same singular point at bifurcation values of different parameters (independently). We need certain definitions as discussed below, cardinality, global stability, dimensionality, and hyperbolicity.

3.4.b.1. Cardinality, (6)

The cardinality of the solution set (x_o), Card (x_o) is the number of distinct characteristic solutions in the set. The set (x_o) splits into two parts, (NOx_o) + (Ox_o) as nonoscillating and oscillating solutions. Thus

$$\text{Card}\ (x_o) = \text{Card}\ (NOx_o) + \text{Card}\ (Ox_o)$$

The Cardinality Rule:

Card (x_o) > Card (NOx_o) implies the existence of (Ox_o).

3.4.b.2. Global Stability, (6)

The Global Stability Theorem:

$$\text{Stab}\ (x_o)_{\text{before peeling}} = \text{Stab}\ (x_o)_{\text{after peeling}}$$

3.4.b.3. Dimensionality, (6)

The Dimensionality Rule:

$$\dim\ (NOx_o) = \dim X - 1$$
$$\dim\ (Ox_o) = 0,1,\ \ldots,\ \dim X - 1.$$

3.4.b.4. Hyperbolicity, (6)

The Hyperbolicity Rule:

If Stab $(NOx_o) = (\dim X)\ .\ w_{m2}$ (unstable), then for the newly created Ox_o,

$$\dim\ (Ox_o) = \dim X - 1.$$

If Stab $(NOx_o) = (\dim X -1-k).\ w_{m1} + (k+1)\ w_{m2}$ (hyperbolic), then,

$$\dim\ (Ox_o) = k.$$

4. ILLUSTRATIVE EXAMPLES

In the following section, we list some of the known examples to illustrate the above classifications and exploding solutions appearing in these systems.

4.1. Example 1. Lorenz Attractor (Cooperative Peeling) (6,3,2)

Bifurcation of Lorenz System: (7, pp.14-20)

$$dx/dt=-mx + my$$
$$dy/dt=-x(z-r)-y$$
$$dz/dt= xy - bz$$

For $b > 0$,

1. $r < 1$ $CLS^o(x) = (S_o$, stable focus)

2. $r > 1$ $CLS(x) = (S_o$, saddle; S_1, S_2, stable foci)

3. $r > r_c$ $CLS'(x) = (S_o$, saddle, S_1, S_2, saddle-foci, LA, Lorenz attractor)

where $r_c = m(A+2)/(2m-A)$, and $A=(b(r-1))^{1/2}$.

Cooperative Peeling: (4, pp.41-42)

$$CLS(x) = (S_o, S_1, S_2)$$

$$M^* = \begin{vmatrix} -m & -m & -m \\ (r-1) & -2 & -2 \\ -b & -A^2 & -A^2 \end{vmatrix}$$

$\Delta = 0$ for all b and r, thus bifurcation (peeling) of CLS(x) into CLS'(x) is cooperative, and results in, (Figure 5)

$$CLS'(x) = (S_o, S_1, S_2, LA)$$

Index of Lorenz attractor shows that this is an exploded point (Figure 6). Moreover the number of possible combinations of this type of exploded point can be determined by the formula

$$N_n = (K_n/2)\ n!$$

Here n is the number of columns in Figure 7, $K_n = K_n + 3$, for $n>5$ or equal to 5 and K1=0,K2=1,K3=2,K4=3.

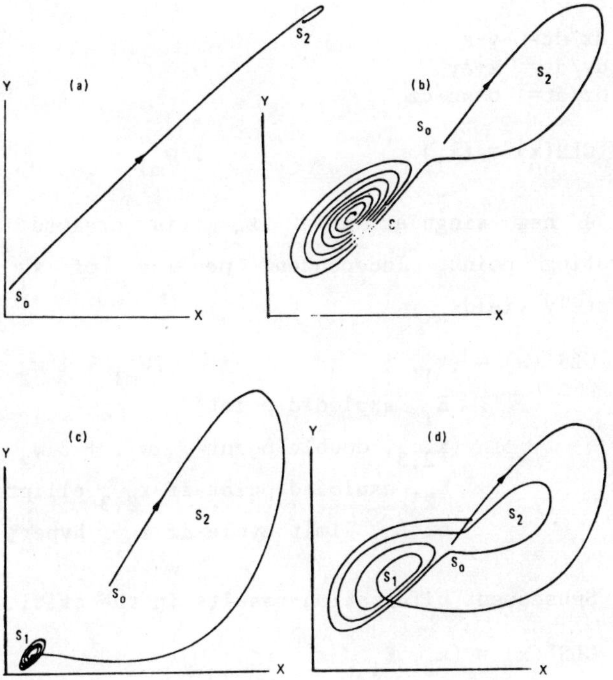

Figure 5. Cooperative peeling to the Lorenz attractor.

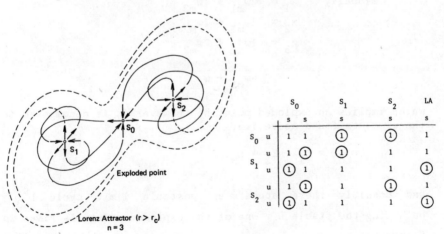

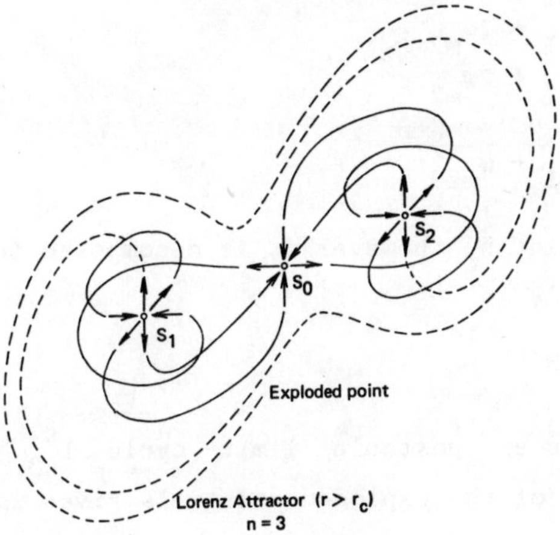

Figure 6. An example of exploded point, Lorenz attractor

		S_0		S_1	S_2	LA
		s	s	s	s	s
S_0	u	1	1	(1)	(1)	1
S_1	u	1	(1)	(1)	1	1
	u	(1)	1	1	1	(1)
S_2	u	1	(1)	1	(1)	1
	u	(1)	1	1	1	(1)

Figure 7. Origin and destination of the half flows.

4.2. Example 2. Rossler Attractor (Independent peeling),(5)

$$dx/dt = -y-z$$
$$dy/dt = x+ay$$
$$dz/dt = b+xz-cz$$

$CLS(x) = (x_1)$ $\qquad 3\ w_{m1}$

A new singular point $x_{2,3}$ is created as a second generating point. Independent peeling of x_1 and $x_{2,3}$ separately yields

$$\begin{aligned}CLS'(x) = (&x_1, && 2W_{m1} + 1\ w_2\\ &E_1, \text{ exploded point})\\ &(x_{2,3}, \text{ double point} && 1w_{m1} + 2\ w_2\\ &E_2, \text{ exploded point if } x_{2,3} \text{ elliptic}\\ &\text{or } L_3, \text{ limit cycle if } x_{2,3} \text{ hyperbolic})\end{aligned}$$

Subsequent bifurcation results in the critical set,

$$\begin{aligned}CLS''(x) = (&x_1, E_1\\ &x_{2,3}, E_2 \text{ (or } (L_2), L_3)\end{aligned}$$

$$\begin{aligned}\text{Globally} \quad x_1 &= 2w_{m1} + 1w_{m2}\\ x_2 &= 1w_{m1} + 2w_{m2}\\ x_3 &= 1w_{m1} + 2w_{m2}\\ \hline & \quad 4w_{m1} + w_{m2}\end{aligned}$$

which implies an exploded point E, however it is decomposed into the elements found above,i.e.

$$E = E_1 + E_2 + L_3$$

and possibly there appears an unstable limit cycle L^u_3 balancing the stable L_3. One of the exploded points is given in Figure 8.

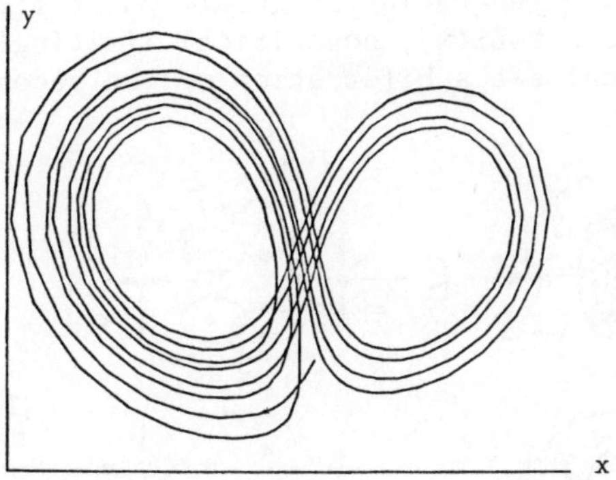

Figure 8. An exploded point in the Rossler attractor.

4.3. Example 3. Toroidal Attractor (Peeling to NCLS), (8)

$$dx/dt = x - ay - xz$$
$$dy/dt = bx + y - yz$$
$$dz/dt = x^2 + y^2 + cz - (z/(z+d))$$

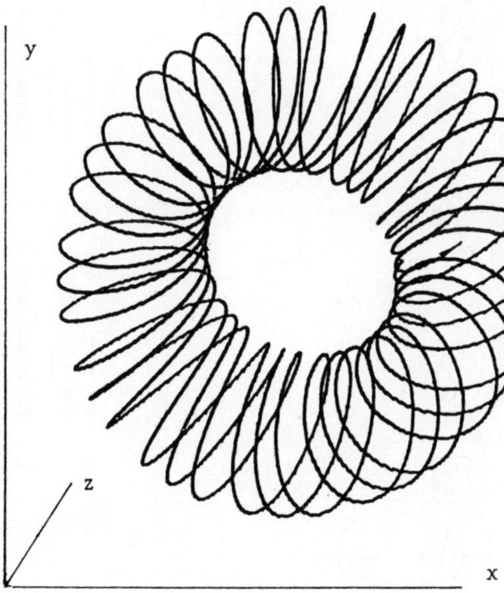

Figure 9. Toroidal attractor.

Bifurcation of the generating singular point first to a CLS(x) subsequently also a NCLS(x), noncritical limiting set is created. In this noncritical set a bifurcation of the second order limit

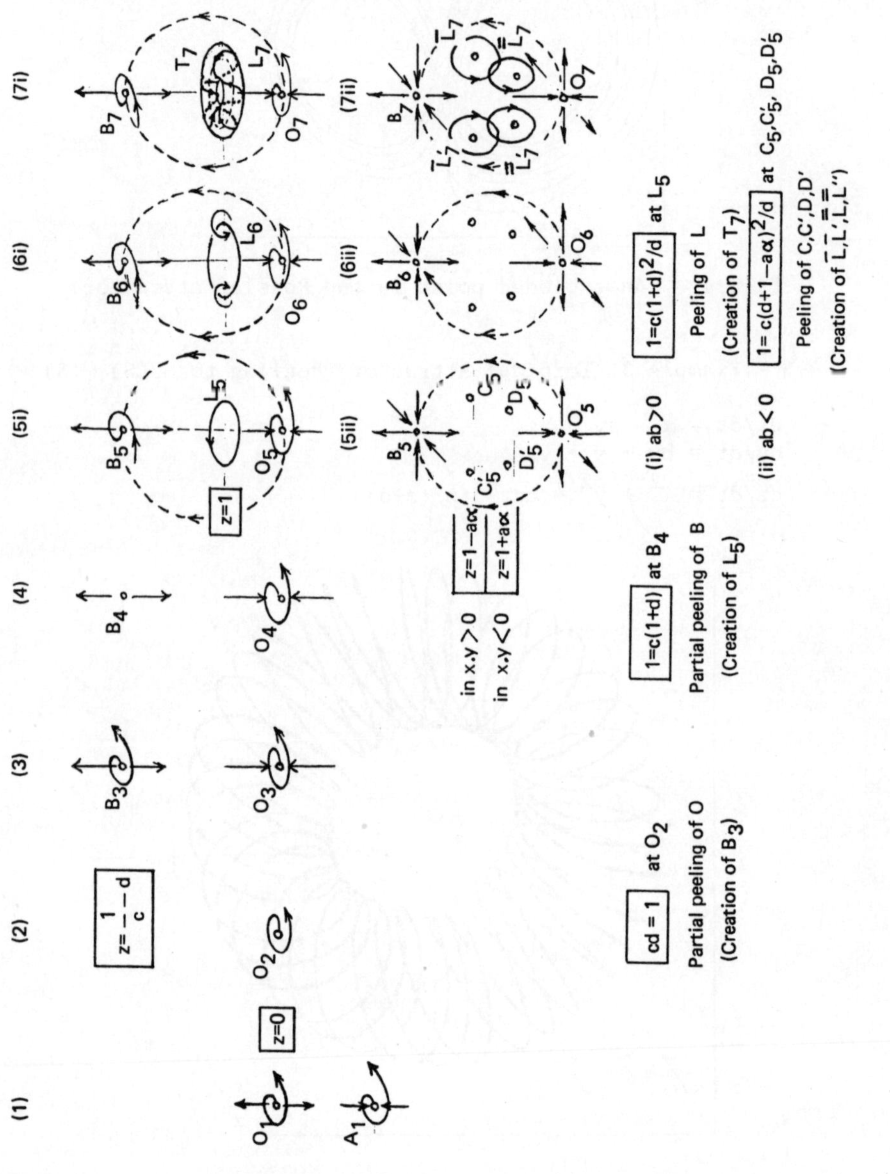

Figure 10. Bifurcations leading to toroidal attractor.

cycle takes place to yield another limit cycle which forms a toroidal surface (Figure 9). Various bifurcation steps are illustrated in Figure 10.

4.4. Example 4. Combined Peeling, (6)

$$dx_1/dt = x_2$$

$$dx_2/dt = -p_2x_1 - p_1x_2 - x_1^3 - x_1^2x_2$$

It should be emphasized that in addition to the above variations in the bifurcation to oscillating or non-oscillating solutions, in the case of multi-parameters, peeling as one parameter varies might be independent from that taking place as another parameter varies. This is illustrated in Figure 11.

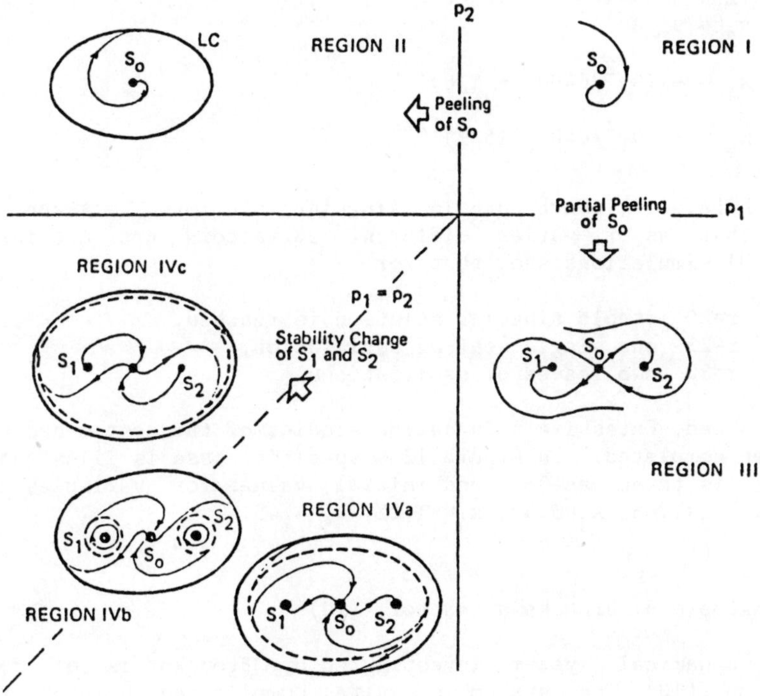

Figure 11. An example of the combined peeling.

4.5. Example 5. Navier-Stokes attractor, (9)

A truncated form of the Navier-Stokes equations is given in (9) as follows:

$$dx_1/dt = -2x_1 + 4x_2x_3 + 4x_4x_5$$
$$dx_2/dt = -9x_2 + 3x_1x_3$$
$$dx_3/dt = -5x_3 - 7x_1x_2 + r$$
$$dx_4/dt = -5x_4 - x_1x_5$$
$$dx_5/dt = -x_5 - 3x_1x_4$$

Singular solutions are found as

$$x_1 = \mp (5/3)^{1/2}$$
$$x_2 = \mp (15)^{1/2} r/80$$
$$x_3 = 9r/80$$
$$x_4 = \mp (9r^2/6400 - 1/6)^{1/2}$$
$$x_5 = \mp (9r^2/640 - 15/2)^{1/2}$$

Therefeore, there are 16 singular solutions. It can be shown that as r varies different attractors are obtained. Numerical simulations show that for

r=20 stable singular solution is reached,
r=25 one single leaf of oscillation,
r=30 two leaves of oscillations

are obtained. Extensive bifurcation studies of the system has not yet been completed. In Figure 12 a specific case is illustrated where r is taken as 33, and initial values for variables are $x_1=1.41$, $x_2=1.68$, $x_3=3.39$, $x_4=-1.22$, $x_5=4.45$.

4.6. Example 6. Broucke attractor, (10)

A dynamical system investigated by Broucke is of four dimension (10). The system is quite complicated, however the equations are given as

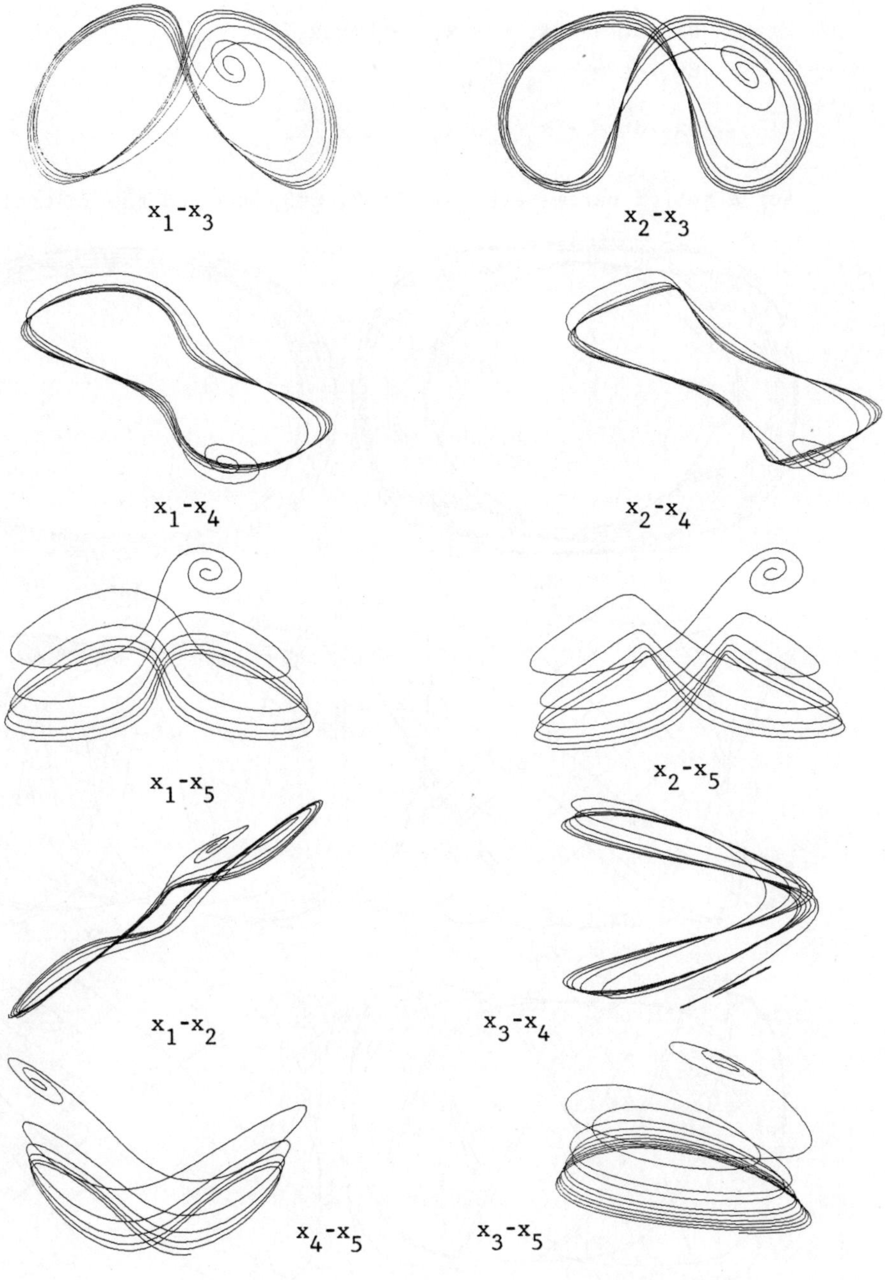

Figure 12. An attractor of Navier-Stokes system.

$$dx_1/dt = x_2$$
$$dx_2/dt = -x_1 - a\,x_1^3 - b\,x_1x_3^2$$
$$dx_3/dt = x_4$$
$$dx_4/dt = -x_3 - c\,x_3^3 - d\,x_1^2x_3$$

For a set of parameters a=1, b=-2, c=3, d=4 and the initial

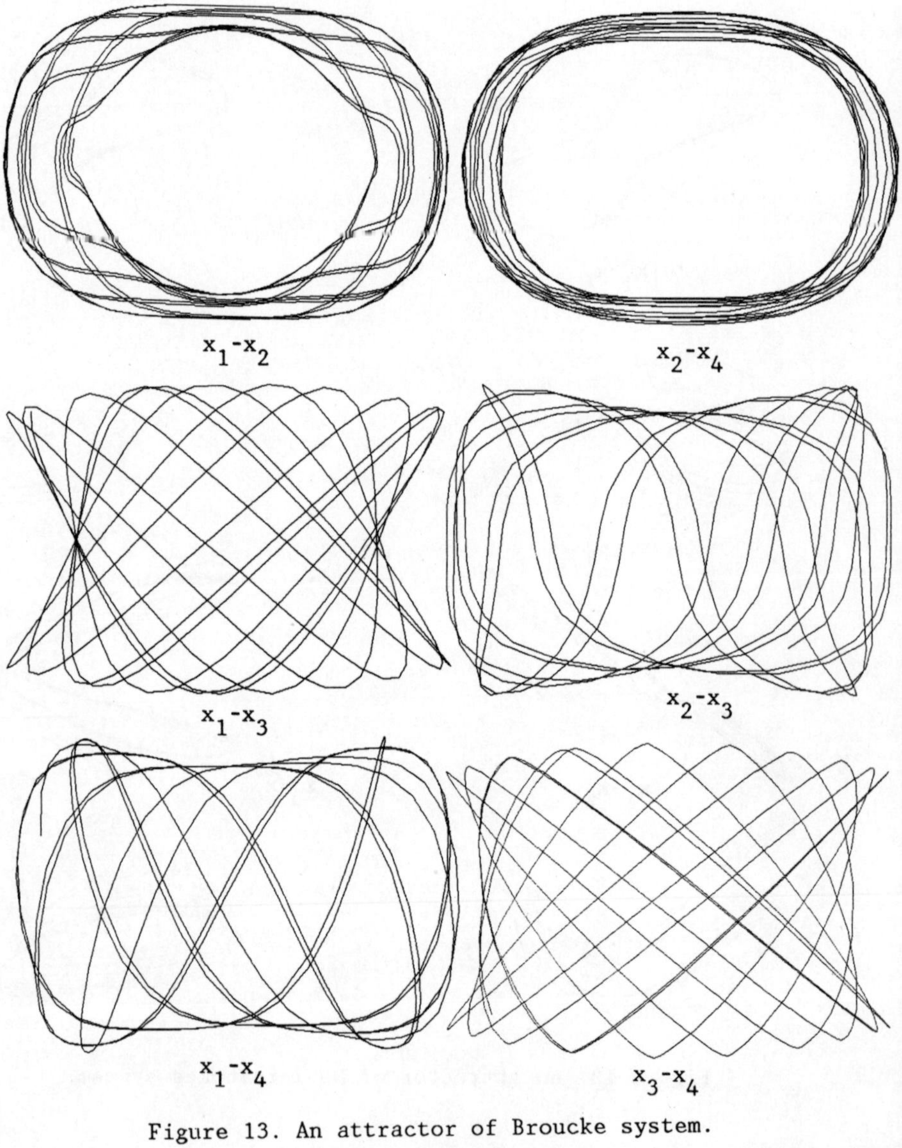

Figure 13. An attractor of Broucke system.

values $x_{10}=10$, $x_{20}=0.02$, $x_{30}=13.0$, $x_{40}=0.0004$ the attractor obtained is illustrated on Figure 13.

Singular solutions can be obtained from the equations

$$x_2 = 0, \quad x_4 = 0,$$
$$1+ax_1^2+bx_3^2 = 0 \quad \text{or } x_1 = 0,$$
$$1+cx_3^2+dx_1^2 = 0 \quad \text{or } x_3 = 0.$$

Therefore the origin is one of the solutions and for a proper set of parameters other points may also be obtained.

5. CONJECTURES THAT MAY BE PLAUSIBLE

In 1971 Stephen Hawking of Cambridge University proposed the possible existence of "mini" black holes wandering throughout the universe as a residue of the "Big Bang" explosion. It is feasible to theorize that at the starting point the whole universe was a single "black hole", a stable singular point from which nothing could escape. The big bang, being a bifurcation, could lead to "peeling" of the entire black hole to a set of characteristic solutions, the critical limiting set being a subset of which,, is a stable object reflecting stability property of the generating black hole.

Furthermore it may be theorized that the critical limiting set would consists of multiple <u>stable</u> objects and perhaps as many <u>unstable</u> objects. All the stable objects by definition are n-dimensional objects (n>1), then they can be referred to as "black n-dimensional objects".

There are two possible cases for the black 3-dimensional objects. These are discussed in Cases A and B below.

CASE A. It can be created by a stability change at the generating point by two of the three eigenvalues crossing the imaginary axis. The outcome is a complex collection of stable and unstable limit cycles. The stable limit cycles are "black limit cycles" while the unstable limit cycles are the planetary orbits.

CASE B. Similarly by a change of stability at the generating point, when only one eigenvalue crossing the imaginary axis, a set of stable and unstable exploded points are created, "black exploded points" interlaced with unstable exploded points on which the planets are placed.

It can be conjectured that in celestial dynamics black exploded points are more plausible objects than the black holes and black limit cycles.

In Figure 14 P_i indicate the unstable orbits, BO_i correspond to stable (black) orbits. In the first case, (A) these are limit cycles while in the second case, (B) they correspond to exploded points.

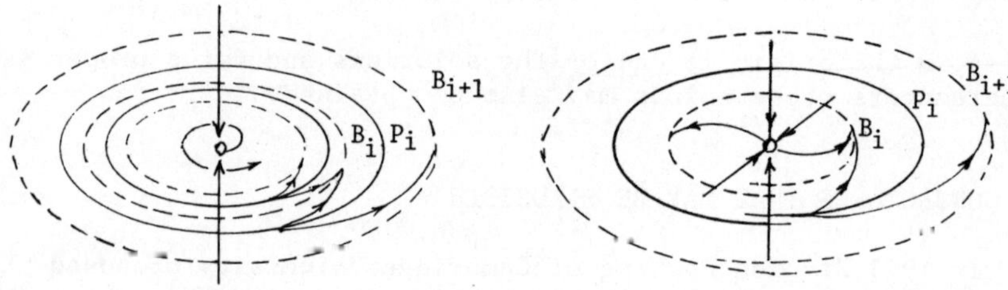

Case A. Case B.

Figure 14. Case A. Black and unstable limit cycles
Case B. Black and unstable exploded points.

6. CONCLUSIONS, (8,9)

Systems appear to explode into bifurcated forms via various types of peeling phenomena. These were collected under X-based and P-based classifications.

Also we showed that the stability index defined may be used in the analysis of oscillating solutions obtained for such bifurcating systems.

We believe that indices based on stochastic consideration, e.g. see (11), may not be as significant as the "deterministic" dimension on which the stability index is based.

A conjecture is also given that the celestial systems may consists of combined stable and unstable oscillating solutions and the unstable orbits may be the real orbits which the celestial bodies might follow.

7. REFERENCES

(1) O. Gurel, Bifurcation Theory and Its Applications, In: Instabilities in Dynamical Systems (V.G. Szebehely, ed.) D. Reidel Publ. Co., (1979)49-60.

(2) Yu. I. Neimark, Invarianthie Mnogoobraziya i Stokhasticheskie Dvijeniya Dinamicheskikh Sistem, (Invariant Manifolds and Stochastic Motion of Dynamical Systems), In: Problemi Asimptoticheskoy Teorii Nelineynikh Kolebaniy (Problems of the Asymptotic Theory of Nonlinear Oscillations), Kiev, Naukova Dumka (1977)160-168.

(3) O. Gurel, Exploded Points, Z. Naturforsch. 36a (1981)72-75.

(4) ________, Necessary & Sufficient Conditions for Cooperative Peeling of Multiple Generating Singular Points, In: Bifurcation Theory and Applications in Scientific Disciplines, The New York Academy of Sciences Annals No.316, (O. Gurel & O. E. Rossler, eds.) (1979)39-42.

(5) ________, Individual Peeling of Multiple Singular Points, Z. Naturforschung 36a (1981)311-316.

(6) ________, On the Cardinality, Stability and Dimensionality of Oscillating Solutions, In: Systems Science and Science (Bela H. Benathy, ed.) Society for General Systems Research, (1980) 266-270.

(7) ________, Poincare's Bifurcation Analysis, In: Bifurcation Theory and Applicats. in Scientific Disciplines, 5-26.

(8) ________ & O. E. Rossler, Bifurcation to Toroidal Surfaces, Mathematica Japonica, vol.23, n.5,(1979)491-507.

(9) V. Franceschini and Claudio Tebaldi, Sequences of Infinite Bifurcations and Turbulence in a Five-Mode Truncation of the Navier-Stokes Equations, J. Stat. Phys. vol. 21, n.6 (1979)707-726.

(10) R. Broucke, On the Construction of a Dynamical System From a Preassigned Family of Solutions, Int. J. Eng. Sci. vol 12 (1979)1151-1162.
(b) Private Communication (1978).

(11) D. A. Russel, J. D. Hanson & E. Ott, Dimension of Strange Attractors, Physical Review Letters,vol.45,n.14 (1980) 1175-1178.

Figures 12 and 13 were produced on IBM 3279 Color Display and printed on IBM 3278 Color Printer. Examples 1 through 6 were solved by using IBM CSMP program.

ON SOME INVARIANT MANIFOLD RESULTS AND THEIR APPLICATIONS

Urs Kirchgraber

ETH Zürich, Switzerland

ABSTRACT. A theory on the existence and the properties of invariant manifolds for a certain class of finite dimensional maps is described, with applications to averaging and to a problem in celestial mechanics.

INTRODUCTION.

The general goal in the qualitative theory of ordinary differential equations is to understand the geometry of the set of orbits, the so-called orbit structure, of a system of ordinary differential equations. This is easy for one-degree of freedom conservative systems. The general plane autonomous case is yet much more difficult and for dimensions higher than 3 the problem is almost intractable under general conditions.

Among the numerous attempts to get partial results at least, the study of distinguished invariant sets has proved to be most fruitful. On the one hand, invariant sets with nice geometrical properties are considered to be of interest in themselves, in particular if they have some stability properties, in addition. Typical examples are equilibrium points, periodic orbits, tori and compact manifolds in general. They quite naturally generalize and replace the study of individual solutions. On the other hand, in many cases the search for invariant sets turns out to be a preliminary step for a more detailed discussion of a dynamical system. The introduction of the so-called center manifold in connection with the Hopf bifurcation problem or the study of the stability of equilibria in the critical case of some purely imaginary or zero eigenvalues is a typical example. The introduction of an invariant

V. Szebehely (ed.), Applications of Modern Dynamics to Celestial Mechanics and Astrodynamics, 301–320.

manifold then facilitates the problem by replacing the given system by a lower dimensional one.

At this stage we make a few historical remarks. The theory of invariant manifolds originates in the work of Poincaré. In the sequence major steps were taken by Hadamard and Perron who introduced the two different concepts which seem to underlie almost all the subsequent work on invariant manifolds. Next we mention the contributions by Bogoliubov, Levinson, Diliberto, Hale which were directed towards applications to the theory of non-linear oscillations. Since about 1960 the number of papers in the field has grown that fast that we do not even make an attempt to provide a comprehensive list of references. We merely mention the work of Fenichel, Hirsch, Pugh, Shub; their treatment covers the most general cases.

The goal of this paper is to describe an approach to the theory of invariant manifolds which is conceptually easy and yet general enough to cover a large number of applications. This is achieved as follows. In a first step the results are derived for a certain class of maps; the proofs in this case involve easy geometrical arguments and are in fact surprisingly simple. Only in a second step the theory is carried over to systems of differential equations; this is basically easy as well, yet occasionally a bit technical. It is emphasized that the approach to the theory of invariant manifolds, as presented in this paper, has been stimulated by the work of Knobloch and Palmer on the same subject.

The organization of the paper is as follows. It consists essentially of two parts. In part one various results on invariant manifolds are described, a few of them are proved. In part two an outline of applications is given.

1.1 Invariant Manifolds for Standard Maps

a) Maps versus Systems of ODE's

Given a system of ordinary differential equations (ODE's), there are usually various ways to attach to this system a map of some Euclidean space into itself. If we consider a periodic system

$$\dot{x} = f(t,x) \qquad f(t+2\pi,x) = f(t,x)$$

and denote by $x(t,\xi)$ its solution corresponding to the initial value ξ for $t = 0$, we may define the period map (cf. Fig. 1)

$$P: \quad \xi \to P(\xi) = x(2\pi,\xi)$$

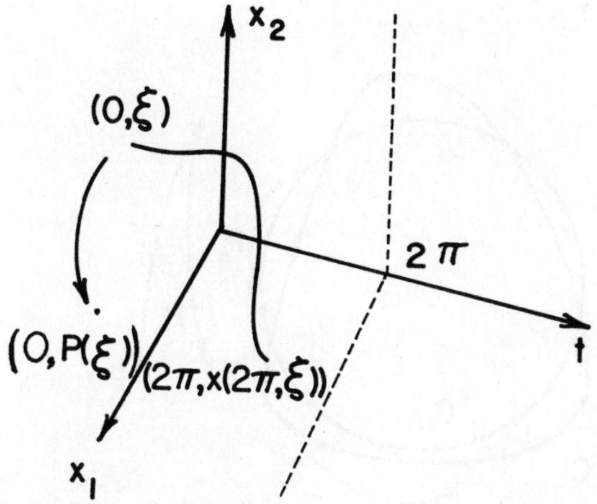

Figure 1

The introduction of the period map is justified essentially by the fact that the long term behaviour of the solutions of the given system of ODE's may well be studied by means of the iterates P^n of P; this is due to the obvious relation

$$x(n2\pi,\xi) = P^n(\xi) \qquad n\in\mathbb{Z}.$$

If we consider an autonomous system $\dot{x} = f(x)$ which admits a closed solution C, we may introduce the classical Poincaré return map (cf. Fig. 2) to study the behaviour of the solutions near to C.

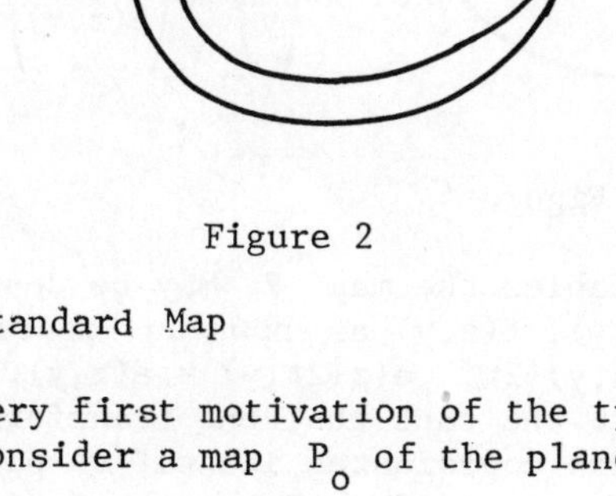

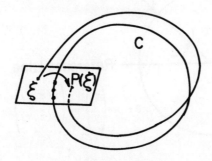

Figure 2

b) The Standard Map

For the very first motivation of the type of maps we are going to study, we consider a map P_o of the plane $\mathbb{R}^2$ into itself, which leaves invariant the unit circle C_o, (cf. Fig. 3). Let P be a perturbation of P_o.

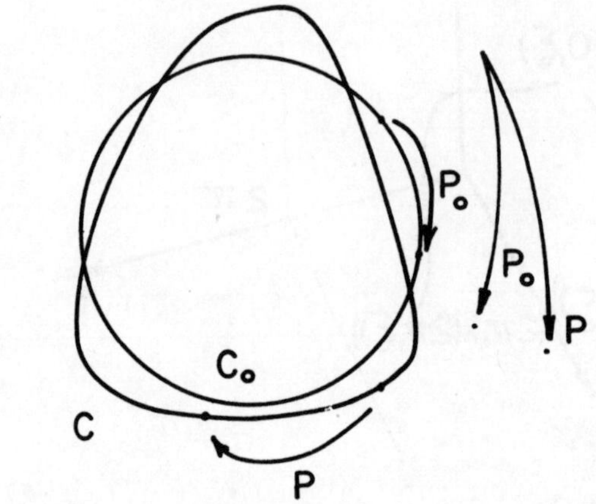

Figure 3

The <u>main problem</u> may be stated now as follows. What kind of conditions do we have to impose on P_o and P such that P admits an invariant closed curve C close to C_o ?

In order to get a simple description of the circle C_o we use polar coordinates, (cf. Fig. 4).

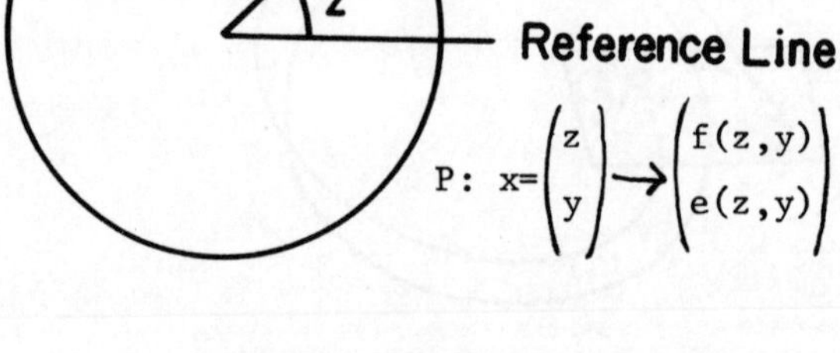

$$P:\ x=\begin{pmatrix} z \\ y \end{pmatrix} \rightarrow \begin{pmatrix} f(z,y) \\ e(z,y) \end{pmatrix} \qquad (1)$$

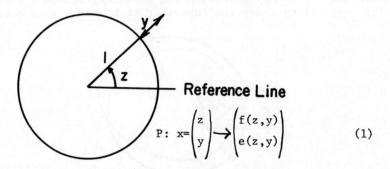

Figure 4

In terms of these variables the map P may be described by means of two functions $e(z,y)$, $f(z,y)$ as shown on Figure 4, where $f(z+2\pi,y) = f(z,y)+2\pi$, $e(z+2\pi,y) = e(z,y)$. These periodicity conditions reflect the fact that the transformation from polar coordinates to Cartesian coordinates identifies points the z-components of which are equal mod 2π. In terms of the polar coordinates the circle C_o is given by y=0. The main problem may now be stated as follows. Give conditions on P such that there is a 2π-periodic function g(z) such that its graph $M = \{ (z,g(z)) | z \in \mathbb{R} \}$ is invariant with respect to P, i.e. $x \in M$ implies $P(x) \in M$, (cf. Fig. 5).

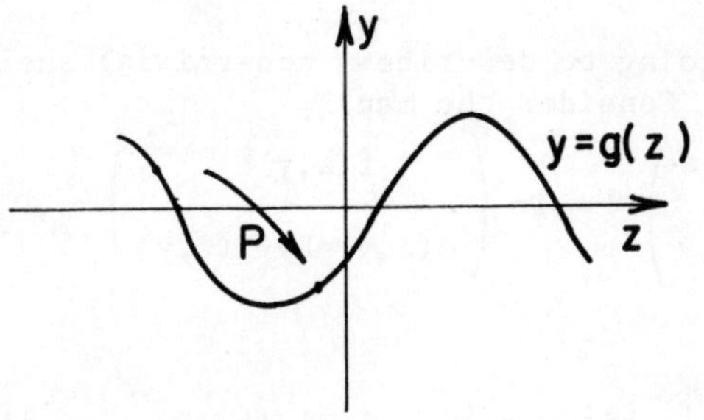

Figure 5

An easy example of a map of type (1) is as follows

$$P: x = \begin{pmatrix} z \\ y \end{pmatrix} \rightarrow \begin{pmatrix} z \\ Ly \end{pmatrix} \tag{2}$$

where we assume $|L^{-1}| \leq \ell < 1$, i.e. the linear transformation $y \rightarrow Ly$ is an expansion. The map (2) obviously admits an inverse $P^{-1} : x=(z,y)^T \rightarrow (z,L^{-1}y)^T$. The geometrical features of the map (2) are easily discussed. Let $M = \{(z,0) | z \in \mathbb{R}\}$. We then have:

(i) The Invariance Property

For $x \in M$ we have $P(x) \in M$ and $P^{-1}(x) \in M$ and therefore $P^n(x) \in M$ for $n \in \mathbb{Z}$.

For d>0 define $U_d = \{(z,y) | z$ arbitrary, $|y| < d\}$.

(ii) The Maximality Property

Given $x \in U_d - M$ there is $n \in \mathbb{N}$ such that

$$P^1(x) \in U_d,\ P^2(x) \in U_d,\ \ldots,\ P^{n-1}(x) \in U_d$$

but

$$P^n(x) \notin U_d.$$

Given $x^o \in U_d$, define $x^i = \begin{pmatrix} z^i \\ y^i \end{pmatrix} = P^i(x^o)$, $i \in \mathbb{N}$. From (i), (ii) we have the following characterization of M: $x^o \in M$ iff $|y^i| < d$ for all $i \in \mathbb{N}$.

(iii) <u>The Property of Stability and Attractivity</u>

<u>Given</u> $\varepsilon>0$ <u>there is</u> $\delta>0$ <u>such that</u> $x\in U_\delta$ <u>implies</u> $P^{-i}(x)\in U_\varepsilon$ <u>for</u> $i\in\mathbb{N}$ <u>and</u> $x\in U_d$ <u>implies</u> $\lim_{i\to\infty} \operatorname{dist}(P^{-i}(x),M)=0$.

Next we are going to describe a non-trivial analogue of the previous example. Consider the map

$$P:\quad x=\begin{pmatrix} z \\ y \end{pmatrix} \longmapsto \begin{pmatrix} f(z,y) \\ e(z,y)=Ly+Y(z,y) \end{pmatrix} \tag{3}$$

where

$$f:\ U_d:=\left\{\begin{pmatrix} z \\ y \end{pmatrix} \middle|\ z\in\mathbb{R}^r,\ y\in\mathbb{R}^t,\ |y|<d\right\}\to\mathbb{R}^r,\quad Y:\ U_d\to\mathbb{R}^t$$

$|.|$ denotes some norm in $\mathbb{R}^{r,t}$, L is a txt-matrix, independent of z,y. Assume that $f(z,y)-z$, $Y(z,y)$ are 2π-periodic with respect to the components of z and that there are constants $K_{zz},K_{zy},K_{yz},K_{yy},M,\ell$ such that

$$|f(z,y)-f(\bar{z},\bar{y})|\le K_{zz}|z-\bar{z}|+K_{zy}|y-\bar{y}|$$

$$|Y(z,y)-Y(\bar{z},\bar{y})|\le K_{yz}|z-\bar{z}|+K_{yy}|y-\bar{y}| \qquad \text{for all } (z,y),\ (\bar{z},\bar{y})\in U_d$$

$$|Y(z,y)|\le M \qquad |L^{-1}y|\le\ell|y|$$

and define $\alpha=K_{zz}$, $\beta=\frac{1}{\ell}-K_{yy}$. Our basic hypotheses (H1), (H2) are as follows.

(H1) <u>There is a constant</u> $\rho\ge 1$ <u>such that</u>

(a) $\beta>\rho>\alpha$

(b) $K_{zy}K_{yz}<(\beta-\rho)(\rho-\alpha)$

(c) $M\ell/(1-\ell)<d$

(H2) (a) P <u>is injective</u>

(b) P <u>is overflowing on</u> U_d i.e. $P(U_d)\supset U_d$.

We claim

<u>Theorem 1. Given a map P satisfying the above mentioned hypotheses there is a map</u> $g:z\in\mathbb{R}^r\to g(z)\in\mathbb{R}^t$, <u>satisfying</u> $|g(z)|\le M\ell/(1-\ell)$,

<u>Lipschitzian</u>, <u>2π-periodic with respect to the components of</u> z, <u>and such that</u> $M=\{(z,g(z))|z\in\mathbb{R}^r\}$ <u>has the property of invariance</u>, <u>maximality, stability and attractivity</u>.

We next give a geometrical interpretation of the requirements (a), (b) of (H1), which are the basic restrictions. To this end we introduce two quantities $\overline{D}_z$, $\underline{D}_y$. $\overline{D}_z$ is what could be called an upper bound on the degree of contraction/expansion of the map P in z-direction and is defined as follows

$$\overline{D}_z = \sup\{\frac{|f(z,y)-f(\overline{z},y)|}{|z-\overline{z}|}\ |z,\overline{z}\in\mathbb{R}^r,\ z\neq\overline{z},\ y\in\mathbb{R}^t,\ |y|<d\}.$$

$\underline{D}_y$ on the other hand is a lower bound on the degree of contraction/expansion of the map P in y-direction:

$$\underline{D}_y=\inf\{\frac{|e(z,y)-e(z,\overline{y})|}{|y-\overline{y}|}|z\in\mathbb{R}^r,\ y,\ \overline{y}\in\mathbb{R}^t,\ y\neq\overline{y},\ |y|,|\overline{y}|<d\}$$

see Fig. 6,7.

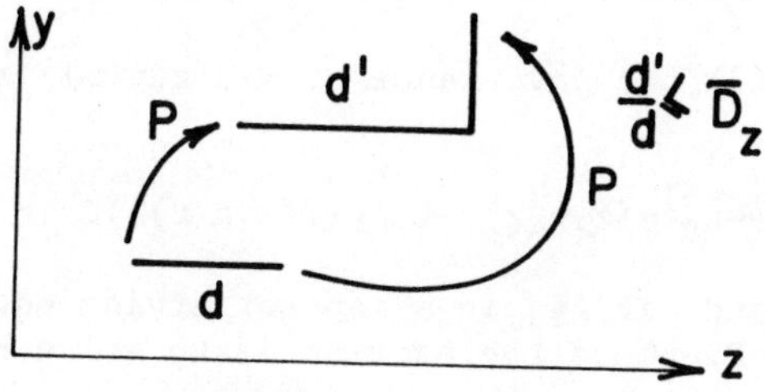

Figure 6

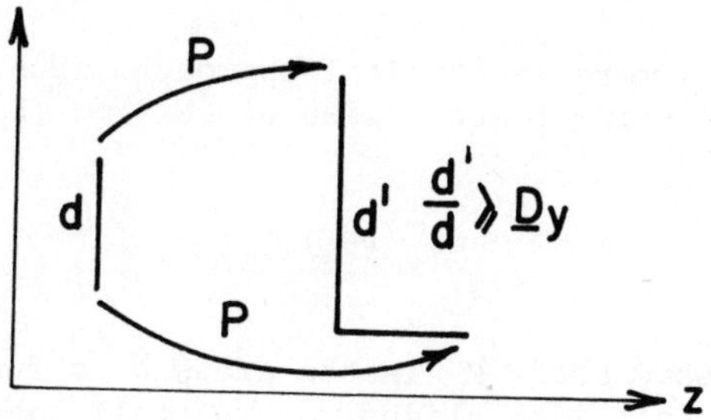

Figure 7

It easily follows that

$$\overline{D}_z \leq \alpha < \beta \qquad \underline{D}_y \geq \beta > 1.$$

Thus part (a) of hypothesis (H1) may be restated as follows: P is an expansion in y-direction. It may or may not be an expansion in z-direction. In any case the y-direction is more expanding than the z-direction.

It remains to comment on hypothesis (b) of (H1). Consider the Lipschitz constants K_{zy}, K_{yz}. If $K_{zy}=0$ f is independent of y. Thus, in this case, the y-component of x does not affect the z-component of the image P(x). Similarly, if $K_{yz}=0$, the z-component of x does not affect the y-component of P(x). If $K_{zy}=K_{yz}=0$ the two components of P are completely decoupled. The product $K_{zy}K_{yz}$ may therefore be considered a measure for the degree of interaction or coupling of the two components. Condition (b) of (H1) thus puts a suitable upper bound on the strength of coupling.

c) Proof of Theorem 1.

Given the map (3), the invariance of $M=\{(z,g(z)) \mid z\in\mathbb{R}^r\}$ with respect to P implies

$$g(z)=-L^{-1}Y(z,g(z))+L^{-1}g(f(z,g(z))). \tag{4}$$

On the other hand: if g is a map satisfying Eq. (4) then M is invariant for P. Thus the problem is to solve Eq. (4).

One way to solve Eq. (4) is as follows. Define the right-hand side of Eq. (4) to be an operator $\mathcal{F}$ acting on the function g. Choosing a suitable Banach space it can be shown that $\mathcal{F}$ in fact is a contraction mapping. This implies (4) to have a solution.

We prefer a more geometrical approach. Due to the fact that U_d is in the domain of attraction of M for P^{-1} one might try to define M by

$$M = \bigcap_{j=0}^{\infty} P^{-j}(U_d).$$

To be able to show that M is the graph of a function we shall have to modify this idea slightly. We shall construct a sequence of manifolds $M_i = \{(z,g_i(z)) \mid z\in\mathbb{R}^r\}$, i=0, 1, ... satisfying $P(M_i)\subset M_{i-1}$ and such that $g_i(z)\to g(z)$ and g(z) being the solution of Eq. (4).

Given $g_{i-1}(z)$ the function $g_i(z)$ will satisfy

$$g_i=-L^{-1}Y(z,g_i)+L^{-1}g_{i-1}(f(z,g_i)), \tag{5}$$

where $g_i=g_i(z)$. While Eq. (4) is to be considered an equation in a function space, z may be kept fixed in Eq. (5) and to solve Eq. (5) for g_i is an easy task in $\mathbb{R}^t$.

Before proceeding we mention that the following easy Lemma holds.

__Lemma 1. Part__ (a), (b) __of hypothesis__ (H1) __imply the existence of a constant__ $\mu>0$ __such that__:

(A) $\ell[K_{yy}+\mu K_{zy}]<1-\ell$

(B) $\frac{1}{\ell}-K_{yy}-\mu K_{zy}=:\chi>\frac{K_{yz}+\mu K_{zz}}{\mu}$.

As to the function g_{i-1} we make the assumption

$$\begin{aligned}&|g_{i-1}(z)|\le M\ell/(1-\ell)\\ &|g_{i-1}(z)-g_{i-1}(\bar{z})|\le\mu|z-\bar{z}|.\end{aligned} \tag{6.i-1}$$

Denote the right-hand side of Eq. (5) by $F_{i-1}(g_i)$. It is easy to see that F_{i-1} maps the ball $B=\{g|g\in\mathbb{R}^t,\ |g|\le M\ell/(1-\ell)\}$ into itself and is a contraction on B with Lipschitz constant $1-\ell$ (thanks to (A) of Lemma 1). Thus there is a unique solution $g_i(z)$ of (5) in B. Moreover, it follows from Eq. (5) and part (B) of Lemma 1 that $g_i(z)$ is Lipschitzian with Lipschitz constant μ. Thus Eq. (6.i-1) implies Eq. (6.i). Moreover Eq. (6.0) holds with $g_o(z)\equiv 0$. Next we show the convergence of the sequence $g_i(z)$.

Using the notation $||s(.)||=\sup\{|s(z)|\ |z\in\mathbb{R}^r\}$ it is not difficult to see that

$$||g_{i+N}(.)-g_i(.)||\le\frac{1}{\chi^i}c\ , \tag{7}$$

where $c = \chi/(\chi-1)||g_1(.)-g_0(.)||$. Note that $\chi > 1$ from part (A) of Lemma 1. It follows from Eq. (7) that $g_i(z)$ is a Cauchy sequence for each $z \in \mathbb{R}^r$. Thus there is $g(z) \in \mathbb{R}^r$ with $\lim_{i\to\infty} g_i(z)=g(z)$.

It follows next from Eq. (6.i) that g(z) is bounded by $M\ell/(1-\ell)$ and μ-Lipschitzian. It is easy to see that all the functions $g_i(z)$ and g(z) are 2π-periodic with respect to the components of z. Going to the limit in Eq. (5), we see that g(z) indeed satisfies Eq. (4). So far we have proved the following. There is a function g(z), bounded by $M\ell/(1-\ell)$, μ-Lipschitzian, periodic and such that $P(M) \subset M$ where $M=\{(z,g(z)) | z \in \mathbb{R}^r\}$.

To prove the remaining properties we proceed as follows. Given $x^o = (z^o,y^o)^T \in U_d$ we define $x^i=(z^i,y^i)=P^i(x^o)$ for those $i \in N$ for which $P^i(x^o)$ is defined. We then claim

$$|y^i-g(z^i)| \geq \chi^i |y^o-g(z^o)|. \tag{8}$$

For i=1 the estimate follows from Eq. (4) and the fact that g is μ-Lipschitzian, for $i > 1$ by induction.

Note that hypothesis (H2) implies that there is a map $P^{-1}:U_d \to U_d$ such that $P(P^{-1}(x))=x$ for $x \in U_d$ and $P^{-1}(P(x))=x$ if $P(x) \subset U_d$. This, together with Eq. (8) for i=1 implies $P^{-1}(M) \subset M$. Thus M is indeed invariant with respect to P and to P^{-1}. The maximality property follows again from Eq. (8) together with the boundedness of g(z) and since $\chi > 1$. To get the stability and attractivity property we define $x^{-j}=P^{-j}(x^o)$ for $x^o \in U_d$, $j \in \mathbb{N}$ and note that $P^k(x^{-j}) \in U_d$ for $k=0,1,\ldots,j$ and $P^j(x^{-j})=x^o$. Thus from Eq. (8)

$$|y^o-g(z^o)| \geq \chi^j |y^{-j}-g(z^{-j})|.$$

This implies the desired results.

1.2 Extensions

a) Our first extension concerns the regularity of the function g(z). We assumed the functions f(z,y), Y(z,y) to be Lipschitzian with respect to z,y and it was possible to prove that g(z) is Lipschitzian with respect to z as well. If we assume that f,Y are differentiable, the derivates $\frac{\partial f}{\partial z}, \frac{\partial f}{\partial y}, \frac{\partial Y}{\partial z}, \frac{\partial Y}{\partial y}$ being uniformly continuous with respect to z,y and satisfying

$$\left|\frac{\partial f}{\partial z}\right| \le K_{zz}, \quad \left|\frac{\partial f}{\partial y}\right| \le K_{zy}, \quad \left|\frac{\partial Y}{\partial z}\right| \le K_{yz}, \quad \left|\frac{\partial Y}{\partial y}\right| \le K_{yy}$$

for $(z,y) \in U_d$, we have

Theorem 2. The map $g(z)$ of Theorem 1 is differentiable with respect to z and its derivative $\frac{\partial g}{\partial z}$ is uniformly continuous.

One might expect that $g(z)$ is as regular as the functions f, Y are. However, to get sharper regularity results, it is necessary to sharpen hypothesis (H1) .

b) We continue to explore the geometrical structure of the map (3). For simplicity we assume $d=\infty$. Consider the z-y-space $\mathbb{R}^{r+t}$. It is possible to assure the existence of two transversal foliations $\mathcal{H}$, $\mathcal{V}$ of this space the structures of which are related to the map P. The sheets of $\mathcal{H}$ are r-dimensional manifolds that are homeomorphic to the z-plane. $\mathcal{H}$ is preserved with respect to P and P^{-1} i.e. the P-image or the P^{-1}-image of a sheet of $\mathcal{H}$ is again a sheet of the foliation. The invariant manifold M of Theorem 1 is a sheet of $\mathcal{H}$, in fact the only sheet that is mapped onto itself. The second foliation $\mathcal{V}$ consists of t-dimensional manifolds that are homeomorphic to the y-plane and is preserved with respect to P and P^{-1} as well, cf. Fig. 8. To arrive at a different characterisation of the two foliations we proceed as follows. For $x=(z,y)^T$ consider $x_n = P^n(x) =: (z_n, y_n)^T, n \in \mathbb{N}$. It is easy to see that the following statement holds.

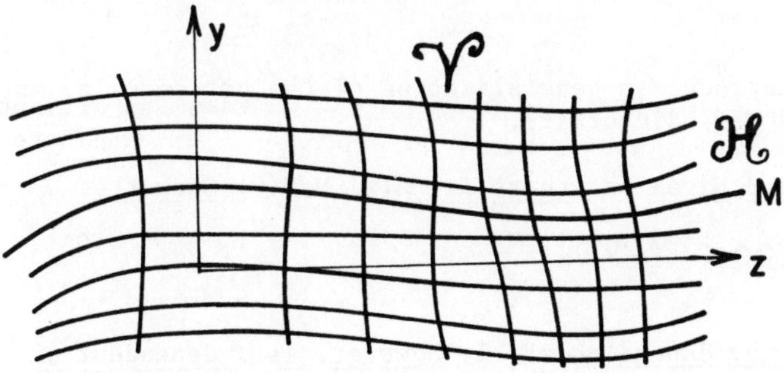

Figure 8

Theorem 3.1. The limit $\lim_{n\to\infty} L^{-n} y_n =: G(x)$ exists.

$G(x)$ may be considered a kind of measure for how fast the projections of $P^n(x)$ into the y-plane diverge. Our next result relates the level sets of $G(x)$ to the foliation $\mathcal{H}$.

Theorem 3.2. *There is a function* $g(z,G)$, 2π-*periodic with respect to the components of* z, *such that* $\mathcal{H} = \bigcup_{G\in\mathbb{R}^t} M(G)$ where

$$M(G) = \{(z,g(z,G))\,|\,z\in\mathbb{R}^r\},\ G\in\mathbb{R}^t, \text{ are the sheets of } \mathcal{H}.$$

Moreover we have $M(G)=\{x\,|\,G(x)=G\}$.

It follows from the maximality property of the invariant manifold M of Theorem 1 that $M(0)=M$.

We briefly describe the construction of $g(z,G)$. From the proof of Theorem 1 the following is known. Given a function $g_2(z)$ which is μ-Lipschitzian and periodic, there is a function $g_1(z)$ with the same properties, and such that $P(M_{g_1})\subset M_{g_2}$ where $M_{g_i}=\{(z,g_i(z))\,|\,z\in\mathbb{R}^r\}$. Symbolically we write $g_1=P^{-1}(g_2)$.

Now $g(z,G)$ is obtained via the following limit

$$g(z,G) = \lim_{n\to\infty} P^{-n}(L^n G). \tag{9}$$

Note that the map $g(z)$ of Theorem 1 was given by

$$g(z)=\lim_{n\to\infty} P^{-n}(0)=\lim_{n\to\infty} P^{-n}(L^n 0).$$

Next we introduce a generalization of the notion of asymptotic phase for periodic orbits.

Theorem 3.3. *Given* x *there is* $A(x)\in M=M(0)$ *such that*

$$|P^{-n}(x)-P^{-n}(A(x))| \le \frac{c}{\chi^n},$$

$n\in\mathbb{N}$, (c *may depend on* x; $\chi>1$, *however, is independent of* x).

Finally we relate the level sets of $A(x)$ to the foliation $\mathcal{V}$.

Theorem 3.4. *There is a function* $x(A,G): M=M(0)\times\mathbb{R}^t\to\mathbb{R}^{r+t}$ *such that*

$$\mathcal{V} = \bigcup_{A\in M} N(A), \text{ where } N(A)=\{x(A,G)\,|\,G\in\mathbb{R}^t\},\ A\in M, \text{ are the sheets}$$

of $\mathcal{V}$. *Moreover we have* $N(A)=\{x\,|\,A(x)=A\}$ *and* $N(A)\cap M(G)=\{x(A,G)\}$.

$x(A,G)$ is constructed as follows. For $n\in\mathbb{N}$ denote the z-component of $P^{-n}(A)$ by z_{-n}. We then have

$$x(A,G) = \lim_{n\to\infty} P^n(z_{-n}, P^{-n}(g(.,G))(z_{-n})), \qquad \text{cf. Fig. 9.}$$

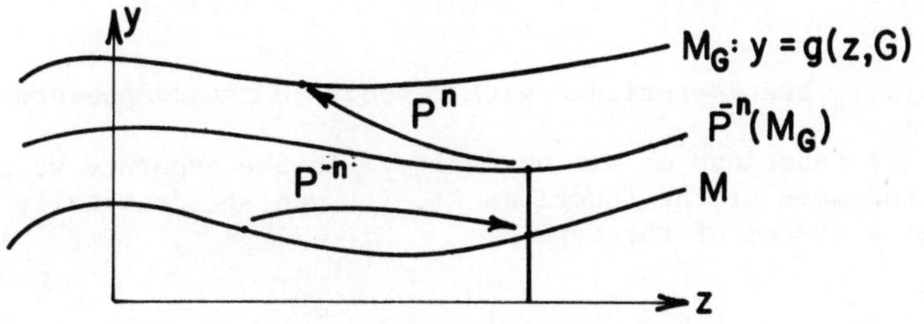

Figure 9

The foliations $\mathcal{H}$, $\mathcal{V}$ may be used to define a new coordinate system $(\bar{z},\bar{y})$, in terms of which the map P is particularly simple.

<u>Theorem 4</u>. <u>The map</u> P <u>is topologically conjugate to</u>

$$\bar{P}: \quad \begin{pmatrix} \bar{z} \\ \bar{y} \end{pmatrix} \to \begin{pmatrix} f(\bar{z}, g(\bar{z})) \\ L\bar{y} \end{pmatrix}.$$

Thus in terms of these coordinates the two components of the map are completely decoupled.

2. Applications

2.1 An Application to the Method of Averaging.

The final problem is to translate the previous results to systems of ODE's. This step is basically simple, yet a bit technical. We will give a result which is motivated by the method of averaging. Given a system of ODE's $\dot{y} = f(y,\varepsilon)$. There sometimes exists a transformation $y = B(\phi,a)$, B 2π-periodic with respect to the components of ϕ, such that the transformed system reads as follows

$$\begin{cases} \dot{\phi} = \nu(a) + \varepsilon R^1(\phi,a) + \varepsilon^2 R^2(\phi,a) + \dots \\ \dot{a} = \qquad \varepsilon T^1(\phi,a) + \varepsilon^2 T^2(\phi,a) + \dots \end{cases}$$

This system may be further simplified by performing a near-identical change of variables, such that the remaining system is as follows:

$$\begin{cases} \dot{\overline{\phi}} = \nu(\overline{a}) + \varepsilon \overline{R}^1(\overline{a}) + \dots + \varepsilon^N \overline{R}^N(\overline{a}) + \varepsilon^{N+1} \rho_\phi(\overline{\phi}, \overline{a}, \varepsilon) \\ \dot{\overline{a}} = \qquad \varepsilon \overline{T}^1(\overline{a}) + \dots + \varepsilon^N \overline{T}^N(\overline{a}) + \varepsilon^{N+1} \rho_a(\phi, a, \varepsilon) \end{cases}$$

where ρ_ϕ, ρ_a are 2π-periodic with respect to the components of $\overline{\phi}$ and nice functions of all variables. In the sequence we shall admit that some of the functions $\overline{R}^i$, $\overline{T}^i$ vanish identically and consider a system of the type

$$\begin{cases} \dot{\phi} = \omega(\varepsilon) + \varepsilon^G R(a,\varepsilon) + \varepsilon^{N+1} \rho_\phi(\phi,a,\varepsilon) \\ \dot{a} = \qquad \varepsilon^J T(a,\varepsilon) + \varepsilon^{N+1} \rho_a(\phi,a,\varepsilon) \end{cases} \tag{10}$$

(for simplicity of notation we have skipped the bar), where

$$\omega(\varepsilon) = \omega^0 + \varepsilon\omega^1 + \dots + \varepsilon^{G-1}\omega^{G-1}$$

$$\varepsilon^G R = \varepsilon^G R^G(a) + \dots + \varepsilon^N R^N(a)$$

$$\varepsilon^J T = \varepsilon^J T^J(a) + \dots + \varepsilon^N T^N(a)$$

$$\rho_\phi(\phi,a,\varepsilon), \quad \rho_a(\phi,a,\varepsilon)$$

are defined on $\mathbb{R}^r \times \Delta \times (-\varepsilon_1, \varepsilon_1)$, Δ being a region in $\mathbb{R}^t$, smooth, 2π-periodic with respect to the components of ϕ. As to the exponents G,J,N we impose the following restrictions

$$0 \le G \le N+1, \quad 1 \le J \le N, \quad G \le J.$$

Note, that if G happens to satisfy $G > J$ it is always possible to write

$$\varepsilon^G R(a,\varepsilon) = \varepsilon^J [R^J + \varepsilon R^{J+1} + \dots + \varepsilon^{G-1-J} R^{G-1} + \varepsilon^{G-J} R(a,\varepsilon)]$$

with $R^J = \dots = R^{G-1} = 0$.

If we skip the remainder terms ρ_ϕ, ρ_a in Eq. (10) for a moment, we see that the two equations in Eq. (10) decouple. We may consider the equation $\dot{a} = \varepsilon^J T(a,\varepsilon)$ or even

$$a' = T(a,\varepsilon) \tag{11}$$

first. Eq. (11) is an autonomous system in $\mathbb{R}^t$ on which we will impose the following conditions:

1) There is $A \in \Delta$ such that $T(A,\varepsilon)=0$ for $\varepsilon \in (-\varepsilon_1, \varepsilon_1)$ i.e. A is an equilibrium solution for Eq. (11).

2) The eigenvalues of the Jacobian $\frac{\partial T}{\partial a}(A,0)$ are distinct and have negative real parts i.e. A is exponentially asymptotically stable for Eq. (11).

3) A is attractive for Δ i.e. any solution of Eq. (11) tends to A as $t \to \infty$.

Let Γ be any compact subset of Δ. We then have:

Theorem 5. Let $N+1>2J-G$. Then there is $\varepsilon_o>0$ and a function $g(\phi,\varepsilon)$: $\mathbb{R}^r x(0,\varepsilon_o) \to \mathbb{R}^t$, continuously differentiable with respect to ϕ and 2π-periodic with respect to the components of ϕ such that

$M_\varepsilon = \{(\phi, A+\varepsilon^{N+1-J} g(\phi,\varepsilon)) \mid \phi \in \mathbb{R}^r\}$ satisfies:

(i) M_ε is invariant for Eq. (10) i.e. any solution with initial value in M_ε is defined for all t and its orbit is contained in M_ε.

(ii) M_ε is maximal i.e. there is a neighborhood U_d of M_ε such that any solution of Eq. (10) with initial value in $U_d - M_\varepsilon$ eventually leaves U_d as $t \to -\infty$.

(iii) M_ε is stable and attractive for $t\to\infty$; moreover $\mathbb{R}^r x\Gamma$ is contained in the domain of attraction of M_ε i.e. any solution of Eq. (10) with initial value in $\mathbb{R}^r x\Gamma$ will tend to M_ε as $t\to\infty$.

2.2 On the Rotation of Mercury

In April 1965, new methods in radar astronomy led to the discovery that the rotation period of Mercury is about 59 days. This was most striking because it had been inferred from earlier optical observations that the planet's rotation period is equal to its orbital revolution period of about 88 days. The publication of their new measurements by Pettengill and Dyce has been followed by a rather controversial discussion and several attempts have been made to explain the 3:2 rotation rate.

Here we restrict ourselves to introduce the simplest possible model and to establish the existence of a certain invariant manifold which seems to be of some importance in the discussion of the Mercury problem.

We shall assume that Mercury is revolving around the Sun in a fixed ellipse and that its spin axis is normal to its orbital plane. The ellipse is specified by its semi-major axis, which is chosen to be the unit of length, and its eccentricity e. r denotes the instantaneous distance from Mercury to the Sun, v the angle between the Mercury-Sun line and a fixed reference line in the orbital plane. The principal moments of interia are A, B, C, $A<B<C$, where C is the moment about the spin axis, Θ denotes the angle between Mercury's long axis and the reference line, cf. Fig. 10.

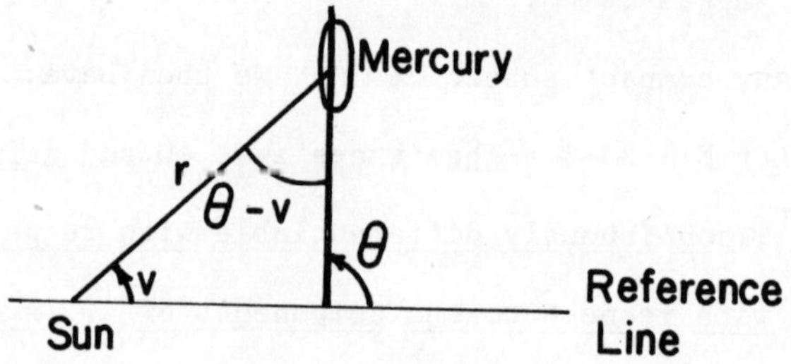

Figure 10

According to Newton's Fundamental Law of Mechanics the rotational motion of Mercury is described by the Equation $C\ddot{\Theta} = T$ where T is the solar torque acting on Mercury. T is assumed to be the sum of a torque T_1 exerted on the permanent axial asymmetry and of a torque T_2 on the tidal bulge raised by the Sun on the planet. T_1 is found to be $T_1 = -\frac{3}{2}(B-A)\mu\frac{1}{r^3}\sin 2(\Theta - v)$, where μ is the gravitational constant times the mass of the Sun. The tidal bulge is due to deformation of the non-rigid planet by the solar gravitational field. Due to the motion of the Sun, relative to Mercury, the tidal bulge follows the Sun's motion. Various models have been proposed for T_2. We shall use the following formula

$$T_2 = -\alpha n C \frac{1}{r^6}(\dot{\Theta} - \dot{v}).$$

In terms of the mean anomaly $M = nt$, $n = \sqrt{\mu}$, the equation of motion becomes

$$\Theta'' = \varepsilon T^1(M, \Theta, e) + \alpha T^2(M, \Theta', e), \qquad (11)$$

where

$$\varepsilon = \frac{3}{2}\,\frac{B-A}{C} \qquad T^1 = \left(\frac{1+e\cos v(M)}{1-e^2}\right)^3 \sin 2(v(M)-\Theta)$$

$$T^2 = \left(\frac{1+e\cos v(M)}{1-e^2}\right)^6 (v'(M)-\Theta'),$$

where prime denotes differentiation with respect to M. (Note the usual relation between the true anomaly v and the mean anomaly M. The order of magnitude of α, ε is 10^{-8}, 10^{-4} respectively. It is therefore reasonable to consider ε a small parameter and to put $\alpha=\varepsilon^2 k$ where $k=O(1)$.

We treat Eq. (11) by means of the method of averaging. We put $\phi_0=M$, $\phi_1=\Theta$, $a=\Theta'$ and rewrite Eq. (11) as follows:

$$\begin{cases} \phi_0' = 1 \\ \phi_1' = a \\ a' = \varepsilon T^1(\phi_0,\phi_1,e)+\varepsilon^2 k T^2(\phi_0,a,e). \end{cases} \tag{12}$$

This system may be simplified by performing a near identical change of variables

$$(\phi_0,\phi_1,a) \leftrightarrow (\bar\phi_0,\bar\phi_1,\bar a),$$

provided the domain of definition of $\bar a$ is restricted to a suitable set $\bar\Delta$, which we do not describe more precisely here. The transformed system is found to read

$$\begin{cases} \bar\phi_0' = 1 \\ \bar\phi_1' = \bar a + O(\varepsilon^5) \\ \bar a' = \varepsilon^2 k\bar T^2(\bar a,e)+\varepsilon^3\bar T^3(\bar a,e)+\varepsilon^4\bar T^4(\bar a,e)+O(\varepsilon^5), \end{cases} \tag{13}$$

where

$$\bar T^2(a,e) = \frac{1}{(1-e^2)^6}\left(1+\frac{15}{2}e^2+\frac{45}{8}e^4+\frac{5}{16}e^6\right) - \frac{a}{(1-e^2)^{9/2}}\left(1+3e^2+\frac{3}{8}e^4\right).$$

By inspection of Eq. (13) we see that $\bar T^2(a,e)$ admits a unique zero $A(e)$ and that $\frac{\partial T^2}{\partial a}(A(e),e)<0$. In order to meet the conditions of Theorem 5 we note that there is a transformation $\bar{a} = \tilde{a} + \varepsilon\tilde{R}^1(e) + \varepsilon^2\tilde{R}^2(e)$ which transforms Eq. (13) into

$$\begin{cases} \bar{\phi}_o' = 1 \\ \bar{\phi}_1' = \tilde{a} + \varepsilon\tilde{R}^1(e) + \varepsilon^2\tilde{R}^2(e) + O(\varepsilon^5) \\ \tilde{a}' = \varepsilon^2 k\bar{T}^2(\tilde{a},e) + \varepsilon^3\tilde{T}^3(\tilde{a},e) + \varepsilon^4\tilde{T}^4(\tilde{a},e) + O(\varepsilon^5) \end{cases} \quad (14)$$

and such that $\tilde{T}^3(A(e),e)=\tilde{T}^4(A(e),e)=0$. The original system (12) and the above system (14) are equivalent in a neighborhood of $\{(\phi_o,\phi_1,A(e))|\phi_o,\phi_1\in\mathbb{R}\}$ provided $A(e)\in\bar{\Delta}$, and for $\varepsilon>0$, sufficiently small. It is obvious that Theorem 5 applies with G=0, J=2, N=4. Eq. (14) admits an invariant manifold with the usual properties, for $\varepsilon>0$, sufficiently small. As to the original system (12) the result is as follows.

<u>Theorem 6</u>. <u>For all</u> $\varepsilon>0$, <u>sufficiently small, there is a function</u> $g^1(\phi_o,\phi_1,e,\varepsilon)$, <u>continuously differentiable and</u> 2π-<u>periodic with respect to</u> ϕ_o,ϕ_1 <u>such that</u> $M_\varepsilon=\{(\phi_o,\phi_1,A(e)+\varepsilon g^1(\phi_o,\phi_1,e,\varepsilon)|\phi_o,\phi_1\in\mathbb{R}\}$ <u>is invariant with respect to</u> Eq. (12). <u>Moreover</u> M_ε <u>possesses the usual properties of maximality, stability and attractivity</u>.

For the fixed eccentricity model the implication of Theorem 6 is as follows. A solution for Eq. (12) either lies above M_ε, on M_ε, or below M_ε. Thus Mercury's spin rate Θ' is for all times either above some value a_{max} or below some value a_{min} or always between a_{min} and a_{max}, where $a_{max}=A(e)+O(\varepsilon)$, $a_{min}=A(e)+O(\varepsilon)$. It should be emphasized however that it is known that Mercury's eccentricity changes within e=0.1 and e=o.25 over a period of some 250,000 years.

REMARKS.

Theorem 1, with a different proof, has been given in Kirchgraber-Stiefel (1978). It is emphasized again, however, that there is a vast literature on the subject, which is to be taken into account as well. The crucial hypothesis (H1) is related to the corresponding assumptions of Knobloch, cf. Knobloch-Kappel (1974).

The extensions in Section 1.2, have been briefly described in Kirchgraber (1979) and were stimulated by Palmer (1975). A detailed version is yet to appear.

Theorem 5 seems to be new and in fact it is an interesting result. On the one hand, it generalizes known invariant manifold results for the averaging method and as far as the global aspect is concerned, it considerably sharpens related results by Banfi (1967), Eckhaus (1975) and Verhulst (1976). The proof of Theorem 5 will be included in the English Edition of the book by Kirchgraber-Stiefel (1978).

The mathematical aspects of the Mercury problem have been described by Kyner (1970), Murdock (1975), Burns (1979). Theorem 6 may be new. Its full proof will appear in the English Edition of the book by Kirchgraber-Stiefel as well. It should be pointed out, however, that there is a similar result due to Burns in a yet unpublished paper.

REFERENCES.

Banfi, C., Sull'approssimazione di processi non stationari in meccanica non lineare., Boll. Un. Mat. Italiana, **22**, (1967) p. 442.

Baumgarte, J., (1973), Asymptotische Stabilisierung von Integralen bei gewöhnlichen Differentialgleichungen 1. Ordnung, ZAMM, **53**, p. 701.

Bogoliubov, N.N. and Y. A. Mitropolski, (1965), Asymptotische Methoden in der Theorie der nicht-linearen Schwingungen.

Burns, T. J., (1979), On the Rotation of Mercury, Celestial Mech., **19**, p. 297.

Burns, T. J., On a Dissipative Model of the Spin-Orbit Resonance of Mercury. To appear.

Diliberto, S. (1960-61), Perturbation Theorems for Periodic Surfaces, Rend. Circ. Mat. Palermo, **9**, **10**, p. 265, 111.

Eckhaus, W., (1975), New Approach to the Asymptotic Theory of Non-linear Oscillations and Wave Propagation. JMAA, **49**, p.575.

Fenichel, N., (1971-72), Persistence and Smoothness of Invariant Manifolds of Flows, Ind. Univ. Math. J., **21**, p. 193.

Fenichel, N., (1974), Asymptotic Stability with Rate Conditions, Ind. Univ. Math. J., **23**, p.1109.

Hadamard, J., (1901), Sur l'itération et les solutions asymptotiques des équations différentielles, Bull. Soc. Math. France, **29**, p. 224.

Hale, J., (1961), Integral Manifolds of Perturbed Differential Systems. Ann. Math., **73**, p. 496.

Hirsch, M.W., C.C. Pugh and M. Shub, (1977), Invariant Manifolds.

Kirchgraber, U., E. Stiefel, (1978), Methoden der analytischen Störungsrechnung.

Kirchgraber, U. (1979), Sur les propriétés géometriques au voisinage d'une variété invariante, C.R.Acad.Sc. Paris, 288, p. 511.

Knobloch, H. W., F. Kappel, Gewöhnliche Differentialgleichungen, (1974).

Kyner, W. T., (1970), Passage through Resonance, in G.E.O. Giacaglia (ed.), Periodic Orbits, Stability and Resonances.

Levinson, N., (1950), Small Periodic Perturbations of an Autonomous System with a Stable Orbit, Ann. Math., 52, p. 727.

Murdock, J. A., (1975), Resonance Capture in Certain Nearly Hamiltonian Systems, JDE, 17, p. 361.

Palmer, K., (1975), Linearization near an Integral Manifold, JMAA, 51, p. 243.

Perron, O., (1930), Die Stabilitätsfrage bei Differentialgleichungen, Math. Z., 32, p. 703.

Verhulst, F., (1976), On the Theory of Averaging in V. Szebehely, B.D. Tapley (eds.), Long-time Predictions in Dynamics.

IS CELESTIAL MECHANICS DETERMINISTIC ?

Victor G. Szebehely, L. B. Meaders Professor

The University of Texas at Austin

ABSTRACT. It is shown that conditions for deterministic systems are neither satisfied by those dynamical systems which are encountered by actual physical situation, nor by problems of interest in celestial mechanics.

1. INTRODUCTION

Our undergraduate dynamics textbooks offer examples which, almost without exceptions, are solvable explicitly. In other words, these mostly artificial problems are not only integrable but even more elementary than the problem of two bodies the solution of which, as it is well known, can not be expressed by simple explicit functions of the time. One purpose of this paper is to correct the erroneous impression created by our elementary dynamics text books and to correct and clarify the misleading teaching of many naive teachers.

2. DETERMINISTIC VERSUS NON-DETERMINISTIC DYNAMICAL SYSTEMS.

Dynamical systems existing in real life usually have the following properties:

1) The systems are not integrable, i.e., the number of globally valid integrals is smaller than necessary to reduce the system to an integrable form.

2) In some very special cases when the system is integrable (such as the problem of two bodies) the solution of the reduced differential equation is not solvable with known or elementary functions of the time without infinite series.

V. Szebehely (ed.), Applications of Modern Dynamics to Celestial Mechanics and Astrodynamics, 321–324.

3) The dynamical system may be represented analytically either by a system of differential equations or by its Hamiltonian or Lagrangian functions. All of these representations are approximate since the actual, real physical systems usually contain either unknown influences or effects which are excluded from the analytical description because of their complexity. In short, our analytical representation is approximate and the effects of neglected terms are unknown, especially for long times.

4) The initial conditions, which in principle determine the solution of the system (satisfying, say, the Lipschitz condition) are known usually only approximately. An excellent example of this is the behavior of a mass-less particle in the restricted problem of three bodies that is placed, initially with zero velocity, at any of the equilibrium points. In general, the initial conditions in the configuration space are not rational numbers. The ordinates of the triangular points contain $\sqrt{3}$, while the abscissae of the collinear points are solutions of fifth order algebraic equations. Numerical integration of the motion of particles placed at the triangular libration points do not start exactly with the proper initial conditions and even in the case of critical stability the integrated orbits are not the expected fixed points , but move away from the equilibrium positions.

At this point it seems to be proper to define problems which belong to deterministic dynamics, as those which do not encounter any of the above listed four difficulties. This leaves us with an empty set of real world problems since deterministic dynamics exists only in our childhood dreams (or in Laplace's fantasy, according to Poincaré). It should be added that one of the serious consequences of the non-deterministic nature of problems in celestial mechanics is that meaningful long-time predictions (especially as $t \to \infty$) are difficult or impossible to obtain.

3. EXAMPLES IN CELESTIAL MECHANICS.

1) Concerning the previously mentioned restricted problem of three bodies the effect of irrational numbers as initial conditions at the equilibrium points easily qualify this problem as one of the simplest examples for non-deterministic dynamics. In addition, the model of the restricted problem is an approximation to all physical problems for a number of reasons: the "third" particle of infinitesimal mass always has finite mass; the primaries seldom move on circles as required by the model; the problem is non-integrable as shown by Poincaré (1896) and finally, regions of the space-phase

show great sensitivity to initial conditions. To emphasize this last point, we attempted to establish regions of the configuration space (Szebehely and McKenzie, 1978) for librational motion and for chaotic (irregular) behavior in the vicinity of the triangular libration points. In these numerical experiments no sharp and clear separation could be established between these two regions because of the high sensitivity to initial condition of the qualitative behavior of the system. Instead of listing the many other existing examples concerning the restricted problem, a few other systems, popular in celestial mechanics are listed shortly.

2) The problem of determining the higher-order gravitational components of the Earth via satellite observations belongs to non-deterministic dynamics. As it is well known, satellites launched in different orbits give different values of the J_{ij} gravitational coefficients of the Earth. The inverse problem of the above, i.e., the long-time determination of satellite orbits, belongs also to non-deterministic dynamics, partially because of our ignorance concerning the model used, especially when drag-effects are of importance.

3) The construction of planetary ephemerides encounters all the above mentioned problems. The initial conditions have been accurately observed for a very short time compared to the life of the system. The gravitational many-body problem is non-integrable, therefore, either generally divergent series solutions or numerical integrations must furnish the results. Both approaches lead to undeterministic dynamics especially for long-time predictions.

4) Predictions concerning the behavior of stellar systems or galaxies also serve as excellent examples for undeterministic dynamics because of the non-integrability, the weakness of the physical model and of the unknown three-dimensional initial conditions. It seems to be clear that if the dynamics of an artificial earth-satellite is undeterministic, then our ability to offer long-time deterministic predictions for the motion of galaxies is non-existent. The same applies to planetary systems.

4. CONCLUSIONS AND FUTURE EXPECTATIONS.

The seriousness that celestial mechanics belongs to the group of problems known as non-deterministic dynamics is not unexpected and it is neither to be underestimated nor to be considered hopeless. Just to mention two examples where significant progresses have been made, attention is directed to the method of regularization used in modern celestial mechanics and to the introduction of (surfaces of section) mapping techniques to help our understanding of the behavior of the system.

The non-deterministic nature of celestial mechanics is to be considered a challenge, rather than a criticism. Criticism must be applied, on the other hand, to those who use self-deception or by misleading statements attempt to justify its deterministic nature. Unquestionably, artificial and non-natural problems may be invented in celestial mechanics (just as in elementary dynamics) which may be "solved" deterministically. These problems and their solutions may result in regression rather than progress in celestial mechanics.

Progress should be expected from the acceptance of the non-deterministic nature and from the developments of new topological statistical techniques to deal with the presently unsolved problems.

ACKNOWLEDGEMENT.

The author gratefully acknowledges the support received in the preparation and presentation of this paper by the N.A.T.O. Scientific Affairs Division, by the Johnson Space Center of N.A.S.A. and by The University of Texas. Discussions with Professor Ilya Prigogine influenced essentially the ideas expressed in this paper.

ON A FAMILY OF CONTINUOUS MAPS OF THE CIRCLE INTO ITSELF RELATED TO THE VAN DER POL EQUATION

Lluís Alsedà, J. Llibre and R. Serra

Facultat de Ciències, Universitat Autònoma de Barcelona
Spain

ABSTRACT. In the study of the equations of the Van der Pol type, M. Levi has proved that the mapping of a certain family f_b of the circle into itself gives essentially all the information on the qualitative behaviour of the motion. On the other hand, L. Block has proved: "Let f be a continuous map of the circle into itself which has a fixed point and a periodic point of period $n > 1$. Then f has periodic points of all periods greater than n in the Sarkovskii's ordering or in the usual ordering".

We study and classify these two orderings for the maps of the family f_b. This is important for a study of the bifurcations.

V. Szebehely (ed.), Applications of Modern Dynamics to Celestial Mechanics and Astrodynamics, 325.

A REVIEW ON SEMICONVERGENT SERIES, PERIODIC SOLUTIONS AND THE VANISHING HESSIAN IN CELESTIAL MECHANICS

Carlos A. Altavista

Observatorio Astronomico, Universidad Nacional de La Plata
Rca. Argentina

ABSTRACT. Highlights of classical research concerning semi-convergent series and periodic solutions of the Three-Body Problem are given in this paper. The problem of the vanishing Hessian is analyzed with regard to both linear and non-linear differential equations, where the case of two degrees of freedom becomes the most important.

V. Szebehely (ed.), Applications of Modern Dynamics to Celestial Mechanics and Astrodynamics, 326.

PERTURBATIVE EFFECTS OF SOLAR RADIATION PRESSURE ON THE ORBITAL MOTION OF HIGH EARTH SATELLITES

L. Anselmo, P. Farinella, A. Milani and A. M. Nobili

Gruppo di Meccanica Spaziale, University of Pisa, Italy

ABSTRACT. From the point of view of the orbit determination, one of the most difficult and important perturbation on the motion of a high Earth satellite is the solar radiation pressure, since it depends on not well known parameters and therefore it is very difficult to model. We discuss the force-model of radiation pressure for spacecrafts of complex shapes and structures.

By general perturbation techniques, orbital effects on the semi-major axis and on the longitude are computed and divided in long and short-periodic types. We show that a) an axially symmetric satellite or a rapidly spinning satellite with constant attidtude does not undergo any long-periodic or secular perturbation both in the semi-major axis and in the longitude, b) even for satellites with an Earth-pointing antenna, for low inclination and eccentricity and for good antenna pointing, long-periodic perturbations in the semi-major axis are small but there are longitude perturbations of secular character. As an example, the perturbations on the orbit of the ESA SIRIO 2 satellite are computed.

We conclude that also spacecrafts of complex shape and structure, including many telecommunication satellites, can be used for high accuracy tracking experiments, provided that: a) high orbits with low eccentricity and inclination are chosen; b) the problems of radiation pressure modelling are taken into account in the design of the antennas and of their pointing system; c) the optical coefficients of the external surface of the spacecraft are known before launch with a reasonable accuracy; d) the maneuvre arcs are long enough to allow the decoupling of the long and short-periodic perturbations in the data analysis. On the other hand, a space craft of complex shape presents no serious problems if it is spin-stabilized.

V. Szebehely (ed.), Applications of Modern Dynamics to Celestial Mechanics and Astrodynamics, 327.

QUASI-RANDOM MOTIONS IN A PERTURBED PENDULUM

Antoni Aubanell

Facultat de Matematiques, Universitat de Barcelona, Spain

ABSTRACT. We consider the equations of the perturbed pendulum $\ddot{x} + \sin x = \varepsilon f(x,t)$ where ε is small. We consider firstly the case $f(x,t) = \sin t$. We show the existence and uniqueness of one hyperbolic periodic orbit in a neighbourhood of the point $(\pi,0)$ in the phase space and we study its invariant manifolds.

Using the results of Holmes and Marsden about the Melnikov function we establish the existence of transversal homoclinic orbits. As a consequence we prove that the Bernoulli shift is included as a subsystem of the Poincaré map in a suitable section of the flow and that the studied motion is therefore of quasi-random kind.

As it is well-known the quasi-randomness allows us to determine some qualitative aspects of the motion for $t \to \pm \infty$. In our case we prove that, choosing suitable initial conditions, it is possible to obtain for the motion of the perturbed pendulum any given sequence of circulations and librations and oscillations in every step, a predetermined number of times around the periodic orbit near $(\pi,0)$.

Finally we generalize these results to larger families of perturbations.

V. Szebehely (ed.), Applications of Modern Dynamics to Celestial Mechanics and Astrodynamics, 328.

SMALL BODIES CAPTURED BY A THIN ANNULAR DISK ORBITING AROUND A PRIMARY BODY

Vittorio Banfi

Observatorio di Pino Torinese, Torino, Italy

ABSTRACT. The disk around the primary body is defined as an annular region composed by a large quantity of particles moving on Keplerian circular orbits. Its equatorial plane is the same as that of the primary. The particles form a "medium" that we will suppose to be continuous. Let us assume that a small body (a grain or a planetesimal), incoming from space, enters the disk. We define capture as the settled motion, after a transient period, within the disk. The orbit of the entering body is assumed to be parabolic, with its axis lying in the equatorial plane. The orbit plane has an inclination i with the preceding one.

It is supposed also that the disk thickness is constant and very small. The disk's resisting medium provides a force which is proportional to the square of the velocity of the entering body.

By straightforward analysis the exit inclination angle i_e (as function of input angle i) is obtained, with also the other motion features. Of course, the first impact transforms the orbit into an elliptic one. For the successive impacts the formulas are obviously to be recalculated. Let us consider the complete results.

(A) Prograde Motion. When $0<i\leq 45°$, i_e decreases monotonically and then the body is "eaten up" sooner or later; when $45°<i<90°$, if the medium is sufficiently dense, the body is captured at the first impact.

(B) Retrograde Motion. If the body's kinetic energy is not completely dissipated (during the crossing time), the body will

V. Szebehely (ed.), Applications of Modern Dynamics to Celestial Mechanics and Astrodynamics, 329–330.

emerge still with retrograde motion. But now i_e will monotonically increase to have prograde motion (see case (A)). If the medium density is sufficiently high, all the preceding kinetic energy is fully dissipated; the body then will emerge with prograde motion (see case (A)).

Therefore, we conclude, by the preceding model analysis, that the capture mechanicsm is highly efficient.

LIE-ALGEBRAIC METHODS IN DYNAMICS AND CELESTIAL MECHANICS

Joachim Baumgarte

Mechanikzentrum der Technischen Universität Braunschweig, F.R.G.

ABSTRACT. The connection between the oscillator problem and the Keplerian motion is considered.

First it is shown that the Lie algebra so(3,2) is characteristic for the two-dimensional oscillator problem. After a point transformation (Levi-Civita) this algebra describes the two-dimensional Kepler problem, provided the eccentric anomaly is used as the independent variable.

Secondly, the algebra so(3,2) can be extended to an algebra so (4,2) which yields the guiding principle for reformulating all the Keplerian formulas in three dimensions.

The concretisation of an abstract isomorphism between the two Lie algebras so(4,2) leads on one side to the KS-transformation and on the other side to the transformation into Delaunay similar elements in the eccentric anomaly.

REFERENCES.

Baumgarte, J., (1978), Das Oszillator-Kepler Problem und die Lie-Algebra, Journal für die reine und angewandte Mathematik, 301, p. 59-76, Walter de Gruyter, Berlin, New York.

Baumgarte, J., (1980), Eine Lie-Algebra, die Delaunay-similar-Elemente in der exzentrischen Anomalie erzeugt, J. Phys. A: Math. Gen., 13, pp. 1145-1158, printed in Great Britain, The Institute of Physics.

V. Szebehely (ed.), Applications of Modern Dynamics to Celestial Mechanics and Astrodynamics, 331–332.

Baumgarte, J., (1979), Eine kanonische Transformation für die Kepler-Bewegung, welche ohne Dimensionserhöhung die Zeit in die exzentrische Anomalie überführt., Zeitschrift für Angewandte Mathematik und Physik (ZAMP), 30.

Baumgarte, J., (1979), Die Invarianzgruppe des Oszillator-Kepler-Problems., ZAMM, 59, pp. 177-187.

PARABOLIC ESCAPE AND CAPTURE IN THE RESTRICTED THREE-BODY PROBLEM FOR LARGER VALUES OF THE JACOBI CONSTANT

Antoni Benseny

Facultat de Matematiques, Universitat de Barcelona, Spain

ABSTRACT. The restricted three-body problem with Jacobi constant C sufficiently large (or with a primary sufficiently small) can be studied as a small perturbation of the two-body problem.

The set of points in the phase space that belong to parabolic orbits is the invariant manifold W of the periodic orbit of infinity.

Using McGehee's variables, the first intersections of W with the plane of points with zero radial velocity have been analyzed for increasing and decreasing times. The greatest terms in the asymptotic development (when C is large) of the Fourier coefficients of the perturbed curve have been determined up to the linear terms in the mass of the smallest primary.

This allows us to detect quasi-integrability if C is large enough, in spite of the non-integrability proved by Poincaré.

A method to find the following intersections has been given. It might help us to determine a lower bound of the stability domain.

Finally, a qualitative study of these intersections gives the topological structure of the sets of capture and escape orbits.

V. Szebehely (ed.), Applications of Modern Dynamics to Celestial Mechanics and Astrodynamics, 333.

COMPATIBILITY CONDITIONS FOR A NON-QUADRATIC INTEGRAL OF MOTION

George Bozis

Department of Theoretical Mechanics, University of Thessaloniki, Greece

ABSTRACT. According to a result due to Bertrand, as mentioned by Whittaker (1937), one can find the forces acting on a dynamical system, if a second integral, besides the energy integral, is known. This integral must satisfy certain conditions. As an application of this idea, Whittaker studies the case of an integral quadratic in the velocities. The corresponding potential $V = V(x,y)$ must then satisfy a second order partial differential equation (solved by Darboux) with coefficients which are functions of the Cartesian position coordinates x and y.

On the other hand, first integrals which include terms in the velocity components, higher than two, have been effectively used both in Celestial Mechanics and in Galactic Dynamics. Conditions are then found which are satisfied by the coefficients of the expression

$$\phi = A\dot{x}^2 + 2B\dot{x}\dot{y} + \Gamma\dot{y}^2 + \Delta\dot{x}^2\dot{y}^2 + E,$$

which is supposed to be a second integral of motion of an autonomous dynamical system with two degrees of freedom. The coefficients are functions of x, y. It is shown that $A = A(y)$, $B = f(x+y) + g(x-y)$, where f and g are arbitrary functions, $\Gamma = \Gamma(x)$, Δ = constant and E can be found from A, B, Γ, Δ. Further, it is shown that if the coefficient $B(x,y)$ is given, then the coefficients $A(y)$ and $\Gamma(x)$ can be found as solutions of ordinary linear differential equations of the first order.

The corresponding potential is given by the formula

$$V = \frac{1}{2\Delta}\left[\, A + \Gamma + 2(f-g)\right].$$

V. Szebehely (ed.), Applications of Modern Dynamics to Celestial Mechanics and Astrodynamics, 334–335.

Depending on the specific case at hand a certain number of arbitrary constants (or arbitrary functions) enter into the potential and the second integral.

CONTACT SYSTEMS AND CELESTIAL MECHANICS

J. Bryant

Besancon, France

ABSTRACT. A brief definition of Contact Systems is given, and their relationships to Hamiltonian Systems are shown. As an application, it is shown that McGehee's equations for Triple Collision are an example of a Contact System.

V. Szebehely (ed.), Applications of Modern Dynamics to Celestial Mechanics and Astrodynamics, 336.

DIMENSIONS OF THE INVARIANT MANIFOLDS ASSOCIATED WITH EQUILIBRIUM POINTS IN THE N-BODY PROBLEM

Josefina Casasayas

Facultat de Ciències, Universitat de Barcelona, Spain

ABSTRACT. According to McGehee (1974) the singularities of total collision in the n-body problem are blown up and an invariant total collision manifold (Δ) is glued in its place. The equilibrium points on Δ are associated with the classes of central configurations: for each class, s_c, there are two hyperbolic equilibrium points s_c^+ (total ejection) and s_c^- (total collision).

We give the dimensions of the stable (W^s) and the unstable (W^u) manifolds of $s_c^{\pm}$ for the planar and spatial n-body problem; the collinear case has been studied by Devaney (1979). The dimensions of $W^{s,u}$ depend on the index of s_c for a convenient restriction of the potential energy, and therefore, we can only express it if we know the corresponding index. For example, we know them in the planar four-body problem with equal masses using the results obtained by Simó (1978).

On the other hand, to each pair of equilibrium points, s_c^+ and s_c^-, we may associate a unique homothetic orbit connecting s_c^+ with s_c^- : $\gamma_h(s_c)$. This orbit can be considered as a heteroclinic orbit. We also prove that a necessary condition in order to have a transversal heteroclinic orbit $\gamma_h(s_c)$ is that the central configuration s_c is a non-degenerate minimum of a convenient restriction of the potential. Recently, Llibre and Simó have proved that this condition is also sufficient.

V. Szebehely (ed.), Applications of Modern Dynamics to Celestial Mechanics and Astrodynamics, 337.

ON THE DEVELOPMENT OF AN ARTIFICIAL SATELLITE THEORY

Shannon Coffey

Naval Research Laboratory, Washington, D.C.

ABSTRACT. An analytic theory for an artificial satellite is developed that can accomodate all of the perturbations due to the earth's gravitational potential. The short period terms are eliminated by two canonical transformations of the Lie type while a third transformation produces the secular Hamiltonian. The literal expressions are explicitly developed by computer in two cases, for the main problem in the theory of an artificial satellite and for several of the tesseral harmonics.

For the main problem, the theory is developed to order three in closed form, without any series expansions in the eccentricity. For the main problem this represents the first closed form extension of Brouwer's theory to order 3. The secular Hamiltonian is developed to order 4 and is identical through order 3 with those given by Brouwer and by Kozai.

For the tesseral harmonics the generator of the short period transformations is produced by expanding the tesserals by the classical series in the eccentricity. The normalization is completed by integrating with respect to Delaunay's elements.

V. Szebehely (ed.), Applications of Modern Dynamics to Celestial Mechanics and Astrodynamics, 338.

MOTION AT THE SECOND ORDER RESONANCES, 3:1 AND 5:3

G. Colombo, University of Padova, Italy and Harvard-Smithsonian Center for Astrophysics, Cambridge, Massachusetts
F. A. Franklin, Harvard-Smithsonian Center for Astrophysics, Cambridge, Massachusetts

ABSTRACT. Studies, both numerical and analytic, of first order resonances, cases for which, for example, the mean motions of an asteroid and Jupiter are in the ratio $(n+1)/n$, are quite extensive. However, relatively little work has been devoted to the second order case in which the ratio is $(n+2)/n$. An example of the former case, the Hilda planets ($n=2$), which form, with respect to the remainder of the asteroidal belt, an isolated group of about 40 bodies centered near 4 A.U. from the Sun, show stable librations that prevent close approaches to Jupiter. Any (hypothetical) non-librating asteroids nearby, not having this "protective" mechanism, would have experienced close approaches. In fact, no bodies are found in these regions. Thus the isolated, stable Hilda group seems to have a clear explanation that has cosmological implications.

The location of the 5:3 resonance near 3.7 A.U. is also somewhat isolated with respect to the main asteroidal belt, which essentially terminates just outside the first order 2:1 resonance at 3.3 A.U. A smaller number (about 20) of non-librating bodies are known between 3.50 and 3.75 A.U., but none are librating. (This small number of objects makes any collisional depopulation extremely unlikely.) The question therefore naturally arises: is the absence of bodies at the 5:3 resonance a case of dynamical instability rather than of evolutionary processes? Inasmuch as few asteroids do librate in apparently stable orbits at the 3:1 resonance, a comparison of these two cases is useful and it provided the motivation for this study. A more detailed analysis of the problem will be published elsewhere. Here we summarize current results as follows.

V. Szebehely (ed.), Applications of Modern Dynamics to Celestial Mechanics and Astrodynamics, 339–340.

A typical orbit in the second order case, when plotted in a frame rotating with Jupiter, is in strong contrast with similar plots for first order resonance--2:1, for example. In the latter the separation between successive aphelia is 180° so that a substantial amplitude of libration of the conjunction between an asteroid and Jupiter about the perihelion that lies between the two aphelia is possible. For the second order cases 3:1 and 5:1, there is an important difference. Orbits (again in the rotating frame) have both aphelia and perihelia lying in the same direction so that any possible pericentric libration necessarily implies apocentric libration as well. The latter are known to be unstable except for very low eccentricities ($e < 0.03$). Stable librations can occur about directions 90° from perihelion in the initial frame, but only when the eccentricity is sufficiently large. We have not yet established accurate limiting values, but it is clearly greater than 0.1 for the 3:1 case and seems likely to be even larger for 5:3. There is, in addition, a quantitative difference between the 3:1 and 5:3 cases that again is most easily apparent in the rotating system. For 3:1 the allowed amplitude of libration is about $\pm 60°$, but it falls to 36° for 5:3. Orbits of low mean eccentricity are associated with large librational amplitudes--which is the reason why we expect a larger eccentricity threshold in the 5:3 case. This fact, when coupled with the larger semi-major axis at 5:3 (hence the increased likelihood of close approaches to Jupiter, as well as increased short period terms) means that stable orbits at this resonance may be quite rare. In fact, although we have found orbits with stable libration at 3:1, we have not yet obtained any at 5:3.

Our numerical integrations of 30 massless bodies injected with random phases and low initial eccentricities near the 5:3 resonance show that no permanent librations are found. The eccentricities are increased (from zero) to an average in excess of 0.1. Some members reach about 0.25, during 800 Jovian periods, which is the current limit of our calculations. (We shall extend the limit somewhat further so as to allow at least one complete apsidal revolution for all bodies.) We also found among the 30 bodies that initially had semi-major axes uniformly spaced near 5:3, a region lying from 0.020 to 0.025 A.U. beyond the location of exact resonance (3.70 A.U.) that is underpopulated. This is shown in plots giving density distribution as a function of time. This question, which might have some relevance to the displaced "Encke-Kepler" gap in Saturn's ring A needs to be further examined for other initial distributions.

The numerical analysis has been carried out by Gian Andrea Bianchini, Silvia Fernandez and Flavio Pellegrini at the University of Padova. The research has been supported in part by the National Research Council of Italy.

ON THE APPLICABILITY OF THE TRANSITION CHAINS MECHANISM

Amadeo Delshams

Facultat de Matematiques, Universitat de Barcelona, Spain

ABSTRACT. A transition torus of a dynamical system is an invariant torus with stable and unstable invariant manifolds, W^s , W^u, which has the property that any neighborhood of every point in W^s is connected by a trajectory of the given dynamical system with any neighborhood of every point in W^u.

Transitions tori appear in the transitions chains mechanism, due to V. Arnold, that is based on transversal intersections of invariant manifolds of different transition tori. The existence of a transition chain in a Hamiltonian system close to an integrable one assures global instability when the dimension of the phase space is more than four.

In this work we study how hyperbolic transition tori under small perturbations in general dynamical systems can be preserved. Some examples are added for clarification. To apply the results to a Hamiltonian system we suppose that there is hyperbolic invariant torus in the non-perturbed case, the flow on the invariant tori being ergodic. The existence of non-degenerate solutions of equations involving integrals that depend only on the non-perturbed flow, provides the transversal intersection required for the applicability of the transition chains mechanism.

V. Szebehely (ed.), Applications of Modern Dynamics to Celestial Mechanics and Astrodynamics, 341.

SMALL DIVISORS IN THE DERIVATIVES OF HANSEN'S COEFFICIENTS

Sebastian Ferrer and Rafael Cid

University of Zaragoza, Spain

ABSTRACT. In an article of Giacaglia (1976) the derivatives of Hansen's coefficients $X_k^{n,m}$ with respect to the eccentricity e are obtained by recurrence. Using this result we calculate the second derivative $d^2X_k^{n,m}/de^2$ and we deduce the first and second partial derivatives of $X_k^{n,m}$ with respect to the variables L, G, of Delaunay. Finally, we analyze the existence of small divisors in the eccentricity of these derivatives.

V. Szebehely (ed.), Applications of Modern Dynamics to Celestial Mechanics and Astrodynamics, 342.

PERIODIC SOLUTIONS OF THE RESTRICTED PROBLEM OF THREE BODIES WITH SMALL VALUES OF THE MASS PARAMETER.

G. Gómez

Secció de Matamàtiques, Universitat Autónoma de Barcelona, Spain

ABSTRACT. Some numerical explorations of periodic solutions of the title problem have already been performed for small mass-ratios by Broucke ($\mu = 0.0121\ldots$) and Markellos ($\mu = 0.00095$). Because the orbits of some of these families become rather complicated, it was not possible to determine their end.

We have computed families (a), (b), (c) of periodic orbits around the collinear equilibrium points and families (f), (h) of retrograde orbits around the primaries for $\mu = 0.01$ and 0.005.

An adequate set of variables is used in order to avoid having the equilibrium points L_1 , L_2 too close to the smallest primary.

A complete picture showing how these families are connected is given.

V. Szebehely (ed.), Applications of Modern Dynamics to Celestial Mechanics and Astrodynamics, 343.

NON-UNIVERSALITY FOR A CLASS OF BIFURCATIONS

Douglas C. Heggie

Department of Mathematics, University of Edinburgh

ABSTRACT. Contopoulos and Zikides have investigated numerically a family of periodic orbits (in a certain potential) which passes from stability to instability infinitely often, at values of the energy, h_k, having a finite limit point, h_{esc}, as $k \to \infty$. From eight such transitions they found numerically that the ratio

$$\frac{h_k - h_{esc}}{h_{k+1} - h_{esc}}$$ tends to a number close to 9.22.

In this paper it is shown analytically that the limiting ratio is $\exp(\pi / \sqrt{2})$. Furthermore, it is shown that the limiting ratio varies from one problem to another, and problems can be constructed in which it takes any value above 1. Thus there is no "quantitative universality" such as has been found for other types of infinite bifurcation by Feigenbaum and others.

V. Szebehely (ed.), Applications of Modern Dynamics to Celestial Mechanics and Astrodynamics, 344.

APPLICATION OF LIE-SERIES TO NUMERICAL INTEGRATION IN CELESTIAL MECHANICS

A. Hanslmeier and R. Dvorak

Institut f. Astronomie, Graz, Austria

ABSTRACT. W. Gröbner gave in his book "Lie-series and Their Applications" the definition of this series. With a linear differential operator D we can evaluate the terms of a convergent series. For testing, this method is applied to the two body problem. It could be shown that the evaluation of the first five terms are sufficient to get good results. Then, Lie series were used to integrate a restricted three body system, a three body system, and a restricted four body system. Because the terms are very complex, much computing time is required. The advantage of solving these problems with Lie series lies in the fact that the same procedure may be used for solving non-conservative systems; for example, a two body system with variable masses. For testing, the method of Lie series is also applied to solve non-autonomous systems.

V. Szebehely (ed.), Applications of Modern Dynamics to Celestial Mechanics and Astrodynamics, 345.

APPLICATIONS OF HAMILTON'S LAW OF VARYING ACTION

Donald L. Hitzl

Lockheed Research Laboratory, Palo Alto, California

ABSTRACT. The Law of Varying Action, originally published by Hamilton in 1834, has recently resurfaced, (Bailey, 1975). He found that "Hamilton's Principle", as it is commonly referred to today, is simply a special case of a more comprehensive theory, Hamilton called the Law of Varying Action. The use of Hamilton's Law to generate approximate series solutions of dynamical problems is described. The approximating series can be constructed either as power series or series in terms of Shifted Legendre Polynomials. As illustrations, numerical results are presented for two problems: (i) a linear harmonic oscillator with small damping; (ii) several periodic orbits of the restricted three-body problem. In both cases, the Legendre Polynomial series gave far superior performance.

REFERENCES.

D. L. Hitzl and D. A. Levinson, (1980), "Application of Hamilton's Law of Varying Action to the Restricted Three-Body Problem", Celestial Mechanics, 22, pp. 255-266.

D. L. Hitzl, (1980), "Implementing Hamilton's Law of Varying Action with Shifted Legendre Polynomials", J. Computational Physics, 38, No. 2, pp. 185-211.

V. Szebehely (ed.), Applications of Modern Dynamics to Celestial Mechanics and Astrodynamics, 346.

SOME CURRENT ASTRODYNAMICS DEVELOPMENTS AT THE NORTH AMERICAN AEROSPACE DEFENSE COMMAND (NORAD)

Felix R. Hoots

Directorate of Astrodynamics Applications, Aerospace Defense Command, Peterson Air Force Base, Colorado

ABSTRACT. The space mission of NORAD requires many basic celestial mechanics calculations. These calculations support such functions as the generation of look angles, ground traces, decay prediction, and calculation of miss distance between objects (COMBO). Because of the large number of satellites (~5,000) and the extensive workload involved, each task must be accomplished with a maximum amount of analytical techniques and a minimum amount of numerical techniques.

The COMBO function is an example of this emphasis on optimized analytical procedures. Here we seek all time points within a given time interval such that the distance between a primary and a secondary satellite reaches a local minimum less than some given threshold D. Since, in general, we consider many different secondary satellites, we use a pair of analytical filters to limit the number of satellites and the intervals of time which must be considered. The first filter determines if the minimum geometric distance between the two orbits is less than D. The second filter then considers respective crossing times for the two satellites of the line of intersection of the two orbit planes. Finally, for sufficiently near crossing times, the local minimum is determined numerically by Newton's method.

V. Szebehely (ed.), Applications of Modern Dynamics to Celestial Mechanics and Astrodynamics, 347.

ABOUT THE TRIPLE COLLISION MANIFOLD IN THE PLANAR THREE-BODY PROBLEM

Maÿlès Irigoyen

Institut Henri Poincaré, Paris, France

ABSTRACT. The formulation of the planar three body problem is founded on the ideas developed in 1974 by R. McGehee for the rectilinear problem, and uses a normalization by the radius of inertia Ω of the system. In this formulation, the positions are defined by the radius Ω, and by three angles. The differential system of eighth order has the property that the radius Ω and one of the angles (which is an ignorable variable) can be separated, so that the problem, for every value of h and C of the integrals of energy and angular momentum, is reduced to a system of sixth order, with two quadratures. Due to this separation, each configuration of the system can be represented on a sphere by two opposite points. The differential system admits two invariant relations; they define an invariant manifold corresponding to the values $h = 0$ and $C = 0$ of the integrals. This invariant manifold F_c is called: "triple collision manifold".

This manifold, of dimension 4, can be regularized at the binary collisions by applying the method of Sundman. It is defined, in a space of dimension 5, by one invariant relation, and can be characterized as follows: at an ordinary point on the configuration sphere, the invariant relation associates with an ellipsoid which degenerates to a cylinder at each binary collision.

On the triple collision manifold, the problem can be reduced to the form of a system of five differental equations, with an invariant relation. The flow thus defined on F_c is gradient-like, and the corresponding Liapunov function is one of the variables of the problem.

V. Szebehely (ed.), Applications of Modern Dynamics to Celestial Mechanics and Astrodynamics, 348–349.

The reduced system admits twenty equlibrium points, corresponding to the solutions of Euler and Lagrange, and it can be shown that, at each of these equilibrium points, the characteristic exponents associated with the stable and unstable invariant manifolds, are related to those given by Siegel in 1941.

PERIODIC ORBITS NEAR HOMOCLINIC ORBITS

Jaume Llibre

Secció de Matemàtiques, Universitat Autònoma de Barcelona
Spain

ABSTRACT. It is known that the orbit-structure of a dynamical system near a homoclinic orbit γ is extremely complicated. However, it is only recently that this complicated structure has begun to be understood. It has been shown (under some hypotheses) that, near γ there are infinitely many long periodic orbits. The flow, near γ, admits a singular Poincaré map $\phi : \Sigma \to \Sigma$. The fixed points of ϕ are called the simple periodic orbits or 1-periodic orbits associated with γ. If γ is a homoclinic orbit to a saddle-focus or to a saddle-center equilibrium point for a Hamiltonian system with two degrees of freedom, then the structure of the 1-periodic orbits is well understood. For these two cases the structure of the 2-periodic orbits (that is, the fixed points of ϕ^2) are described.

V. Szebehely (ed.), Applications of Modern Dynamics to Celestial Mechanics and Astrodynamics, 350.

SOME NUMERICAL RESULTS OF A SEMI-ANALYTIC ORBIT THEORY USING OBSERVED DATA

Joseph J. F. Liu

Directorate of Astrodynamic Applications, Aerospace Defense Command, Peterson Air Force Base, Colorado

ABSTRACT. The application of a semianalytic orbit theory in a real-world environment using observed data is examined. This examination is an extension of a previous paper based on simulated perfect observations. The analysis is intended to compare the adequacy of the mathematical modeling versus observations which are unevenly distributed in time and space and contain various uncertainties and inaccuracies. The examination includes orbit determination, short-term prediction and decay estimation. Parallel results for a special perturbations program are also generated to provide a meaningful reference.

The investigation shows that the numerical results obtained from both the special perturbations program and the semianalytic theory are markedly comparable in accuracy. It is also evident that the differences in the mathematical modeling and the method of integration between the special perturbations and the semianalytic solutions are relatively small and may be within the noise level of the available data. Additionally, due to the significant improvement in efficiency over that of the special perturbations program and the compactness of the computer code, the semianalytic theory is not only feasible for applications in a real-world environment but also readily suitable for a mini-computer and/or a spacecraft on-board system.

REFERENCES.

Liu, J.J.F., France, R.G. and Hujsak, R.S. (1981), "Application of a Semianalytic Orbit Theory Using Observed Data", AAS Paper 81-179.

Liu, J.J.F. and Alford, R.L. (1980), J. Guidance and Control, 3, p. 304.

V. Szebehely (ed.), Applications of Modern Dynamics to Celestial Mechanics and Astrodynamics, 351.

ORBITAL BEHAVIOUR IN THE VICINITY OF UNSTABLE PERIODIC ORBITS IN DYNAMICAL SYSTEMS WITH THREE DEGREES OF FREEDOM

P. Magnenat

Geneva Observatory, Switzerland

ABSTRACT. In a dynamical system with three degrees of freedom, periodic orbits may undergo three main types of linear instability: a) there is only one dilating direction emanating from the representative point in the four-dimensional space of sections; b) there are two such dilating directions; c) complex instability. The structure of the neighbourhood of periodic orbits in the space of section is described for some examples of orbits in each of these three cases.

In the first case, the evolution of an asymptotic curve of the bi-dimensional case, when a perturbation is added in the third dimension, shows a progressive diminution of the oscillations and an orientation of the latter in the approximate direction of the third dimension.

In the second case, where the degree of dissolution is expected to be maximum, the aspect of the set of consequents determined by the two dilating eigenvectors in the 4-D space of section is still more intricate and difficult. The presence of two dilating directions leads to the notion of asymptotic surface.

The neighbourhood of orbits suffering from complex instability (3rd case), which is a new feature with respect to bi-dimensional problems, is studied in the space of sections by means of the maximum Lyapunov Characteristic Number technique. It is shown that the motion can deviate far from the vicinity of the representative point as soon as the orbit is of complex instability. When the perturbation is large enough, the stochasticity produced by this type of instability can be very important.

V. Szebehely (ed.), Applications of Modern Dynamics to Celestial Mechanics and Astrodynamics, 352.

ON THE CONVERGENCE OF FORMAL INTEGRALS IN FINITE TIME

Jose Martinez

Universidad de Valencia, Spain

ABSTRACT. Consider a differential system: $x = f(x) + \varepsilon g(x)$, $x \in R^n$. Let $h(x) = h_o(x) + \varepsilon h_1(x) \ldots$ a "third" integral. For finite time t, I obtain an ε_o such that the series $h(x)$ converges if $\varepsilon < \varepsilon_o$. When t tends to infinite, ε_o tends to zero.

V. Szebehely (ed.), Applications of Modern Dynamics to Celestial Mechanics and Astrodynamics, 353.

A NUMERICAL STUDY OF THE ASYMPTOTIC SOLUTIONS TO THE FAMILY (C) OF PERIODIC ORBITS IN THE RESTRICTED PROBLEM OF THREE BODIES

Regina Martinez, Jaime Llibre and Carles Simo

Universitat Autonoma de Barcelona, Spain

ABSTRACT. If the Jacobian constant C is equal to C_2 the invariant manifolds of asymptotic orbits to the equilibrium point L_2 are one-dimensional. We call W_L^s (W_L^u) the stable (unstable) branch in the part of the Hill's region which contains the primary with greatest mass. The behavior of W_L^s and W_L^u is described in terms of the parameter of masses m. There exist a countable set of values of m $(m_i)_{i\in N}$ which accumulate to zero, such that for every m_i, W_L^s and W_L^u coincide.

When C_2-C is positive and sufficiently small, an unstable periodic solution (the Liapounov periodic orbit) near the equilibrium point L_2 exists. We define $W_{P.O.}^s$ and $W_{P.O.}^u$ in the analogous form to W_L^s and W_L^u. The intersections of $W_{P.O.}^s$ and $W_{P.O.}^u$ with a suitable surface of section permit to decide if there exist homoclinic orbits which turn once around the greatest primary. We have computed, for values of m ranging from m_1 to m_3, the values of C for which these homoclinic orbits exist.

V. Szebehely (ed.), Applications of Modern Dynamics to Celestial Mechanics and Astrodynamics, 354.

THE EQUIVALENCE OF THE GENERATORS OF DEPRIT'S AND GIORGELLI-GALGANI'S CHANGE OF VARIABLES IN DIFFERENTIAL SYSTEMS.

Michel Rapaport

Floirac, France

ABSTRACT. After having extended to the non-canonical case an algorithm proposed by Giorgelli-Galgani (1978), we show that the generator of their transformation is similar to the generator of Deprit's algorithm.

If $\vec{T} = \sum_{i=1} \varepsilon^i \ \vec{T}_i$ and $\vec{W} = \sum n\varepsilon^{n-1} \ \vec{W}^{(n)}$ are

respectively their expressions, we have

$$\vec{T}_i = \vec{W}^i \quad \text{for any } n.$$

V. Szebehely (ed.), Applications of Modern Dynamics to Celestial Mechanics and Astrodynamics, 355.

STABILIZATION OF SPIRAL DENSITY WAVES IN FLAT GALAXIES FOR A HYDRODYNAMICAL MODEL

Wolfgang Renz

Institut für Theoretische Physik, RWTH Aachen, F.R.G.

ABSTRACT. The disk component of flat galaxies is described by a 2-dimensional isothermal hydrodynamical model with the approximation of tightly wounded perturbations. We treat the growth and stabilization of spiral density waves (a problem already considered by Toomre) by taking nonlinearities into account and use an ε-expansion, as in bifurcation theory, which is valid near the critical point.

Our model contains

1) damping of the perturbations due to induced shock waves in the interstellar material, which is introduced into the equations by a relaxation time approximation; this works well whenever the damping is caused by a weak coupling of the system under consideration with a reservoir (in this case the interstellar material).

2) excitation by Jeans instabilities at the corotation region, as proposed by several authors (for discussion see e.g. Lau and Bertin).

We end by demonstrating the compatibility of a stationary spiral wave with the damping process and obtain a relationship between the absolute amplitude of the spiral density wave (σ_1/σ_0) and the damping constant γ. $(\sigma_1/\sigma_0) = 5\%$ corresponds to $\gamma^{-1} < 10^9$ yrs which coincides with the usual accepted values.

Furthermore, the model can be extended to other kinds of instabilities and to an adiabatic closure of the system of equations.

V. Szebehely (ed.), Applications of Modern Dynamics to Celestial Mechanics and Astrodynamics, 356.

STABILITY OF PERIODIC ORBITS NEAR A HOMOCLINIC ORBIT FOR ANALYTICAL HAMILTONIANS WITH TWO DEGREES OF FREEDOM

Carles Simó

Facultat de Matematiques, Universitat de Barcelona, Spain

ABSTRACT. Analytical Hamiltonian systems with two degrees of freedom can have homoclinic orbits, asymptotic to a fixed point of complex saddle or saddle center type. Near this homoclinic orbit some families of periodic orbits exist. This is in fact a structurally stable situation. The changes of stability along the different families are studied, allowing for predictions of bifurcations.

The main tools are an analytic approximation and the implicit function theorem. The approximation is obtained through composition of linear maps (the contribution to the Poincaré map far away from the fixed point) with suitable twist maps (the contribution near the fixed point). For several cases the twist is singular in the sense that the rotation goes to infinity when approaching one of the boundaries of the annulus where the map is defined.

Several applications to the Hénon-Heiles and Contopoulos potentials, to the restricted three body problem, etc. are given.

V. Szebehely (ed.), Applications of Modern Dynamics to Celestial Mechanics and Astrodynamics, 357.

A METHOD FOR THE INVESTIGATION OF THE INTEGRABILITY OF DYNAMICAL SYSTEMS

Anand Sivaramakrishnan

The University of Texas at Austin

ABSTRACT. Conventional approaches to theories of motion of dynamical systems generally involve using integrable approximations (or truncated Birkhoff normal forms), and often the equations of motion are modified by an averaging procedure that eliminates one or more variables. The averaged system has no rigorous connection with the original system.

Liapunov characteristic exponents provide a means of comparing the non-integrability of the averaged system with the original one. Such a method is suggested here in the hope that researchers will attempt such comparisons and the usefulness of Liapunov characteristic exponents can be evaluated from the results of different averaging methods in a variety of dynamical problems.

V. Szebehely (ed.), Applications of Modern Dynamics to Celestial Mechanics and Astrodynamics, 358.

REAPPEARANCE OF ORDERED MOTION IN CLASSICAL NON-INTEGRABLE HAMILTONIAN SYSTEMS

R. L. Somorjai and M. K. Ali

Division of Chemistry, National Research Council of Canada

ABSTRACT. We present numerical examples of two coupled non-linear oscillators which demonstrate that the degree of stochasticity does not always increase monotonically with increasing energy. This phenomenon is by no means pathological in non-integrable Hamiltonians that model interacting physical systems. Both strongly and weakly coupled sets of interacting units show the phenomenon. These results also indicate that the concept of a single stochastic transition energy requires extension.

Some necessary conditions on the total system potentials are given that aid in the selection of coupled oscillator systems which have only limited and local stochasticity regions and only in narrow energy ranges. The examples are for "linear combinations of exponentials" type of pair of potentials. The implications of confined stochasticity on soliton-supporting properties of linear chains of oscillators that couple via such potentials are studied numerically.

V. Szebehely (ed.), Applications of Modern Dynamics to Celestial Mechanics and Astrodynamics, 359.

HOPF-BIFURCATION IN A NEARLY HAMILTONIAN SYSTEM

Franz Spirig

E.T.H., Zürich

ABSTRACT. Consider a plane autonomous system of ordinary differential equations depending on a parameter μ

$$\dot{x} = f(x,\mu). \tag{1}$$

Let this system satisfy the Hopf conditions; i.e.,

1) 0 is an equilibrium point: $f(0,\mu) \equiv 0$ for all μ.

2) The Jacobian matrix $\frac{\partial f}{\partial x}(0,\mu)$ admits a pair of conjugate complex eigenvalues $\alpha(\mu) \pm i\beta(\mu)$, which cross the imaginary axis for $\mu = 0$ with non-zero velocity from left to the right: $\alpha(0) = 0$, $\alpha'(0) > 0$. According to Hopf's Theorem (1) admits a family of periodic solutions. The direction of bifurcation and the stability behaviour is deduced from a certain quantity χ, provided χ does not vanish. Now assume that

3) $f(x,0)$ is Hamiltonian.

Then $\chi = 0$. In order to remedy this difficulty an additional perturbation term is introduced, driven by a small parameter ε, i.e., we replace (1) by

$$\dot{x} = f(x,\mu) + \varepsilon g(x,\mu,\varepsilon). \tag{2}$$

The crucial quantity χ now depends on ε and is found to be

$$\chi(\varepsilon) = \varepsilon\, \chi'(0) + O(\varepsilon^2)\ , \text{ where}$$

V. Szebehely (ed.), Applications of Modern Dynamics to Celestial Mechanics and Astrodynamics, 360–361.

$$\chi'(0) = -(F_1^{11} + F_1^{22})(L_1^{12} + L_2^{22}) + (F_2^{11} + F_2^{22})(L_1^{11} + L_2^{12})$$

$$+ L_1^{111} + L_1^{122} + L_2^{112} + L_2^{222}$$

and $$F^{ij} = \frac{1}{\beta(0)} \frac{\partial^2 f}{\partial x_i \partial x_j}(0,0)$$

$$L^{ij} = -\frac{1}{2\alpha'(0)\beta(0)}\left(\frac{\partial g_1}{\partial x_1}(0,0,0) + \frac{\partial g_2}{\partial x_2}(0,0,0)\right)\frac{\partial^3 f}{\partial\mu\partial x_i \partial x_j}(0,0)$$

$$+ \frac{1}{\beta(0)} \frac{\partial^2 g}{\partial x_i \partial x_j}(0,0,0)$$

$$L^{ijk} = -\frac{1}{2\alpha'(0)\beta(0)}\left(\frac{\partial g_1}{\partial x_1}(0,0,0) + \frac{\partial g_2}{\partial x_2}(0,0,0)\right)\frac{\partial^4 f}{\partial\mu\partial x_i \partial x_j \partial x_k}(0,0)$$

$$+ \frac{1}{\beta(0)} \frac{\partial^3 g}{\partial x_i \partial x_j \partial x_k}(0,0,0).$$

The result is as follows:

For $\varepsilon > 0$ and sufficiently small, there is a $\mu_0(\varepsilon)$ such that the bifurcating solutions exist for $\mu \in (0,\mu_0(\varepsilon))$ if $\chi'(0) < 0$ or for $\mu \in (-\mu_0(\varepsilon),0)$ if $\chi'(0) > 0$. In the first case they are stable and unstable in the second case.

The above problem is of importance in connection with the bifurcation of an invariant torus in a neighbourhood of an equilibrium point with two purely imaginary and one zero eigenvalue.

A REALIZATION OF DYNAMICAL SYSTEMS IN BOOLEAN ALGEBRAS

Mehmet Ülküdas

Ege University, Izmir, Turkey

ABSTRACT. Simple multiplicative equation $a\,x = c$ represents all the equations in a single unknown in Boolean algebra. The solution $x(t) = c + (1-a)t$ of the ordinary differential equation $\dot{x} = 1-a$ with the initial condition $x(0) = c$ is, formally, also the solution of the fundamental Boolean equation $ax = c$. Under that striking analogy, there arises the question: "Which of the two outwardly distinct appearances are the same ?"

The unification visualized as possible in topological dynamics.

The motion of a dynamical system of classical mechanics can be represented by a single point moving in a representative space. Generously endowing a proper structure to the representative space, and further modifying the time concept, the motion of the mobile (representative point) had been previously established as 'the inertial motion', i.e., as the straight-path motion of a particle with uniform velocity. This served once to be an ultra-generalization of the Galilean principle of inertia for any but "real" motion.

In the unifying picture of topological dynamics, a remarkable "Boolean"replica for the inertial motion is provided.

V. Szebehely (ed.), Applications of Modern Dynamics to Celestial Mechanics and Astrodynamics, 362.

RELATIVISTIC ASTRODYNAMICS: PROBLEMS IN INTERSTELLAR FLIGHT

Giovanni Vulpetti

Department of Space Technology, Telespazio S.P.A. per le Telecomunicazioni Spaziali, Rome, Italy

ABSTRACT. After the definition of suitable frames of references, the relativistic rocket equation is examined. A selection is made among the most energetic nuclear reactions which could be employed for generating a high velocity of ejection and a suitable thrust. One of the best candidates is the matter-anti-matter annihilation which is illustrated. In such a type of propulsion there are several enormous problems which present theoretical difficulties. One of these consists of converting and directionalizing the momentum/energy of a great deal of gamma rays produced in the proton-anti-proton annihilation. A possible solution is discussed. This would exploit the polarization of the vacuum by means of supercharged nuclei. A model of propulsion system is presented and both advantages and disadvantages are pointed out through numerical examples. Finally, an interplanetary "softer" version is also discussed.

V. Szebehely (ed.), Applications of Modern Dynamics to Celestial Mechanics and Astrodynamics, 363.

THE ORIGIN OF THE KIRKWOOD GAPS: A MAPPING FOR ASTEROIDAL MOTION NEAR THE 3/1 COMMENSURABILITY

Jack Wisdom

California Institute of Technology, Pasadena, California

ABSTRACT. A mapping of the phase space onto itself with the same low order resonance structure as the 3/1 commensurability in the planar elliptic three-body problem is derived. This mapping is approximately one thousand (1000) times faster than the usual method of numerical integration of the averaged equations of motion (as used by Schubart, Froeschlé and Scholl in their studies of the asteroid belt). This mapping exhibits some very surprising behavior that might provide the key to the origin of the gaps. A test asteroid placed in the gap may evolve for a million years with low eccentricity (< 0.05) and then suddenly may jump to large eccentricity (> 0.3) becoming a Mars-crosser. The asteroid can then be removed by a close encounter with Mars. To test this hypothesis a distribution of 300 test asteroids in the neighborhood of the 3/1 commensurability was evolved for two million years. When the Mars-crossers are removed, the distribution of initial conditions displays a gap at the location of the 3/1 Kirkwood gap. The planar elliptic mapping is then extended to include the inclinations and the secular perturbations of Jupiter's orbit. The two million year evolution of the 300 test asteroids is repeated using the full mapping. The resulting gap is somewhat larger yet still too small. Finally, the possibility that over longer times more asteroids will become Mars-crossers is tested by studying the evolution of one test asteroid near the border of the gap for a much longer time. A jump in its eccentricity occurs after 18 million years indicating that indeed it may simply be a matter of time for the full width of the gap to open.

V. Szebehely (ed.), Applications of Modern Dynamics to Celestial Mechanics and Astrodynamics, 364.

COMPLETELY INTEGRABLE SYSTEMS AND SINGULARITY IN THE COMPLEX t - PLANE

H. Yoshida

Department of Astronomy, University of Tokyo

ABSTRACT. For certain Hamiltonian dynamical systems with several parameters, we always may obtain integrable cases through a postulate that the solution should have Laurent expansion with sufficient number of arbitrary constants near a singularity in the complex t-plane. Famous classical examples are tops of Euler, Lagrange, and Kovalevskaja. For a Hamiltonian system

$$H = \frac{1}{2}(p_1^2 + p_2^2) + \frac{1}{2}(q_1^2 + q_2^2) + q_1^2 q_2 + \frac{1}{3}(1 - 2\varepsilon)\, q_2^3 ,$$

this postulate (procedure) determines the value of parameter as $\varepsilon = 0$, which is shown to be integrable. Hénon-Heiles case ($\varepsilon = 1$) is rejected. Another interesting example is a series of Generalized Toda Lattice systems, which can be written in a Lax-type differential equation.

V. Szebehely (ed.), Applications of Modern Dynamics to Celestial Mechanics and Astrodynamics, 365.

INDEX OF NAMES

INDEX OF SUBJECTS